KB048273

미국수학을 요리하다!
미국 현지 20년 경력 Brian 선생님의 미국수학 만점 레시피!

미국 아마존 스테디셀러

RECIPE

이연욱(Brian Rhee) 지음

SAT SUBJECT TEST MATH LEVEL 2

최신 SAT Math Level 2 시험에
나오는 모든 문제유형 포함

최신 출제문제 경향분석 및
빠르고 정확한 문제풀이법 공개

Preface

미국 버지니아 수학학원 원장 및 강사로서 현장에서 20년 이상 학생들을 지도하는 동안 수많은 교재를 연구했고 분석하여 AP Calculus AB/BC를 수강하는 학생들이 좋은 성적을 받도록 도움을 주었습니다. 하지만, 항상 안타깝게 생각하는 있던 것은 유명한 출판사의 교재들이라 할지라도 실제 시험 출제 경향이나 스타일과 차이가 있는 경우들도 간혹 보이고 턱없이 부족한 개념설명과 문제 난이도로 인해서 진정으로 학생들의 수학실력 향상에 도움을 주지 못한다는 것입니다. 그럼에도 불구하고 적절한 대안이 없어서 그 교재들을 사용하여 학생들을 지도해야 한다는 답답함이 있었습니다. 마음속에 있던 이러한 안타까움은 좋은 교재를 만들어 보자는 용기로 바뀌어 집필을 시작하게 했고, 온/오프라인의 많은 지도경험을 통해 얻은 이해하기 쉬운 개념설명과 핵심포인트를 교재에 소개한다는 목표를 가지고 노력한 결과 집필을 마칠 수 있었습니다.

이 책은 방대한 AP Calculus AB/BC의 토픽들을 45개의 레슨으로 세분화한 것입니다. 처음 28개의 레슨은 AP Calculus AB와 AP Calculus BC의 공통 토픽들이고, 나머지 17개의 레슨은 AP Calculus BC의 토픽들로 나뉩니다. 각각의 레슨에는 쉽게 이해할 수 있는 개념설명과 5~10개의 연습문제를 통해서 학생들이 핵심포인트를 이해할 수 있도록 구성했습니다.

이 책은 미국 및 한국의 학원교재와 1:1 개인지도 교재로 기획되어 미국에서 먼저 출판하였으며, 미국 아마존닷컴에서는 스테디셀러로 학생들의 인기를 얻고 있습니다. 동시에 유학전문 인터넷 강의 사이트인 마스터프렙(www.masterprep.net)에서 AP Calculus AB 와 AP Calculus BC의 교재로 사용되고 있으며 학생들의 좋은 반응으로 수학 교육자의 한 사람으로서 보람을 느끼고 있습니다. 책의 기획 의도상 교재에는 Answer keys만 포함되어 있고 상세한 풀이는 포함되지 않았습니다. 자세한 풀이를 원하는 학생들은 지도하시는 선생님께 도움을 요청하시거나 마스터프렙의 인터넷 강의를 듣는 것을 추천합니다.

한국에서 교재를 출판할 수 있도록 도와주신 헤르몬하우스와 항상 저에게 용기와 격려를 주시는 마스터프렙의 권주근 대표님께 진심으로 감사드리고, AP Calculus AB/BC를 공부하는 학생들에게 도움을 많이 주는 교재로 남기를 기대합니다.

2020년 11월
이연욱

이 책의 특징

AP Calculus AB/BC의 토픽들을
45개의 레슨으로 세분화

1부 28개 레슨
- AP Calculus AB
  AP Calculus BC의 공통 토픽

2부 17개 레슨
- AP Calculus BC의 토픽

개념 설명 5~10개의 개념 설명과
연습 문제를 통해 핵심포인트 구성

저자 소개

이연욱 (Brian Rhee)

미국 New York University
미국 Columbia University
미 연방정부 노동통계청 근무

마스터프렙 수학영역 대표강사
미국 버지니아의 No.1 수학전문 학원
Solomon Academy 대표

미국 버지니아의 No.1 수학전문 학원인 솔로몬 학원(Solomon Academy)의 대표이자 소위 말하는 1타 수학 강사이다. 버지니아에 위치한 명문 토마스 제퍼슨 과학고(Thomas Jefferson High School for Science & Technology) 및 버지니아 주의 유명한 사립, 공립학교의 수많은 학생을 지도하면서 명성을 쌓았고 좋은 결과로 입소문이 나 있다. 그의 수많은 제자들이 Harvard, Yale, Princeton, MIT, Columbia, Stanford와 같은 아이비리그 및 여러 명문 대학교에 입학하였을 뿐만 아니라, 중학교 수학경시대회인 MathCounts에서는 버지니아 주 대표 5명 중에 3명이 바로 선생님의 제자라는 점과, 지도한 다수의 학생들이 미국 고교 수학경시대회인 AMC, AIME를 거쳐 USAMO에 입상한 사실들은 선생님의 지도방식과 능력을 입증하고 있다.

현재 미국과 한국을 오가면서 강의하고 있으며, 한국에서는 SAT 전문 학원에서의 강의와 더불어 No.1 유학전문 인터넷 강의 사이트인 마스터프렙(www.masterprep.net)에서 SAT 2 Math Level 2 강의를 시작으로 AP Calculus와 그 외의 다른 수학 과목을 영어 버전과 한국어 버전으로 강의하고 있다.

아마존닷컴(www.amazon.com)에서 미국수학전문 교재의 스테디셀러 저자이기도 한 선생님은 SAT 2 Math Level 2, SAT 1 Math, SHSAT/TJHSST Math workbook, IAAT와 AP Calculus AB & BC 등 다수의 책을 출판하였고, 지금도 여러 수학책을 집필 중이며 한국에서도 지속적으로 선생님의 책이 시리즈로 소개될 예정이다.

고등학교 때 이민을 가서 New York University에서 수학학사, Columbia University에서 통계학석사로 졸업한 후, 미 연방정부 노동통계청에서 통계학자로 근무했으며, 미국수학과 한국수학 모두에 정통한 강점을 가지고 있을 뿐만 아니라 Pre-Algebra에서부터 AP Calculus, Multivariable Calculus, Linear Algebra까지 지도할 수 있는 실력자이다. 또한 미국 수학경시대회인 MathCounts, AMC 8/10/12와 AIME까지의 고급 수학을 모두 영어와 한국어로 강의를 하는 20년 경력의 진정한 미국수학 전문가로 탄탄한 이론은 물론, 경험과 실력과 검증된 결과를 모두 갖춘 보기 드문 선생님이다.

Contents

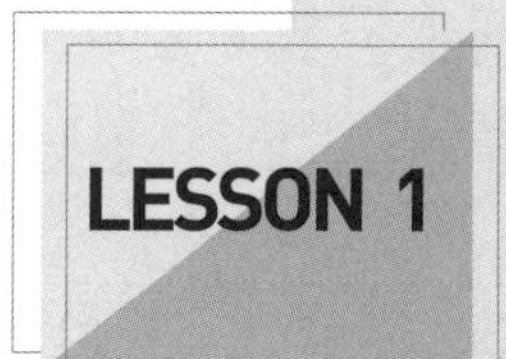

LESSON 1 Functions

Definition of Function

A relation is a set of ordered pairs. A function is a special relation where each x-value is related to exactly one corresponding y-value, or where x-values are not repeated. For instance, a set A=$\{(-1,-2),(0,-1),(2,1),(4,3)\}$ is a relation because set A is a set of ordered pairs. Also, set A is a function because the x-values, -1, 0, 2, and 4, are not repeated. The set of the x-values and the y-values in set A are called the **domain** and the **range** of the function, respectively. Thus, the domain of the function is $\{-1,0,2,4\}$ and the range of the function is $\{-2,-1,1,3\}$.

If set $B=\{(1,3),(2,5),(1,2),(3,5)\}$, set B is a relation, but not a function because the x-value, 1, is repeated twice.

Function Notation and Value of Function

Functions are often denoted by f or g. The value of function f at x is denoted by $f(x)$. In order to evaluate $f(2)$, the value of function f at $x=2$, substitute 2 for x. For instance, if $f(x)=x^2+1$, $f(2)=2^2+1=5$.

1. When substituting a negative numerical value in a function, make sure to use parentheses to avoid a mistake. If $f(x)=x^2+1$, $f(-2)=(-2)^2+1=5$.

2. To evaluate $f(x+1)$ when $f(x)=x^2+1$, substitute $x+1$ for x in $f(x)$.

$$f(x+1)=(x+1)^2+1=x^2+2x+2$$

Vertical Line Test

The vertical line test is a graphical way to determine if a graph represents a function or not. If all vertical lines intersect the graph at most one point, the graph in Figure 1 represents a function. Otherwise, the graph does not represent a function as shown in Figure 2.

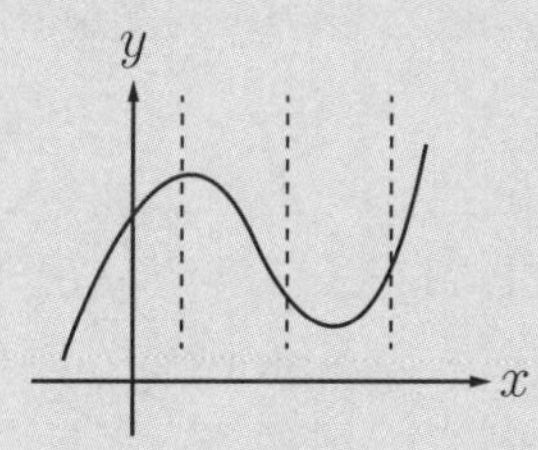

Figure 1

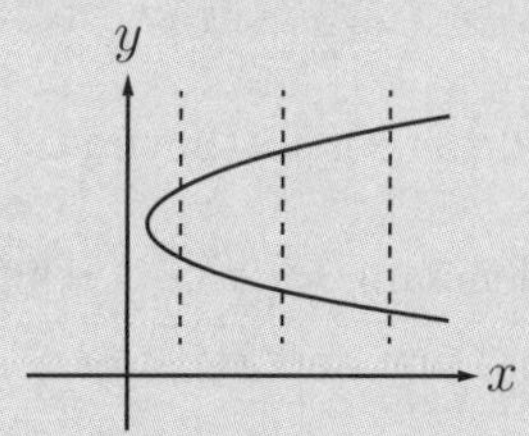

Figure 2

Operations on Functions

Functions can be added, subtracted, multiplied, divided, or combined to create a new function. Below are types of operations on functions. Let $f(x) = x^2 + 1$ and $g(x) = x - 3$.

- The sum of $f(x)$ and $g(x)$.

$$f(x) + g(x) = x^2 + 1 + x - 3 = x^2 + x - 2$$

- The difference of $f(x)$ and $g(x)$.

$$f(x) - g(x) = x^2 + 1 - (x - 3) = x^2 - x + 4$$

- The product of $f(x)$ and $g(x)$: use the distributive property.

$$f(x) \cdot g(x) = (x^2 + 1)(x - 3) = x^3 - 3x^2 + x - 3$$

- The quotient of $f(x)$ and $g(x)$.

$$\frac{f(x)}{g(x)} = \frac{x^2 + 1}{x - 3}, \text{ where } x - 3 \neq 0$$

- The composition function: $(f \circ g)(x)$ or $f\big(g(x)\big)$.

$$\begin{aligned}(f \circ g)(x) &= f\big(g(x)\big) = f(x - 3)\\ &= (x - 3)^2 + 1 = x^2 - 6x + 9 + 1\\ &= x^2 - 6x + 10\end{aligned}$$

1. The quotient of $f(x)$ and $g(x)$, or $\frac{f(x)}{g(x)} = \frac{x^2+1}{x-3}$ is a rational function. In order for the rational function to be defined, the denominator, $g(x) = x - 3$, cannot be zero. Otherwise, it will be undefined. The rational function $\frac{x^2+1}{x-3}$ has a vertical asymptote at $x = 3$.
2. $(f \circ g)(x)$ or $f\big(g(x)\big)$ is read as f composed with g. This operation substitutes $g(x)$ for x in $f(x)$.

Example 1 Finding vertical asymptotes

If $f(x) = \dfrac{x - 3}{x^2 - 4}$, find the vertical asymptotes, if any.

Solution Set the denominator $x^2 - 4$ equal to zero and solve for x: $x^2 - 4 = 0$. Thus, $x = \pm 2$. Therefore, the rational function $f(x) = \dfrac{x - 3}{x^2 - 4}$ has two vertical asymptotes at $x = 2$ and $x = -2$.

Example 2 Evaluating a composition function

If $f(x) = x^2 - 3$ and $g(x) = 3x - 2$, evaluate $f\big(g(2)\big)$.

Solution Since $g(2) = 3(2) - 2 = 4$, substitute $g(2) = 4$ for x in $f(x)$.

$$\begin{aligned} f\big(g(2)\big) &= f(4) && \text{Substitute } g(2) = 4 \text{ for } x \text{ in } f(x) \\ &= (4)^2 - 3 \\ &= 13 \end{aligned}$$

Therefore, the value of $f(g(2))$ is 13.

Odd and Even Functions

Let $f(x)$ be a function. $f(x)$ is an odd function when $f(-x) = -f(x)$ for all values of x. Any odd function is symmetric with respect to the origin. Whereas, $f(x)$ is an even function when $f(-x) = f(x)$ for all values of x. Any even function is symmetric with respect to the y-axis.

	Odd functions	Even functions
Definition	$f(-x) = -f(x)$	$f(-x) = f(x)$
Graph	Symmetric with respect to the origin	Symmetric with respect to the y-axis
Example	$x^3,\ \sin x,\ \tan x$	$x^2,\ \cos x$

Tip Not all functions are either odd or even. Some functions are neither.

Example 3 Determining odd and even functions

If $f(x) = x^3 - 2x + 3$, determine whether function $f(x)$ is odd, even, or neither.

Solution Evaluate $f(-x)$ and $-f(x)$.

$$\begin{aligned} f(x) &= x^3 - 2x + 3 \\ f(-x) &= (-x)^3 - 2(-x) + 3 = -x^3 + 2x + 3 \\ -f(x) &= -(x^3 - 2x + 3) = -x^3 + 2x - 3 \end{aligned}$$

Since $f(-x) \neq -f(x)$ and $f(-x) \neq f(x)$, $f(x)$ does not satisfy the definition of an odd function or an even function. Therefore, $f(x)$ is neither.

Piecewise Functions

A piecewise function is defined by multiple functions with different domains. For instance, the piecewise function $f(x)$ below

$$f(x) = \begin{cases} 3x - 3, & x < 2 \\ -2x + 8, & x \geq 2 \end{cases}$$

is defined by two different functions: $g(x) = 3x - 3$, $x < 2$ in Figure 3, and $h(x) = -2x + 8$, $x \geq 2$ in Figure 4.

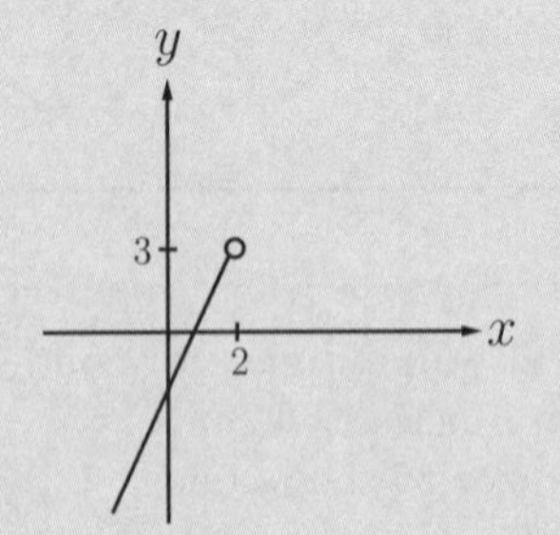

Fig. 3: $g(x) = 3x - 3$, $x < 2$

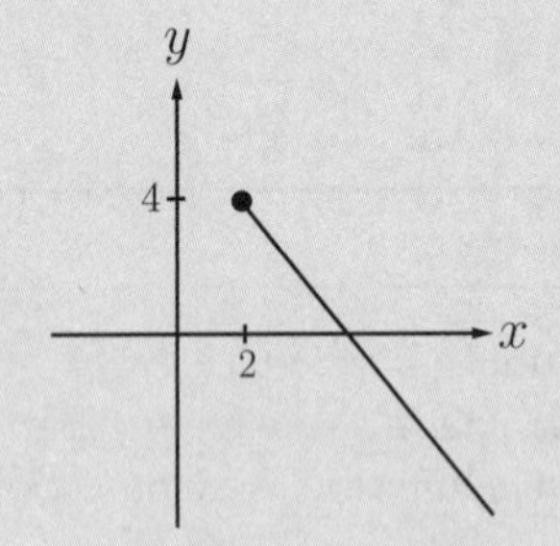

Fig. 4: $h(x) = -2x + 8$, $x \geq 2$

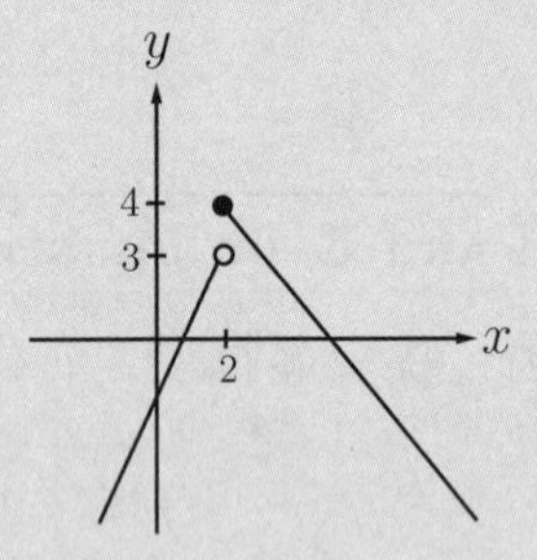

Fig. 5: Piecewise function $f(x)$

As shown in Figure 5, the graph of the piecewise function, $f(x)$, consists of the graphs of $g(x)$ and $h(x)$. The domain and range of the piecewise function $f(x)$ are (∞, ∞) and $(-\infty, 4]$, respectively.

Tip: In order to evaluate the piecewise function at $x = a$, or $f(a)$, check the conditions on the right side of the piecewise function to see where $x = a$ belongs to. For instance, to evaluate $f(1)$, use the equation $3x - 3$ for f because $x = 1$ belongs to $x < 2$. Thus, $f(1) = 3(1) - 3 = 0$.

Example 4 Evaluating piecewise functions

$$f(x) = \begin{cases} x^2 - 3, & x < -1 \\ x + 4, & x \geq -1 \end{cases}$$

For the following function f above, evaluate $f(-2)$ and $f(3)$.

Solution In order to evaluate the piecewise function at $x = -2$ and $x = 3$, check the conditions on the right side of the piecewise function to see where $x = -2$ and $x = 3$ belong. Since $x = -2$ satisfies the condition $x < -1$, the equation for f is $x^2 - 3$. Thus, $f(-2) = (-2)^2 - 3 = 1$. Likewise, $x = 3$ satisfies the condition $x \geq -1$, the equation for f is $x + 4$. Thus, $f(3) = 3 + 4 = 7$.

Inverse Functions

Addition and subtraction can be considered as an example of do and undo processes. For instance, you start with a number, x. If you add 2 to x, then subtract 2 from the result, you will get the original number x as shown below.

$$x \xrightarrow{\text{Add 2}} x+2 \xrightarrow{\text{Subtract 2}} x$$

In mathematics, a function, f, and its inverse function, f^{-1}, are considered as do and undo functions. Thus, when a function, f, is composed with its inverse, f^{-1}, or vice versa, the result is x.

$$f(f^{-1}(x)) = x, \qquad f^{-1}(f(x)) = x$$

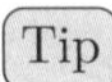

1. The inverse of f is denoted by f^{-1}. Note that the -1 used in f^{-1} is not an exponent. Thus, the inverse function of $f(x)$, $f^{-1}(x)$, is not the reciprocal of $f(x)$. In other words, $f^{-1}(x) \neq \dfrac{1}{f(x)}$. For instance, if $f(x) = x^3 + 5$, $f^{-1}(x) \neq \dfrac{1}{x^3+5}$.
2. The domain of f is the range of f^{-1}. Whereas, the range of f is the domain of f^{-1}.

Horizontal Line Test

The horizontal line test is a graphical way to determine if a function has an inverse function or not. If all horizontal lines intersect the graph at most one point, the function has an inverse function as shown in Figure 6. Otherwise, the function does not have an inverse function as shown in Figure 7.

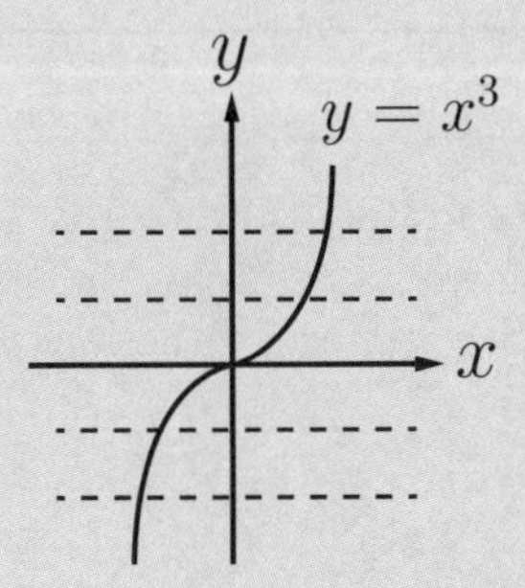

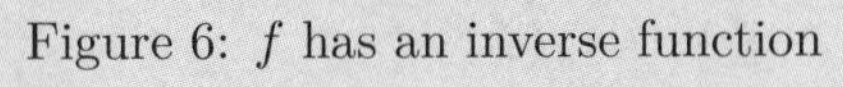
Figure 6: f has an inverse function

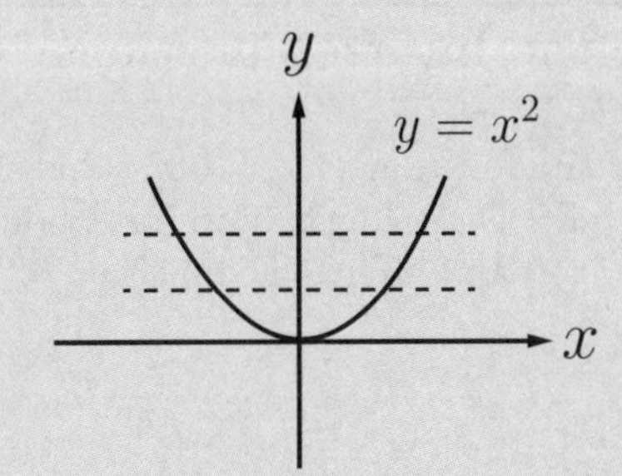

Figure 7: f does not have an inverse function

Tip As shown in Figure 6, if every horizontal line intersects the graph of the function f at most one point, f is said to be **one-to-one**. All one-to-one functions have inverse functions.

Finding Inverse Functions

There are four steps to find the inverse of a function.

Step 1: Replace $f(x)$ with the y variable.

Step 2: Switch the x and y variables.

Step 3: Solve for y.

Step 4: [Optional] Replace the y variable with $f^{-1}(x)$.

Perform the horizontal line test to determine if a function has an inverse function. If the function passes the horizontal line test, follow the four steps shown above to find the inverse function.

Example 5 Finding the inverse function

Find the inverse function of $f(x) = x^3 + 5$.

Solution Replace $f(x)$ with the y variable so that $f(x) = x^3 + 5$ becomes $y = x^3 + 5$. Next, switch the x and y variables and solve for y.

$$\begin{aligned} y &= x^3 + 5 && \text{Switch the } x \text{ and } y \text{ variables} \\ x &= y^3 + 5 && \text{Subtract 5 from each side} \\ x - 5 &= y^3 && \text{Take the cube root of each side} \\ y &= \sqrt[3]{x-5} \end{aligned}$$

Therefore, the inverse function of $f(x) = x^3 + 5$ is $y = \sqrt[3]{x-5}$.

Graphing the Inverse Functions

The graph of the function $f(x) = \sqrt{x} + 2$ is shown in Figure 8. The graph of the inverse function is obtained by reflecting the graph of $f(x)$ about the line $y = x$ as shown in Figure 9. In other words, the graph of $f(x)$ and its inverse function $f^{-1}(x)$ are symmetric with respect to the line $y = x$.

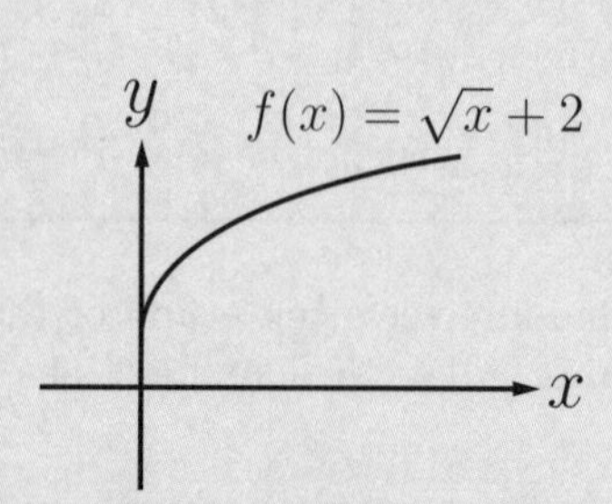

Figure 8: The graph of $f(x)$

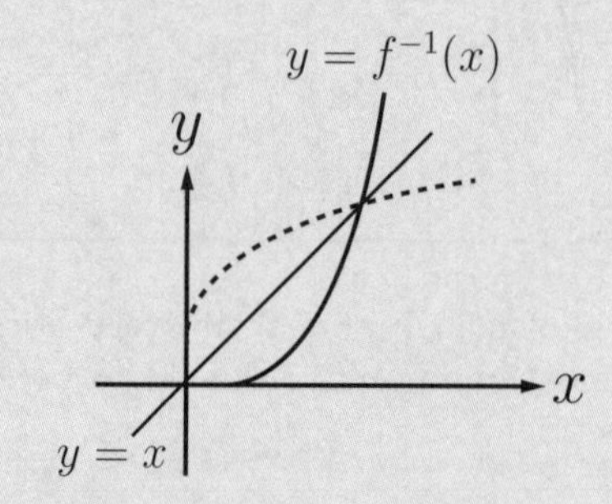

Figure 9: The graph of $f^{-1}(x)$

Example 6 Finding the inverse function

Find the inverse function of $f(x) = \sqrt{x} + 2$ and state the domain and range of the inverse function.

Solution The graph of $f(x) = \sqrt{x} + 2$ shown in Figure 8 suggests that the domain and range of the function are $x \geq 0$ and $y \geq 2$, respectively. Since the domain of a function is the range of the inverse function, and the range of a function is the domain of the inverse function, the domain and range of the inverse function are $x \geq 2$ and $y \geq 0$, respectively. The graph of the inverse function is shown in Figure 9.

In order to find the inverse function algebraically, switch the x and y variables and solve for y.

$y = \sqrt{x} + 2$	Switch the x and y variables
$x = \sqrt{y} + 2$	Subtract 2 from each side
$x - 2 = \sqrt{y}$	Square both sides
$y = (x-2)^2$	Replace y with $f^{-1}(x)$
$f^{-1}(x) = (x-2)^2, \quad x \geq 2$	Since the domain of $f^{-1}(x)$ is $x \geq 2$

Therefore, the inverse function is $f^{-1}(x) = (x-2)^2$, $x \geq 2$.

EXERCISES

1. Which of the following function is an even function?

(A) $y = \cos x$

(B) $y = \tan x$

(C) $y = \sin x$

(D) $y = 2x^3 + 4x + 2$

(E) $y = x^2 - 2x + 1$

2. What is the range of the piecewise function defined below?

$$f(x) = \begin{cases} e^x, & x \geq 0 \\ -x-2, & x < 0 \end{cases}$$

(A) $y > -2$

(B) $y \geq -2$

(C) $y < 1$

(D) $y \leq 1$

(E) $(-\infty, \infty)$

3. What is the inverse function of $y = 2x^3 - 1$?

(A) $y = 2x^3 + 1$

(B) $y = \frac{x+1}{2}$

(C) $y = \sqrt[3]{\frac{x+1}{2}}$

(D) $y = 3\sqrt{2x+1}$

(E) $y = \left(\frac{x+1}{2}\right)^3$

4. At what values of x is $\frac{x+2}{x^2-1}$ undefined?

(A) $x = -2$

(B) $x = 1$ or $x = -1$.

(C) $x = -1$ only

(D) $x = 1$ only

(E) $x = 0$

5. If $f(x) = x^2 - 1$ and $g(x) = \sqrt{x+1}$, what is $g(f(x))$?

(A) 1

(B) x

(C) $-x$

(D) $|x|$

(E) x^2

6. If $f(x) = x^2 + 4x$, what is $f(x-2)$?

(A) $x^2 + 12$

(B) $x^2 - 4$

(C) $x^2 - 8x - 4$

(D) $x^2 - 8x + 12$

(E) $x^2 + 2x - 4$

7. If $f(x) = \frac{x+3}{x-1}$ and $g(x) = \sqrt{x-1}$, what is the value of $f(g(5))$?

(A) 3

(B) 4

(C) 5

(D) 6

(E) 7

8. Which of the following function satisfies $f(x) = f^{-1}(x)$?

(A) $y = x + 1$

(B) $y = 2x$

(C) $y = x^2$

(D) $y = \sqrt{x}$

(E) $y = \frac{1}{x}$

9. If $f(x) = \left(\frac{x+1}{2}\right)^3$, evaluate $f\left(f^{-1}(2)\right)$.

(A) 2

(B) 3

(C) 4

(D) 6

(E) 8

10. What is the range of the inverse function of $f(x) = \sqrt{2x-1}$?

(A) $y < 0$

(B) $y \leq 0$

(C) $y \leq \frac{1}{2}$

(D) $y > \frac{1}{2}$

(E) $y \geq \frac{1}{2}$

ANSWERS AND SOLUTIONS

1. (A)

Tip
1. Definition of odd functions: $f(-x) = -f(x)$
2. Definition of even functions: $f(-x) = f(x)$

$\tan x$ and $\sin x$ are odd functions. As shown below, $f(x) = 2x^3 + 4x + 2$ and $f(x) = x^2 - 2x + 1$ do not satisfy the definition of odd and even functions.

$f(x) = 2x^3 + 4x + 2$	$f(-x) = -2x^3 - 4x + 2$	$-f(x) = -2x^3 - 4x - 2$
$f(x) = x^2 - 2x + 1$	$f(-x) = x^2 + 2x + 1$	$-f(x) = -x^2 + 2x - 1$

Therefore, $f(x) = 2x^3 + 4x + 2$ and $f(x) = x^2 - 2x + 1$ are neither odd nor even functions which leaves (A) $y = \cos x$ as the correct answer.

2. (A)

The graph of the piecewise function is given below.

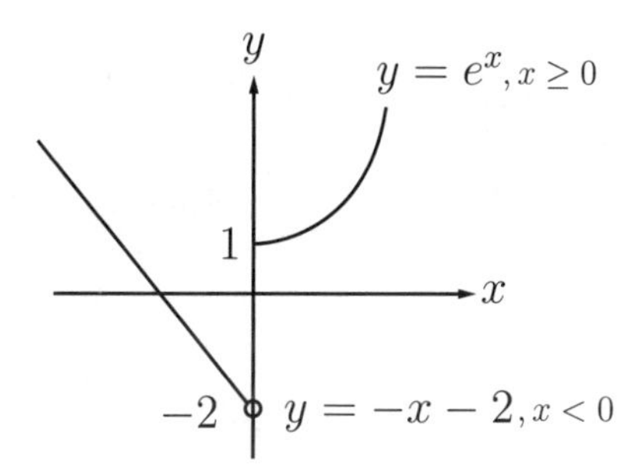

Therefore, the range of the piecewise function is $y > -2$.

3. (C)

Switch the x and y variables and solve for y.

$y = 2x^3 - 1$	Switch the x and y variables
$x = 2y^3 - 1$	Add 1 to each side
$x + 1 = 2y^3$	Divide each side by 2
$\frac{x+1}{2} = y^3$	Take the cube root of each side
$y = \sqrt[3]{\frac{x+1}{2}}$	

Therefore, the inverse function of $f(x) = 2x^3 - 1$ is $y = \sqrt[3]{\frac{x+1}{2}}$.

4. (B)

Set the denominator $x^2 - 1$ equal to zero and solve for x: $x^2 - 1 = 0$. Thus, $x = \pm 1$. Therefore, $\frac{x+1}{x^2-1}$ is undefined at $x = \pm 1$.

5. (D)

> Tip
> $$\sqrt{x^2} = |x| = \begin{cases} x, & x \geq 0 \\ -x, & x < 0 \end{cases}$$

In order to find $g(f(x))$, substitute $x^2 - 1$ for x in $g(x)$.

$g(f(x)) = g(x^2 - 1)$	Substitute $x^2 - 1$ for x in $g(x)$
$= \sqrt{(x^2 - 1) + 1}$	Simplify
$= \sqrt{x^2}$	
$= \lvert x \rvert$	

Thus, $g(f(x)) = |x|$. Therefore, (D) is the correct answer.

6. (B)

In order to find $f(x-2)$, substitute $x - 2$ for x in $x^2 + 4x$.

$f(x) = x^2 + 4x$	Substitute $x - 2$ for x
$f(x-2) = (x-2)^2 + 4(x-2)$	
$= x^2 - 4x + 4 + 4x - 8$	Since $(x-2)^2 = x^2 - 4x + 4$
$= x^2 - 4$	

Therefore, $f(x-2) = x^2 - 4$.

7. (C)

In order to evaluate $g(5)$, substitute 5 for x in $\sqrt{x-1}$. Thus, $g(5) = \sqrt{5-1} = 2$. Likewise, in order to evaluate $f(g(5))$, substitute 2 for x in $\frac{x+3}{x-1}$.

$f(g(5)) = f(2)$	Since $g(5) = 2$
$f(2) = \dfrac{2+3}{2-1}$	Substitute 2 for x in $\dfrac{x+3}{x-1}$
$= 5$	

Therefore, $f(g(5)) = 5$.

8. (E)

> Tip
> $$y = x \qquad\qquad y = \frac{1}{x}$$
> The two functions shown above have inverse functions that are the same as the original functions. This can be denoted by $f(x) = f^{-1}(x)$.

Therefore, (E) is the correct answer.

9. (A)

> **Tip** When a function, f, is composed with its inverse, f^{-1}, or vice versa, the result is x. In other words, cancel out f with f^{-1} so that the result is x.
>
> $$f(f^{-1}(x)) = x, \qquad f^{-1}(f(x)) = x$$

Therefore, $f(f^{-1}(2)) = 2$.

10. (E)

> **Tip**
> 1. The domain of the inverse function is the range of the original function.
> 2. The range of the inverse function is the domain of the original function.

In order to find the domain of $f(x) = \sqrt{2x-1}$, set the expression $2x - 1$ greater than or equal to zero: $2x - 1 \geq 0$. Thus, $x \geq \frac{1}{2}$. Since the range of the inverse function is the domain of the function $f(x) = \sqrt{2x-1}$, the range of the inverse function is $y \geq \frac{1}{2}$. Therefore, (E) is the correct answer.

LESSON 2 Graphs of Parent Functions

Graphs of Parent Functions

A family of functions is a group of functions that all have a similar shape. A parent function is the simplest function and is used as a reference to graph more complicated functions in the family. The table below summarizes the graph, domain, and range of each parent function.

Parent function	Graph
Constant $y = k$ Domain: $(-\infty, \infty)$ Range: $y = k$ 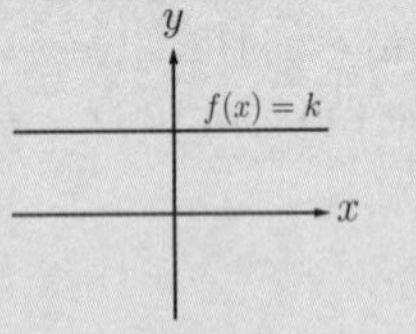 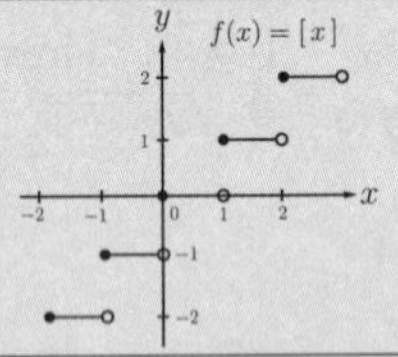	$f(x) = k$
Linear $y = x$ Domain: $(-\infty, \infty)$ Range: $(-\infty, \infty)$ 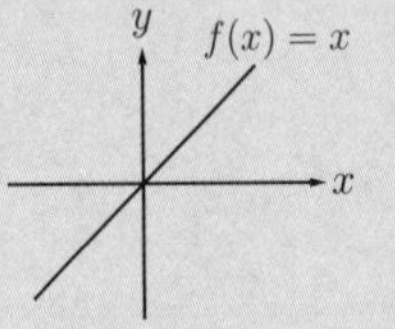 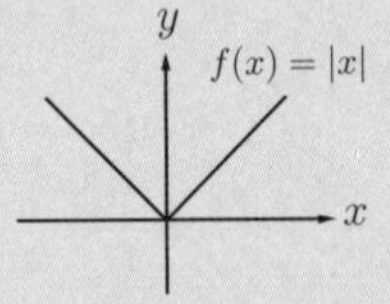	$f(x) = x$
Quadratic $y = x^2$ Domain: $(-\infty, \infty)$ Range: $[0, \infty)$	$f(x) = x^2$
Cubic $y = x^3$ Domain: $(-\infty, \infty)$ Range: $(-\infty, \infty)$ 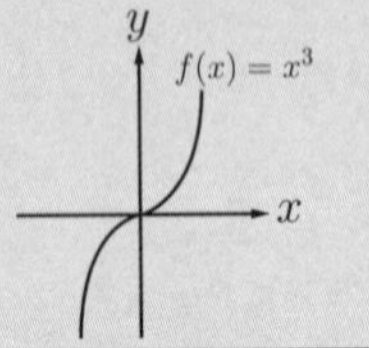 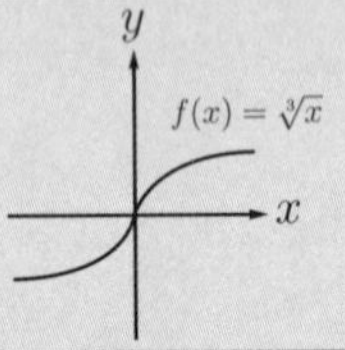	$f(x) = x^3$
Rational 1 $y = \frac{1}{x}$ Domain: $x \neq 0$ Range: $y \neq 0$	$f(x) = \frac{1}{x}$
Exponential $y = a^x$, $a > 1$ Domain: $(-\infty, \infty)$ Range: $(0, \infty)$	$(0, 1)$ $f(x) = a^x$, $a > 1$

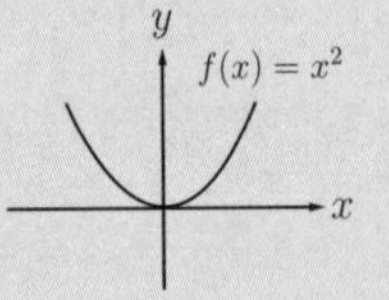

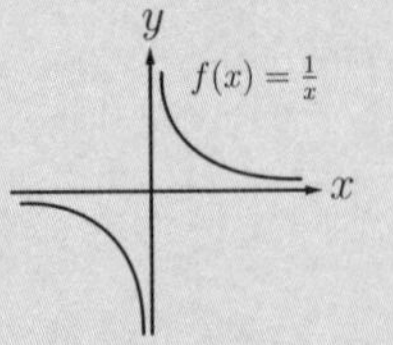

Parent function	Graph
Greatest integer $y = [x]$ Domain: $(-\infty, \infty)$ Range: All integers	$f(x) = [x]$
Absolute value $y = \lvert x \rvert$ Domain: $(-\infty, \infty)$ Range: $[0, \infty)$	$f(x) = \lvert x \rvert$
Square root $y = \sqrt{x}$ Domain: $[0, \infty)$ Range: $[0, \infty)$	$f(x) = \sqrt{x}$
Cube root $y = \sqrt[3]{x}$ Domain: $(-\infty, \infty)$ Range: $(-\infty, \infty)$	$f(x) = \sqrt[3]{x}$
Rational 2 $y = \frac{1}{x^2}$ Domain: $x \neq 0$ Range: $(0, \infty)$	$f(x) = \frac{1}{x^2}$
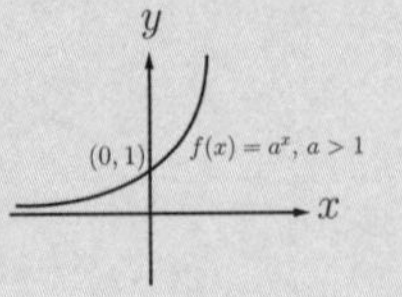 Logarithmic $y = \log_a x$, $a > 1$ Domain: $(0, \infty)$ Range: $(-\infty, \infty)$	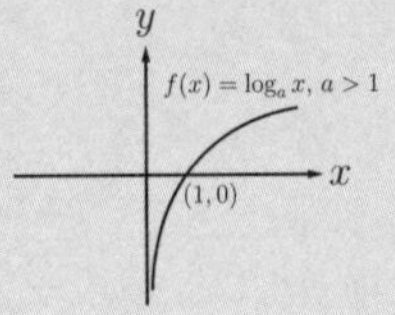 $f(x) = \log_a x$, $a > 1$; $(1, 0)$

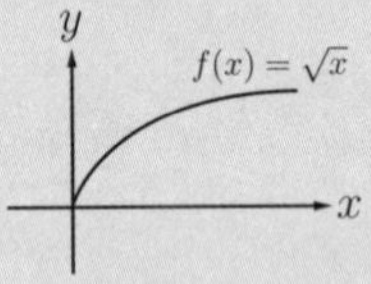

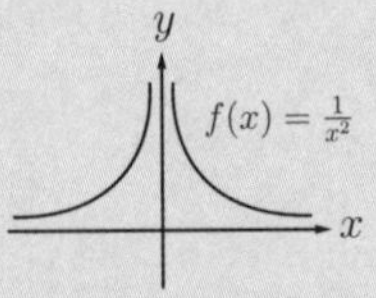

Transformations

The general shape of each parent function can be moved or resized by transformations. For instance, the three functions shown below

$$y = x^2, \qquad y = (x-1)^2, \qquad y = \frac{1}{2}x^2 + 3$$

are in the family of quadratic functions and have the same shape. After moving or resizing the graph of the parent function, $y = x^2$, we can obtain the graphs of $y = (x-1)^2$ and $y = \frac{1}{2}x^2 + 3$.

Transformations consist of horizontal and vertical shifts, horizontal and vertical stretches and compressions, and reflections about the x-axis and y-axis. The table below summarizes the transformations.

Transformation	Function Notation	Effect on the graph of $f(x)$
Horizontal shift	$y = f(x-1)$	Move the graph of $f(x)$ right 1 unit
	$y = f(x+2)$	Move the graph of $f(x)$ left 2 units
Vertical shift	$y = f(x) + 3$	Move the graph of $f(x)$ up 3 units
	$y = f(x) - 4$	Move the graph of $f(x)$ down 4 units
Horizontal stretch and compression	$y = f(2x)$	Horizontal compression of the graph of $f(x)$ by a factor of $\frac{1}{2}$
	$y = f(\frac{1}{3}x)$	Horizontal stretch of the graph of $f(x)$ by a factor of 3
Vertical stretch and compression	$y = 3f(x)$	Vertical stretch of the graph of $f(x)$ by a factor of 3
	$y = \frac{1}{2}f(x)$	Vertical compression of the graph of $f(x)$ by a factor of $\frac{1}{2}$
Reflection about the x-axis, y-axis, and origin	$y = -f(x)$	Reflect the graph of $f(x)$ about the x-axis
	$y = f(-x)$	Reflect the graph of $f(x)$ about the y-axis
	$y = -f(-x)$	Reflect the graph of $f(x)$ about the origin

1. Translation means moving right, left, up, or down. Sometimes, horizontal shifts or vertical shifts are referred to as translations.

2. Horizontal shifts, written in the form $f(x-h)$, do the opposite of what they look like they should do. $f(x-1)$ means to move the graph of $f(x)$ right 1 unit. Whereas, $f(x+2)$ means to move the graph of $f(x)$ left 2 units.

3. Horizontal stretches and compressions, written in the form $f(cx)$, also do the opposite of what they look like they should do. $f(2x)$ means a horizontal compression of the graph of $f(x)$ by a factor of $\frac{1}{2}$. Whereas, $f(\frac{1}{3}x)$ means a horizontal stretch of the graph of $f(x)$ by a factor of 3.

Order of Transformations

In order to graph a function involving more than one transformation, use the order of transformations.

1. Horizontal shift
2. Horizontal and vertical stretch and compression
3. Reflection about x-axis and y-axis
4. Vertical shift

The order of transformations suggests to first perform the horizontal shift. Afterwards, perform the horizontal and vertical stretch and compression. Next, perform the reflection about the x-axis and y-axis. Finally, perform the vertical shift.

Example 1 Graphing the function by transformations

Graph $f(x) = -\sqrt{x-1}+2$ and state the domain and range of the function.

Solution If $f(x) = \sqrt{x}$, $-f(x-1)+2$ in function notation represents $-\sqrt{x-1}+2$. The function notation $-f(x-1)+2$ suggests that the graph of $-\sqrt{x-1}+2$ involves a horizontal shift, reflection about the x-axis, and a vertical shift from the graph of the parent function $y = \sqrt{x}$. Thus, perform the horizontal shift first. Next, perform the reflection about the x-axis, and lastly, perform the vertical shift.

Let's start with the graph of $f(x) = \sqrt{x}$.

$f(x) = \sqrt{x}$	The graph of the parent function as shown in Fig. 1
$f(x-1) = \sqrt{x-1}$	Move the graph of $\sqrt{x}$ right 1 unit as shown in Fig. 2
$-f(x-1) = -\sqrt{x-1}$	Reflect the graph of $\sqrt{x-1}$ about the x-axis as shown in Fig. 3
$-f(x-1)+2 = -\sqrt{x-1}+2$	Move the graph of $-\sqrt{x-1}$ up 2 units as shown in Fig. 4

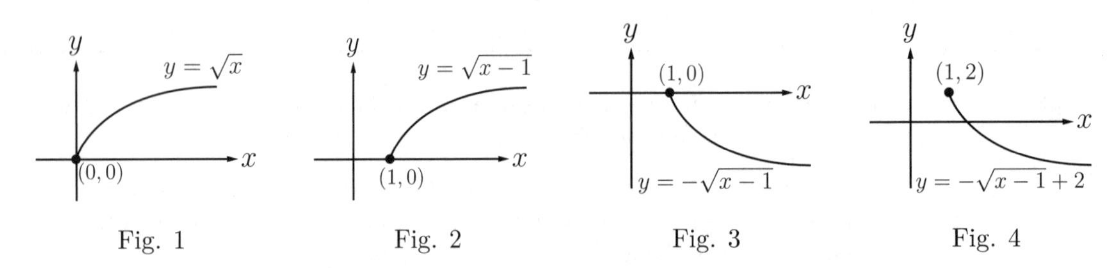

Fig. 1 Fig. 2 Fig. 3 Fig. 4

As shown in Fig. 4, the domain of the function is $[1, \infty)$, and the range of the function is $(-\infty, 2]$.

Graph of $y = |f(x)|$

In order to graph the function $y = |f(x)|$, reflect the part of the graph of $y = f(x)$ that lies below the x-axis about the x-axis. For instance, in order to graph $y = |x^2 - 3|$, start with the graph of $y = x^2 - 3$ as shown in Figure 5. Determine the part of the graph that lies below the x-axis as shown in Figure 6. Lastly, reflect the part of the graph that lies below the x-axis about the x-axis as shown in Figure 7.

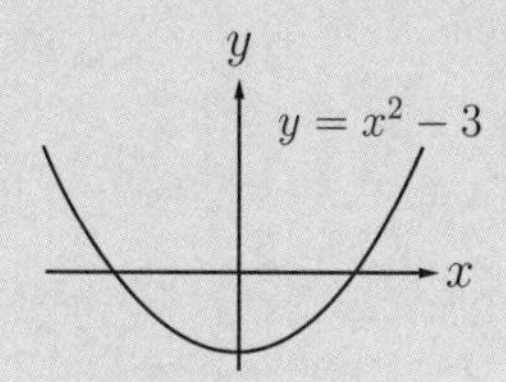

Fig 5: $y = x^2 - 3$

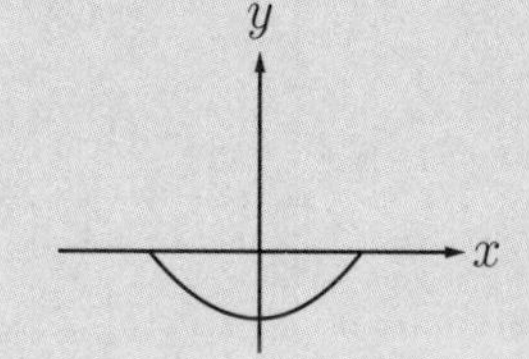

Fig 6: Part below the x-axis

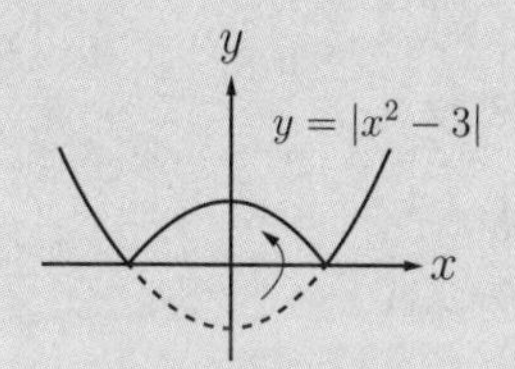

Fig 7: $y = |x^2 - 3|$

Tip In general, the graph of $y = |f(x)|$ is **NOT** a reflection of the entire graph of $y = f(x)$ about the x-axis nor about the y-axis.

Example 2 Graphing $y = |f(x)|$

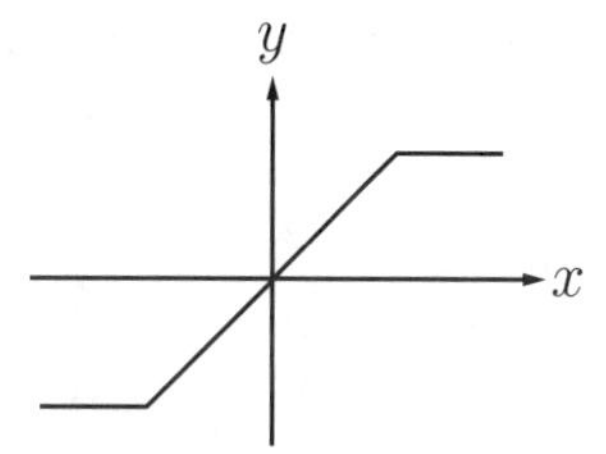

The graph of $y = f(x)$ is given above. Graph $y = |f(x)|$.

Solution The function $f(x)$ is an odd function since the graph of $y = f(x)$ is symmetric with respect to the origin. To graph $y = |f(x)|$, reflect the part of the graph of $y = f(x)$ that lies below the x-axis about the x-axis. The graph of $y = |f(x)|$ is shown below.

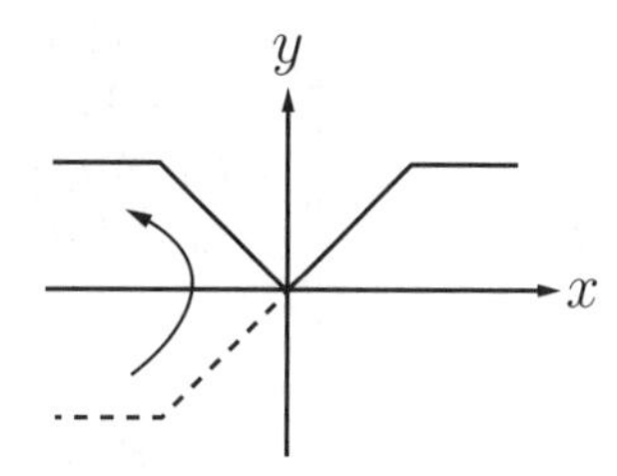

Graph of $y = |f(x)|$

Since the graph of $y = |f(x)|$ is symmetric with respect to the y-axis, the function $y = |f(x)|$ is an even function.

EXERCISES

1. If $f(x) = (x-1)^2 - 4$ and $g(x) = f(x+1)$, what is the vertex of $g(x)$?

   (A) $(2, -3)$

   (B) $(0, -3)$

   (C) $(0, -4)$

   (D) $(-1, -4)$

   (E) $(-2, -4)$

2. Let $f(x) = x^3 + 1$. If $g(x)$ is obtained by translating $f(x)$ 2 units to the right and 3 units up, what is $g(x)$?

   (A) $g(x) = (x-2)^3 + 3$

   (B) $g(x) = (x-2)^3 + 4$

   (C) $g(x) = (x-3)^2 + 3$

   (D) $g(x) = (x+2)^3 + 3$

   (E) $g(x) = (x+2)^3 + 4$

3. If the line $y = 2x - 3$ is reflected about the x-axis, what is the equation of the new line?

   (A) $y = 3x - 2$

   (B) $y = 3x + 2$

   (C) $y = 2x + 3$

   (D) $y = -2x - 3$

   (E) $y = -2x + 3$

4. What is the range of $y = \sqrt{x+1} - 2$?

   (A) $y > 1$

   (B) $y < 1$

   (C) $y < -1$

   (D) $y \geq -2$

   (E) $y \leq -2$

5. What is the x-intercept of $y = |x-2| + 4$?

   (A) No x-intercept.

   (B) $x = -2$

   (C) $x = 4$

   (D) $x = 6$

   (E) $x = -2$ or $x = 6$

6. If $f(x) = \frac{1}{x^2-4}$ and $g(x) = f(x-3)$, what are the vertical asymptotes of $g(x)$?

   (A) $x = -5$ and $x = -1$

   (B) $x = -2$ and $x = 2$

   (C) $x = 1$ and $x = 5$

   (D) $y = -5$ and $y = -1$

   (E) $y = 1$ only $y = 5$

7. If point $(2, 4)$ is on the graph of $f(x)$, which of the following point must be on the graph of $f(2x)$?

   (A) $(1, 2)$

   (B) $(1, 4)$

   (C) $(1, 6)$

   (D) $(3, 2)$

   (E) $(3, 6)$

8. The graph of e^x is translated 2 units to the right and 1 unit down. If the resulting graph represents $g(x)$, what is the value of $g(2.2)$?

   (A) -0.18

   (B) 0.22

   (C) 2.32

   (D) 8.03

   (E) 23.53

9. If the graph of $f(x)$ is translated 3 units to the right, reflected about the y-axis, vertically stretched by a factor of 2, and shifted 4 units down, which of the following function represents the resulting graph?

    (A) $f(2x+3)-4$

    (B) $f(\frac{1}{2}x+3)-4$

    (C) $f(\frac{1}{2}x-3)-4$

    (D) $2f(-x+3)-4$

    (E) $2f(-x-3)-4$

10. If $f(x)=|x^2-4|$, $-1\le x\le 3$, what is the range of $f(x)$?

    (A) $0\le y\le 5$

    (B) $0\le y\le 4$

    (C) $3\le y\le 4$

    (D) $3\le y\le 5$

    (E) $-3\le y\le 5$

ANSWERS AND SOLUTIONS

1. (C)

> Tip $f(x+1)$ means moving $f(x)$ 1 unit to the left.

The quadratic function $f(x)=(x-1)^2-4$ has its vertex at $(1,-4)$. Since $f(x+1)$ means moving the graph of $f(x)$, including its vertex, 1 unit to the left, the new x and y coordinates of the vertex is $(0,-4)$. Therefore, (C) is the correct answer.

2. (B)

Since $g(x)$ is obtained by translating $f(x)$ 2 units to the right and 3 units up, $g(x)=f(x-2)+3$.

$$f(x)=x^3+1$$

$$f(x-2)=(x-2)^3+1 \qquad \text{Substitute } x-2 \text{ for } x \text{ in } x^3+1$$

$$f(x-2)+3=(x-2)^3+4$$

Therefore, $g(x)=f(x-2)+3=(x-2)^3+4$.

3. (E)

> Tip $y=-f(x)$ means reflecting the graph of $f(x)$ about the x-axis.

The line $y=2x-3$ is reflected about the x-axis. Thus, the new equation of the line is $y=-(2x-3)$ or $y=-2x+3$. Therefore, (E) is the correct answer.

4. (D)

Let $f(x) = \sqrt{x}$. $f(x+1)-2$ represents $y = \sqrt{x+1}-2$ and suggests that the graph of $y = \sqrt{x+1}-2$ is obtained by translating the graph of $y = \sqrt{x}$ 1 unit to the left and 2 units down as shown below.

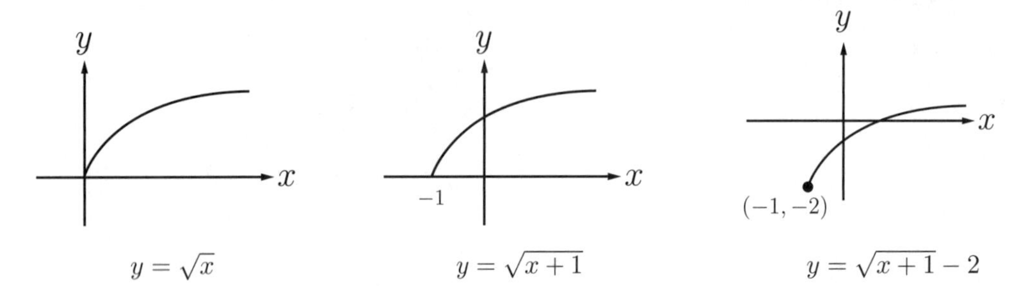

Therefore, the range of $y = \sqrt{x+1} - 2$ is $y \geq -2$.

5. (A)

Let $f(x) = |x|$. $f(x-1)+2$ represents $y = |x-1|+2$ and suggests that the graph of $y = |x-1|+2$ is obtained by translating the graph of $y = |x|$ 1 unit to the right and 2 units up as shown below.

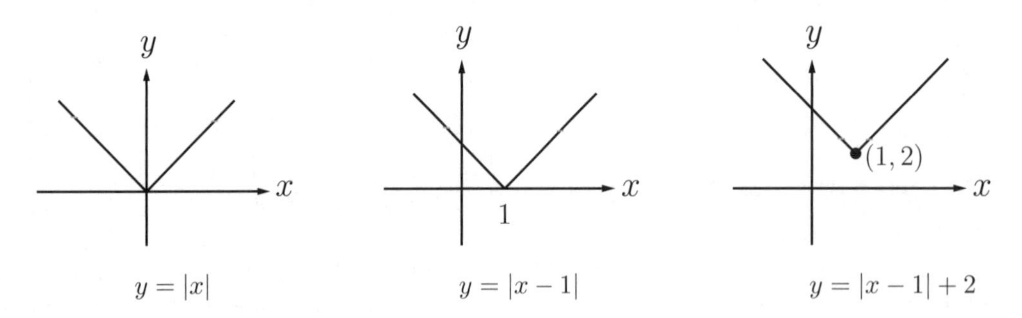

Since the graph of $y = |x-1| + 2$ never touches or crosses the x-axis, there is no x-intercept.

6. (C)

Set the denominator of $f(x)$ equal to zero and solve for x: $x^2 - 4 = 0$. Thus, $x = \pm 2$. Since $g(x) = f(x-3)$, the graph of $g(x)$ is obtained by translating the graph of $f(x)$ including the vertical asymptotes 3 units to the right. Therefore, the vertical asymptotes of $g(x)$ are $x = -2+3 = 1$ and $x = 2+3 = 5$.

7. (B)

> **Tip** $f(2x)$ means the graph, $f(x)$, will be horizontally compressed by a factor of $\frac{1}{2}$. This implies that the x-coordinate of any point on the graph is multiplied by $\frac{1}{2}$ and the y-coordinate of any point on the graph remains the same.

Since $(2, 4)$ is on the graph of $f(x)$, the graph of $f(2x)$ must contain the point $(1, 4)$. Therefore, (B) is the correct answer.

8. (B)

Let $f(x) = e^x$. Since the graph of e^x is translated 2 units to the right and 1 unit down, $g(x)$ can be expressed as $g(x) = f(x-2) - 1 = e^{x-2} - 1$. Thus,

$$g(x) = e^{x-2} - 1 \qquad \text{Substitute 2.2 for } x$$
$$g(2.2) = e^{2.2-2} - 1 = 0.22$$

Therefore, $g(2.2) = 0.22$.

9. (E)

> **Tip** The order of transformations suggests to perform the horizontal shift first. Afterwards, perform the horizontal and vertical stretch and compression. Next, perform the reflection about the x-axis and y-axis. Finally, perform the vertical shift.

1. Horizontal shift of 3 units to the right: $f(x) \Longrightarrow f(x-3)$
2. Reflection about the y-axis: $f(x-3) \Longrightarrow f(-x-3)$
3. Vertical stretch by a factor of 2: $f(-x-3) \Longrightarrow 2f(-x-3)$
4. Vertical shift of 4 units down: $2f(-x-3) \Longrightarrow 2f(-x-3) - 4$

Therefore, $2f(-x-3) - 4$ represents the resulting graph.

10. (A)

The graphs of $y = x^2 - 4$ and $y = |x^2 - 4|$ are shown below.

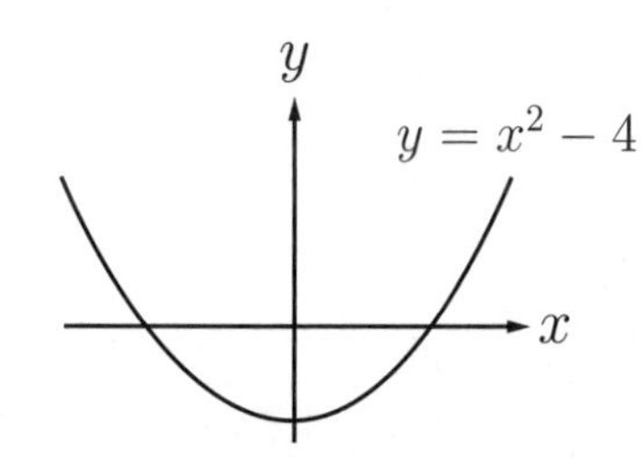

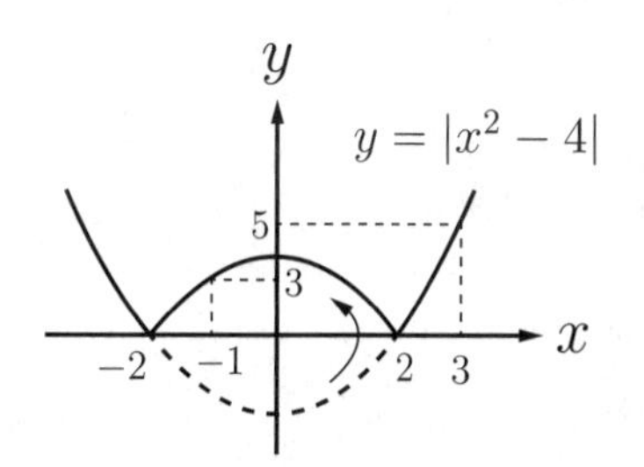

In order to find the range of $y = |x^2 - 4|$, $-1 \le x \le 3$, find the largest and smallest values of y. The largest value of y is 5 when $x = 3$, and the smallest value of y is 0 when $x = 2$. Therefore, the range of $y = |x^2 - 4|$, $-1 \le x \le 3$ is $0 \le y \le 5$.

LESSON 3 Linear and Quadratic Functions

Linear Equations (Linear Functions)

The **slope** of a line, often denoted by m, is a number that describes the steepness of the line. The larger the absolute value of the slope, $|m|$, the steeper the line is (closer to y-axis). If a line passes through the points (x_1, y_1) and (x_2, y_2), the slope m is defined as

$$m = \frac{\text{Rise}}{\text{Run}} = \frac{y_2 - y_1}{x_2 - x_1}$$

If the points (x_1, y_1) and (x_2, y_2) are given, the following formulas are useful in solving graph problems.

Midpoint Formula:	$\left(\frac{x_1 + x_2}{2}, \frac{y_1 + y_2}{2}\right)$
Distance Formula:	$D = \sqrt{(x_2 - x_1)^2 + (y_2 - y_1)^2}$

A linear equation or a linear function represents a line in the xy-plane. A line can be expressed in three different forms.

1. **Slope-intercept form:** $y = mx + b$, where m is slope and b is y-intercept.
2. **Point-slope form:** If the slope of a line is m and the line passes through the point (x_1, y_1),
$$y - y_1 = m(x - x_1)$$
3. **Standard form:** $Ax + By = C$, where A, B, and C are integers.

Below are classifications of the lines by slope:

- Lines that rise from left to right have positive slope.
- Lines that fall from left to right have negative slope.
- Horizontal lines have zero slope (example: $y = 2$).
- Vertical lines have undefined slope (example: $x = 2$).
- Parallel lines have the same slope.
- Perpendicular lines have negative reciprocal slopes (product of the slopes equals -1).

The x-intercept of a line is the point where the line crosses the x-axis. **The y-intercept** of a line is the point where the line crosses the y-axis.

To find the x-intercept of a line	$\Longrightarrow$	Substitute 0 for y and solve for x
To find the y-intercept of a line	$\Longrightarrow$	Substitute 0 for x and solve for y

Solving Systems of Linear Equations

A system means more than one. A linear equation represents a line. Thus, a **system of linear equations** represents more than one line. Below is an example of a system of linear equations.

$$2x - y = 5$$
$$3x + y = 10$$

A solution to a system of linear equations is an ordered pair (x, y) that satisfies all equations in the system. In other words, a solution to a system of linear equation is an intersection point that lies on both lines. In the figure above, $(3, 1)$ is an ordered pair that satisfies both equations,

$$2x - y = 5 \implies 2(3) - 1 = 5$$
$$3x + y = 10 \implies 3(3) + 1 = 10$$

and is the intersection point of both lines.

Solving a system of linear equations means finding the x and y coordinates of the intersection point of the lines. There are two methods to solve a system of linear equations: **substitution** and **linear combinations.**

1. Substitution method
In the example above, write y in terms of x in the first equation. $2x - y = 5 \implies y = 2x - 5$. Substitute $2x - 5$ for y in the second equation.

$$3x + y = 10 \implies 3x + (2x - 5) = 10$$
$$5x - 5 = 10$$
$$x = 3 \implies y = 2x - 5 = 2(3) - 5 = 1$$

The solution to the system using the substitute method is $(3, 1)$.

2. Linear combinations method
In the example above, the coefficient of the y variable in each equation is the opposite. Thus, adding the two equations eliminates the y variables. Then, solve for x.

$$2x - y = 15$$
$$3x + y = 10 \quad \text{Add two equations}$$
$$5x \quad = 15$$
$$x = 3$$

Substitute 3 for x in the first equation and solve for y. Thus, $y = 5$. Therefore, the solution to the system using the linear combinations method is $(3, 1)$.

Number of Solutions in a System of Linear Equations

The number of solutions in a system of linear equations is equal to the number of intersection points that lie on both equations. The table below summarizes the relationship between the graph of a system of linear equations and the number of solutions.

Graph	Description	Number of intersection points	Number of solutions
y, x, $2x-y=5$, $(3,1)$, $3x+y=10$	Two lines intersect in one point.	1	One solution
y, x, $y=x+3$, $y=x-2$	Two lines are parallel. There is no intersection point.	0	No solutions
y, x, $2x-y=5$, $4x-2y=10$	Two lines are the same line.	Infinitely many	Infinite solutions

Tip

$$ax+by=c$$
$$dx+ey=f$$

If the system of linear equations is given above, algebraically, the number of solutions can be determined by the following rules.

Case 1: If $\dfrac{a}{d} \neq \dfrac{b}{e}$, the two lines intersect in one point. Thus, there is one solution.

Case 2: If $\dfrac{a}{d} = \dfrac{b}{e} \neq \dfrac{c}{f}$, the two lines are parallel. Thus, there are no solutions.

Case 3: If $\dfrac{a}{d} = \dfrac{b}{e} = \dfrac{c}{f}$, the two lines are the same line. Thus, there are an infinite number of solutions.

Quadratic Functions and Quadratic Equations

A quadratic function is a polynomial function of degree 2 and can be expressed in three forms.

Standard form:	$y = ax^2 + bx + c$	
Vertex form:	$y = a(x-h)^2 + k,$	Vertex (h, k)
Factored form:	$y = a(x-p)(x-q),$	where p and q are x-intercepts.

The graph of a quadratic function is a parabola. Depending on the value of the **leading coefficient**, $a\,(a \neq 0)$, the graph of a quadratic function either opens up or opens down.

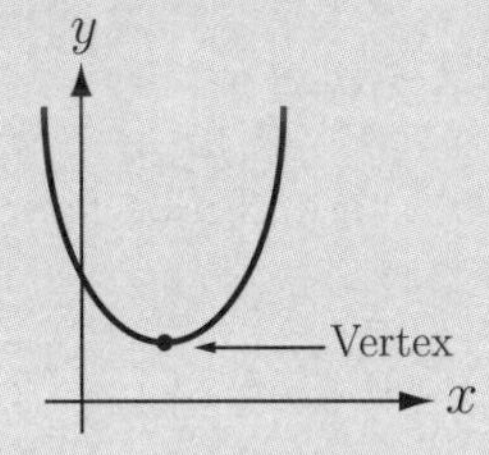

Figure 1: If $a > 0$, opens up

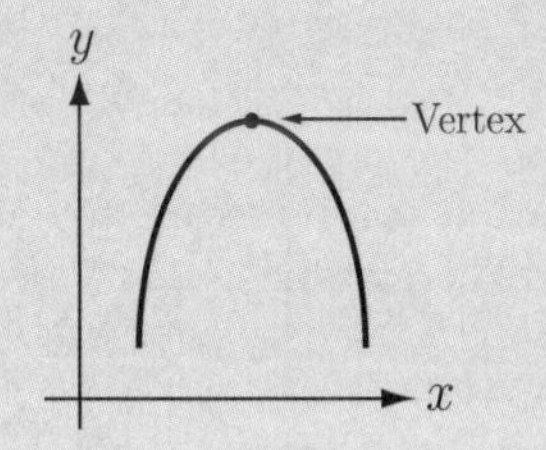

Figure 2: If $a < 0$, opens down

The vertex is the minimum point when the parabola opens up as shown in Figure 1 and the maximum point when the parabola opens down as shown in Figure 2.

Below are ways to find the x-coordinate of the vertex from each form.

$$\text{Standard form: } y = ax^2 + bx + c \implies x = -\frac{b}{2a}$$

$$\text{Vertex form: } y = a(x-h)^2 + k \implies x = h$$

$$\text{Factored form: } y = a(x-p)(x-q) \implies x = \frac{p+q}{2}$$

To evaluate the y-coordinate of the vertex, substitute the value of the x-coordinate in each form.

The x-intercept and y-intercept of quadratic functions

- To solve for the x-intercept: substitute 0 for y. Then, quadratic functions become quadratic equations as shown below.

$$\text{Standard form: } y = ax^2 + bx + c \implies ax^2 + bx + c = 0$$

$$\text{Vertex form: } y = a(x-h)^2 + k \implies a(x-h)^2 + k = 0$$

$$\text{Factored form: } y = a(x-p)(x-q) \implies a(x-p)(x-q) = 0$$

- To solve for the y-intercept: substitute 0 for x and evaluate the y-intercept.

Solving Quadratic Equations

Solving a quadratic equation is finding the x-intercept(s) of the quadratic function. There are two common methods to solve a quadratic equation: **Factoring** and **Quadratic formula**.

Factoring
Factoring is an important tool that is required for solving a quadratic equation. Factoring is the opposite of expanding. Factoring a quadratic expression is to write the expression as a product of two linear terms. Below is an example.

$$(x-2)(x-3) \xrightarrow{\text{Expanding}} x^2-5x+6$$
$$x^2-5x+6 \xrightarrow{\text{Factoring}} (x-2)(x-3)$$

If a quadratic equation can be expressed as $x^2+(p+q)x+pq=0$, it can be factored as $(x+p)(x+q)=0$. For instance,

$$x^2-5x+6=0 \implies x^2+(-2+-3)x+(-2)(-3)=0 \implies (x-2)(x-3)=0$$

Once a quadratic equation is written in a factored form, use the **zero product property** to solve the equation.

Zero product property: If $ab=0$, then $a=0$ or $b=0$

Thus, the solutions to $(x-2)(x-3)=0$ is

$$(x-2)(x-3)=0 \implies (x-2)=0 \text{ or } (x-3)=0 \implies x=2 \text{ or } x=3$$

The Quadratic Formula
The quadratic formula is a general formula for solving quadratic equations. The solutions to the quadratic equation $ax^2+bx+c=0$ are as follows:

$$x=\frac{-b\pm\sqrt{b^2-4ac}}{2a}$$

Example 1 Finding solutions of a quadratic equation

Solve the equation: $x^2-x-1=0$

Solution Since the quadratic equation cannot be factored, use the quadratic formula to solve the equation.

$$\begin{aligned} x &= \frac{-b\pm\sqrt{b^2-4ac}}{2a} \\ &= \frac{-(-1)\pm\sqrt{(-1)^2-4(1)(-1)}}{2(1)} \\ &= \frac{1\pm\sqrt{5}}{2} \end{aligned}$$

Substitute 1 for a, -1 for b, and -1 for c

Solving Quadratic Inequalities

Solving a quadratic inequality means finding the x-values for which the graph of a quadratic function lies above or below the x-axis. A quadratic inequality can be solved algebraically. However, solving a quadratic inequality **graphically** is highly recommended.

In order to solve $ax^2 + bx + c > 0$ (or $ax^2 + bx + c \geq 0$) graphically,

Step 1 Find the x-intercepts of $y = ax^2 + bx + c$: Let $y = 0$ and solve for x using factoring or the quadratic formula.

Step 2 Graph $y = ax^2 + bx + c$.

Step 3 From the graph in step 2, find the x-values for which the graph lies **above** (or on and above) the x-axis.

In order to solve $ax^2 + bx + c < 0$ (or $ax^2 + bx + c \leq 0$) graphically,

Step 1 Find the x-intercepts of $y = ax^2 + bx + c$: Let $y = 0$ and solve for x using factoring or the quadratic formula.

Step 2 Graph $y = ax^2 + bx + c$.

Step 3 From the graph in step 2, find the x-values for which the graph lies **below** (or on and below) the x-axis.

Example 2 Solving a quadratic inequality graphically

Solve $x^2 - 5x + 6 \geq 0$

Solution Substitute 0 for y in $y = x^2 - 5x + 6$ and solve $x^2 - 5x + 6 = 0$ using factoring.

$$(x-2)(x-3) = 0 \implies (x-2) = 0 \text{ or } (x-3) = 0 \implies x = 2 \text{ or } x = 3$$

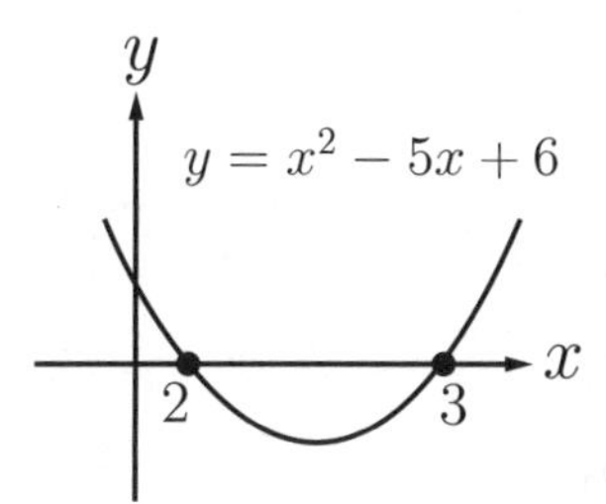

Figure 3

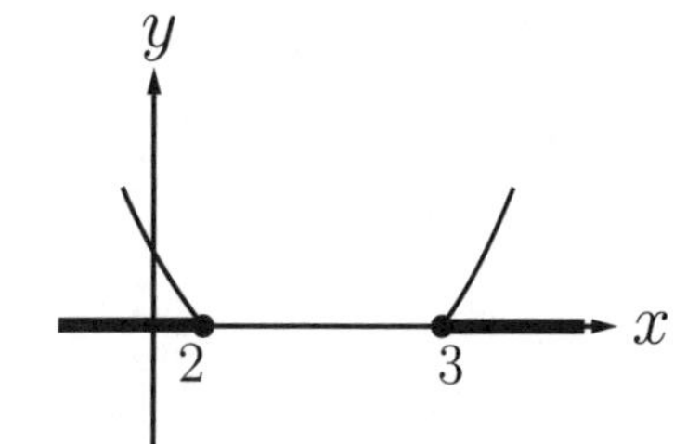

Figure 4

As shown in Figure 3, the x-intercepts of $y = x^2 - 5x + 6$ are 2, and 3. Since the graph lies on and above the x-axis to the left of $x = 2$ and to the right of $x = 3$ as shown in Figure 4, the solution to $x^2 - 5x + 6 \geq 0$ is $x \leq 2$ or $x \geq 3$.

EXERCISES

1. If the slope of a line that passes through the points $(2, 3)$ and $(6, y)$ is $\frac{1}{2}$, which of the following is the value of y ?

   (A) 5

   (B) 4

   (C) 3

   (D) 2

   (E) 1

2. What is the maximum value of the quadratic function $y = -(x+2)^2 + 7$?

   (A) -7

   (B) -5

   (C) 2

   (D) 5

   (E) 7

3. If the line $y = mx - 3$ passes through quadrants II, III, and IV, what is a possible value of m ?

   (A) $m = 0$

   (B) $m < 0$

   (C) $m \leq 0$

   (D) $m > 0$

   (E) $m \geq 0$

4. If 5 is a solution to $y = x^2 - 3x + b$, what is the value of the other solution?

   (A) -2

   (B) -1

   (C) 1

   (D) 2

   (E) 3

5. If the number of solutions to the system of linear equations given below is zero, what is the value of a ?

$$2x + ay = 4$$
$$6x + 9y = -1$$

   (A) 1

   (B) 2

   (C) 3

   (D) 6

   (E) 9

6. What are the x-intercepts of the quadratic function $y = x^2 - x - 1$?

   (A) $x = -2.37$ or $x = 3.53$

   (B) $x = -1.67$ or $x = 2.67$

   (C) $x = -1.35$ or $x = 0.79$

   (D) $x = -0.62$ or $x = 1.62$

   (E) $x = -0.38$ or $x = 3.71$

7. What is the vertex of the quadratic function $y = 2(x+2)(x-4)$?

(A) $(-2, -16)$

(B) $(-1, 18)$

(C) $(0, -16)$

(D) $(0, -8)$

(E) $(1, -18)$

8. Which of the following equation is the perpendicular bisector of the line segment connected by $(0, 2)$ and $(4, 0)$?

(A) $y = -\frac{1}{2}x + 2$

(B) $y = \frac{1}{2}x$

(C) $y = -2x + 5$

(D) $y = 2x + 3$

(E) $y = 2x - 3$

9. If the area of a square with side length x is equal to the perimeter of an equilateral triangle with side length $x + 6$, what is the area of the equilateral triangle?

(A) 6

(B) $18\sqrt{2}$

(C) 36

(D) $36\sqrt{3}$

(E) 72

10. Solve: $2x^2 - 5x + 2 < 0$

(A) $-2 < x < -\frac{1}{2}$

(B) $\frac{1}{2} < x < 2$

(C) $x < -2$ or $x > -\frac{1}{2}$

(D) $x < -\frac{1}{2}$ or $x > 2$

(E) $x < \frac{1}{2}$ or $x > 2$

ANSWERS AND SOLUTIONS

1. (A)

The slope of the line that passes through $(2, 3)$ and $(6, y)$ is $\frac{1}{2}$.

$$\text{Slope} = \frac{y_2 - y_1}{x_2 - x_1}$$

$$\frac{1}{2} = \frac{y-3}{6-2} \qquad \text{Cross multiply}$$

$$2(y-3) = 4 \qquad \text{Solve for } y$$

$$y = 5$$

Therefore, the value of y is 5.

2. (E)

The vertex of the quadratic function is $(-2, 7)$. This is the maximum point because the quadratic function opens down $(a < 0)$. The maximum value of the quadratic function is same as the y-coordinate of the vertex. Therefore, the maximum value of the quadratic function is 7.

3. (B)

Since the line $y = mx - 3$ passes through quadrants II, III, and IV as shown below,

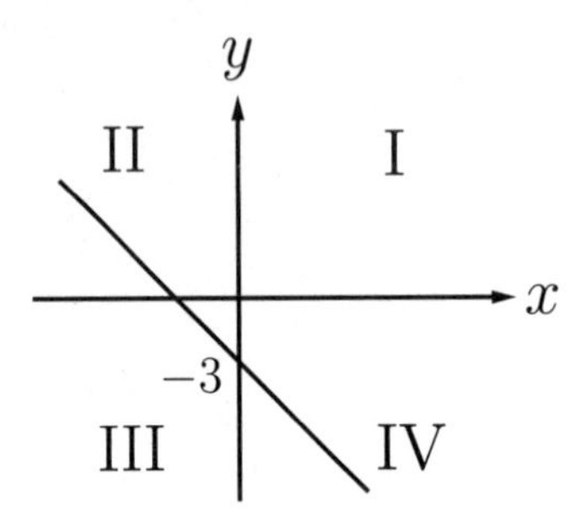

the slope of the line, m, must be negative, or, $m < 0$. Therefore, (B) is the correct answer.

4. (A)

Since 5, or the ordered pair $(5, 0)$, is a solution to the quadratic equation, substitute 5 for x and 0 for y in the equation and solve for b.

$x^2 - 3x + b = y$	Substitute 5 for x and 0 for y
$(5)^2 - 3(5) + b = 0$	Solve for b
$b = -10$	

$x^2 - 3x + b = 0$ becomes $x^2 - 3x - 10 = 0$. Use the factoring method to solve for the other solution.

$x^2 - 3x - 10 = 0$	Use the factoring method
$(x + 2)(x - 5) = 0$	Use the zero product property
$x = -2$ or $x = 5$	

Therefore, the other solution to the quadratic equation is -2.

5. (C)

Tip

In the system of linear equations given below,

$$ax + by = c$$
$$dx + ey = f$$

the number of solutions to the system is zero if $\frac{a}{d} = \frac{b}{e} \neq \frac{c}{f}$.

The number of solutions to the system of linear equations below is zero if $\frac{2}{6} = \frac{a}{9}$. Thus, $a = 3$.

$$2x + ay = 4$$
$$6x + 9y = -1$$

Therefore, (C) is the correct answer.

6. (D)

Tip

Use the quadratic formula to find the solutions to the quadratic equation $ax^2 + bx + c = 0$.

$$\text{Quadratic formula:} \quad x = \frac{-b \pm \sqrt{b^2 - 4ac}}{2a}$$

In order to find the x-intercepts, substitute 0 for y and solve the quadratic equation $x^2 - x - 1 = 0$. Since $x^2 - x - 1$ is not factorable, use the quadratic formula.

$$x = \frac{-b \pm \sqrt{b^2 - 4ac}}{2a} = \frac{-(-1) \pm \sqrt{(-1)^2 - 4(1)(-1)}}{2} = \frac{1 \pm \sqrt{5}}{2}$$

Therefore, the solutions to $x^2 - x - 1 = 0$ are $x = \frac{1+\sqrt{5}}{2} = 1.62$ or $x = \frac{1-\sqrt{5}}{2} = -0.62$.

7. (E)

The quadratic function $y = 2(x+2)(x-4)$ is expressed in factored form. The x-coordinate of the vertex is the mean of the x-intercepts of the quadratic function. Since the x-intercepts are -2 and 4, the x-coordinate of the vertex is $x = \frac{-2+4}{2} = 1$. To find the y-intercept, substitute 1 for x in $y = 2(x+2)(x-4)$. Thus, the y-intercept is $2(1+2)(1-4) = -18$. Therefore, the vertex of the quadratic function is $(1, -18)$.

8. (E)

The slope of the line segment connected by $(0,2)$ and $(4,0)$ is:

$$\text{Slope} = \frac{0-2}{4-0} = -\frac{1}{2}$$

The midpoint between $(0,2)$ and $(4,0)$ is:

$$\text{Midpoint} = \left(\frac{0+4}{2}, \frac{2+0}{2}\right) = (2,1)$$

The slope of the perpendicular bisector is the negative reciprocal of $-\frac{1}{2}$, or 2. The equation of the perpendicular bisector in slope-intercept form is $y = 2x + b$. Since the perpendicular bisector passes through the midpoint of the line segment, $(2,1)$ is the solution to the equation $y = 2x + b$.

$$\begin{aligned} y &= 2x + b && \text{Substitute 2 for } x \text{ and 1 for } y \\ 1 &= 2(2) + b \\ b &= -3 \end{aligned}$$

Therefore, the equation of the perpendicular bisector is $y = 2x - 3$.

9. (D)

> Tip The area of an equilateral triangle with side length s is $\frac{\sqrt{3}}{4}s^2$

The area of the square with side length x is x^2. The perimeter of the equilateral triangle with side length $x+6$ is $3(x+6)$. Set the area and the perimeter equal to each other and solve for x. Choose the positive value of x since the length of the square is positive.

$$\begin{aligned}(x^2 &= 3(x+6)\\ x^2 &= 3x+18\\ x^2-3x-18 &= 0\\ (x+3)(x-6) &= 0\\ x = -3 \quad &\text{or} \quad x = 6\end{aligned}$$

Since $x = 6$, the side length of the equilateral triangle is $x + 6 = 12$. Therefore, the area of the equilateral triangle is $\frac{\sqrt{3}}{4}s^2 = \frac{\sqrt{3}}{4}(12)^2 = 36\sqrt{3}$.

10. (B)

Substitute 0 for y in $y = 2x^2 - 5x + 2$ and solve $2x^2 - 5x + 2 = 0$ by factoring.

$$2x^2 - 5x + 2 = 0 \implies (2x-1)(x-2) = 0 \implies x = \frac{1}{2} \quad \text{or} \quad x = 2$$

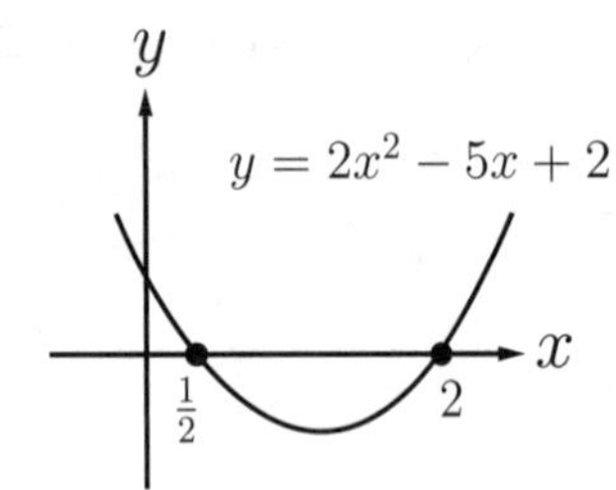

Figure 5

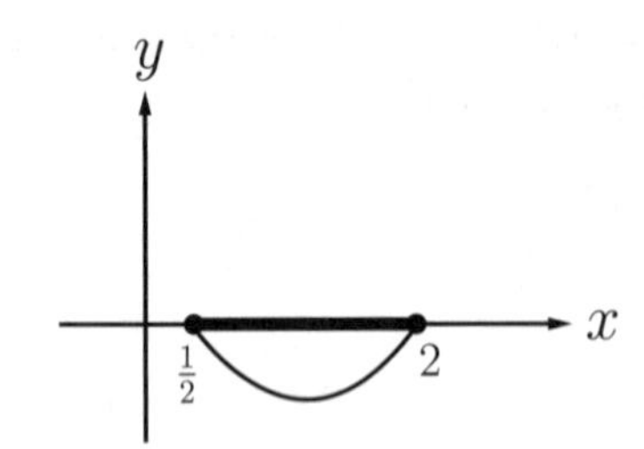

Figure 6

As shown in Figure 5, the x-intercepts of $y = 2x^2 - 5x + 2$ are $\frac{1}{2}$ and 2. Since the graph lies below the x-axis when $\frac{1}{2} < x < 2$ in Figure 6, the solution to $2x^2 - 5x + 2 < 0$ is $\frac{1}{2} < x < 2$.

Complex Numbers

Definition of Complex Numbers

The equation $x^2 = -1$ does not have real solutions because the square of a real number x is either 0 or positive, never negative. In order to provide a solution to $x^2 = -1$, complex numbers are created.

A complex number is written in the form $a + bi$, where a and b are real numbers, and i is an imaginary unit which satisfies $i^2 = -1$. The number a is the real part of the complex number, and the number bi is the imaginary part of the complex number.

1. The imaginary unit, denoted as i, is defined as $i = \sqrt{-1}$. Thus, $i^2 = -1$.
2. The property of the square root of a negative number is as follows:

$$\sqrt{-3} = \sqrt{-1 \cdot 3} = \sqrt{-1}\sqrt{3} = i\sqrt{3}$$

Example 1 Solving a quadratic equation

Solve $(x+1)^2 = -3$

Solution The equation $(x+1)^2 = -3$ means that the square of a number $x+1$ is equal to -3. Since the square of a real number cannot be negative, the solutions to $(x+1)^2 = -3$ involve complex numbers.

$$\begin{aligned} (x+1)^2 &= -1 \cdot 3 && \text{Since } i^2 = -1 \\ (x+1)^2 &= 3i^2 && \text{Take the square root of both sides} \\ x+1 &= \pm i\sqrt{3} && \text{Subtract 1 from each side} \\ x &= -1 \pm i\sqrt{3} \end{aligned}$$

Therefore, the solutions to $(x+1)^2 = -3$ is $x = -1 + i\sqrt{3}$ or $x = -1 - i\sqrt{3}$.

Powers of i

The table below shows the powers of i, which repeat in a pattern: i, -1, $-i$, and 1.

Powers of i	i	i^2	i^3	i^4	i^5	i^6	i^7	i^8	$\cdots$	i^{12}	$\cdots$	i^{4n}
Value	i	-1	$-i$	1	i	-1	$-i$	1	$\cdots$	1	$\cdots$	1

1. $i^3 = i^2 \cdot i = -1 \cdot i = -i$. $\qquad i^7 = i^4 \cdot i^3 = 1 \cdot i^3 = 1 \cdot -i = -i$.
2. $i^{4n} = 1$ means that if the power of i is a multiple of 4, the value is always equal to 1. For instance, $i^4 = 1$, $i^8 = 1$, $i^{12} = 1$, and $i^{1000} = 1$.

Example 2 Evaluating powers of i

Evaluate i^{123}.

Solution The closest multiple of 4 which is smaller than 123 is 120. Since $i^{4n} = i^{120} = 1$,

$$i^{123} = i^{120} \cdot i^3 = 1 \cdot i^3 = -i$$

Therefore, the value of i^{123} is $-i$.

Absolute Value of Complex Numbers

A complex number can be plotted in the complex plane, where the horizontal axis is the real axis and the vertical axis is the imaginary axis. Two complex numbers, $z_1 = 3 + 4i$, and $z_2 = -2 - 3i$, are plotted in the complex plane as shown in Figure 1.

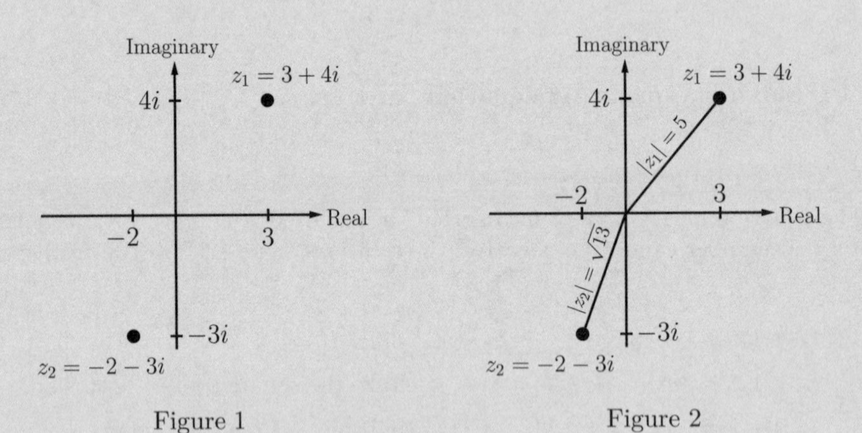

Figure 1 Figure 2

The absolute value of a complex number $a + bi$, denoted by $|z|$, is the distance from the origin to the complex number in the complex plane. The formula for finding the absolute value of a complex number $a + bi$ is as follows:

$$\text{If } z = a + bi, \qquad |z| = \sqrt{a^2 + b^2}$$

The absolute values of $z_1 = 3 + 4i$ and $z_2 = -2 - 3i$ are shown in Figure 2 and are calculated below.

$$|z_1| = |3 + 4i| = \sqrt{3^2 + 4^2} = 5$$
$$|z_2| = |-2 - 3i| = \sqrt{(-2)^2 + (-3)^2} = \sqrt{13}$$

Tip Note that $|a + bi| \neq \sqrt{a^2 + (bi)^2}$. For instance, $|3 + 4i| \neq \sqrt{3^2 + (4i)^2} = \sqrt{-7}$.

Operations on Complex Numbers

There are five operations for complex numbers: Equality, addition, subtraction, multiplication, and division.

1. Equality of complex numbers: Two complex numbers are equal if and only if their real parts are equal and their imaginary parts are equal.

$$a + bi = c + di \qquad \text{if and only if } a = c \text{ and } b = d$$

2. Addition of complex numbers: Add real parts and imaginary parts of the complex numbers.

$$4 + 3i + 1 + 2i = (4 + 1) + (3i + 2i) = 5 + 5i$$

3. Subtraction of complex numbers: Subtract real parts and imaginary parts of the complex numbers.

$$4 + 3i - (1 + 2i) = (4 - 1) + (3i - 2i) = 3 + i$$

4. Multiplication of complex numbers: Use FOIL (First, Outer, Inner, Last) method.

$$\begin{aligned}(4 + 3i) \cdot (1 + 2i) &= 4 + 8i + 3i + 6i^2 && \text{Since } i^2 = -1\\ &= (4 - 6) + (8i + 3i)\\ &= -2 + 11i\end{aligned}$$

5. Division of complex numbers: Multiply the numerator and denominator of the quotient by the conjugate of the denominator. If $z = x + yi$, the conjugate of z, denoted by $\overline{z}$, is $\overline{z} = x - yi$.

$$\begin{aligned}\frac{4 + 3i}{1 - 2i} &= \frac{4 + 3i}{1 - 2i} \cdot \frac{1 + 2i}{1 + 2i} && \text{Since the conjugate of } 1 - 2i \text{ is } 1 + 2i\\ &= \frac{(4 + 3i)(1 + 2i)}{(1 - 2i)(1 + 2i)} && \text{Use FOIL method}\\ &= \frac{-2 + 11i}{1 - 4i^2} && \text{Since } i^2 = -1\\ &= \frac{-2 + 11i}{5}\end{aligned}$$

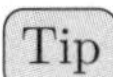

1. The product of a complex number and its conjugate is a nonnegative real number. Let $z = x + yi$ and its conjugate $\overline{z} = x - yi$.

$$\begin{aligned}z\overline{z} &= (x + yi)(x - yi) && \text{Use FOIL method}\\ &= x^2 + y^2\end{aligned}$$

2. The conjugate of bi is $-bi$. For instance, the conjugate of $3i$ is $-3i$.

$$z = 3i = 0 + 3i \implies \overline{z} = 0 - 3i = -3i$$

EXERCISES

1. Which of the following expression cannot be equal to $-i$?

(A) i^3

(B) i^{15}

(C) i^{35}

(D) i^{45}

(E) i^{55}

2. Solve the equation: $2x^2 + 5 = 1$

(A) $x = i\sqrt{2}$ or $x = -i\sqrt{2}$

(B) $x = 2i$ or $x = -2i$

(C) $x = 2i\sqrt{2}$ or $x = -2i\sqrt{2}$

(D) $x = \sqrt{2}$ or $x = -\sqrt{2}$

(E) $x = 2\sqrt{2}$ or $x = -2\sqrt{2}$

3. $-2i(3 + 4i) =$

(A) $-6 + 8i$

(B) $-8 - 6i$

(C) $-8 + 6i$

(D) $8 + 6i$

(E) $8 - 6i$

4. If $z_1 = 1 - 3i$ and $z_2 = 3 - 2i$, what is $2z_2 - 3z_1$?

(A) $5 + 3i$

(B) $3 + 8i$

(C) $3 + 5i$

(D) $-7 - 12i$

(E) $-12 - 7i$

5. If $z = 4 - 5i$, what is the absolute value of z ?

(A) $-3i$

(B) $3i$

(C) 3

(D) $\sqrt{41}$

(E) $\sqrt{51}$

6. If $\dfrac{4 + 2i}{1 + i} = a + bi$, what is $a + b$?

(A) 5

(B) 4

(C) 3

(D) 2

(E) 1

7. $(5 - 3i)^2 =$

(A) 16

(B) 34

(C) $34 - 30i$

(D) $16 + 30i$

(E) $16 - 30i$

8. Solve: $x^2 + 2x + 5 = 0$

(A) $x = -1 + 2i$ or $x = -1 - 2i$

(B) $x = -2 + 4i$ or $x = -2 - 4i$

(C) $x = 2 + i$ or $x = 2 - i$

(D) $x = 1$ or $x = -3$

(E) $x = 3$ or $x = -1$

9. If z is a solution of $x^2 - 6x + 25 = 0$, what is $|z|$?

(A) 2.83

(B) 5

(C) 7

(D) 8.83

(E) 9

10. If the two complex numbers $z_1 = (a-8)+bi$ and $z_2 = b+(4-a)i$ are equal, what is the sum of a and b ?

(A) 2

(B) 3

(C) 4

(D) 5

(E) 6

ANSWERS AND SOLUTIONS

1. (D)

Tip: If the power of i is a multiple of 4, the value is always equal to 1; that is, $i^{4n} = 1$.

(A) $i^3 = -i$

(B) $i^{15} = i^{12} \cdot i^3 = i^3 = -i$

(C) $i^{35} = i^{32} \cdot i^3 = i^3 = -i$

(D) $i^{45} = i^{44} \cdot i = i$

(E) $i^{55} = i^{52} \cdot i^3 = i^3 = -i$

Therefore, (D) is the correct answer.

2. (A)

$2x^2 + 5 = 1$ is equal to $x^2 = -2$. Since the square of a real number cannot be negative, the solutions to $x^2 = -2$ involve complex numbers.

$2x^2 + 5 = 1$	Subtract 5 from each side
$2x^2 = -4$	Divide each side by 2
$x^2 = -2$	Since $i^2 = -1$
$x^2 = 2i^2$	Take the square root of both sides
$x = \pm i\sqrt{2}$	

Therefore, the solutions to $2x^2 + 5 = 1$ are $x = i\sqrt{2}$ or $x = -i\sqrt{2}$.

3. (E)

$$-2i(3+4i) = -6i - 8i^2 = 8 - 6i \qquad \text{Since } i^2 = -1$$

4. (C)

$$\begin{aligned} 2z_2 - 3z_1 &= 2(3-2i) - 3(1-3i) \\ &= 6 - 4i - 3 + 9i \\ &= (6-3) + (9-4)i \\ &= 3 + 5i \end{aligned}$$

Therefore, $2z_2 - 3z_1 = 3 + 5i$.

5. (D)

> Tip If $z = a + bi$, $|z| = \sqrt{a^2 + b^2}$.

Since $z = 4 - 5i$, the absolute value of z, $|z|$, is $\sqrt{4^2 + (-5)^2} = \sqrt{41}$.

6. (D)

> Tip In order to divide two complex numbers, multiply the numerator and denominator of the quotient by the conjugate of the denominator.

$$\begin{aligned} \frac{4+2i}{1+i} &= \frac{4+2i}{1+i} \cdot \frac{1-i}{1-i} && \text{Since the conjugate of } 1+i \text{ is } 1-i \\ &= \frac{(4+2i)(1+i)}{(1+i)(1-i)} && \text{Use the FOIL method} \\ &= \frac{6-2i}{1-i^2} && \text{Since } i^2 = -1 \\ &= \frac{6-2i}{2} && \text{Simplify} \\ &= 3 - i \end{aligned}$$

Since $\frac{4+2i}{1+i} = 3 - i$, $a = 3$ and $b = -1$. Therefore, the value of $a + b = 2$.

7. (E)

$$\begin{aligned} (5-3i)^2 &= (5-3i)(5-3i) && \text{Use the FOIL method} \\ &= 25 - 15i - 15i + 9i^2 && \text{Since } i^2 = -1 \\ &= (25-9) - 30i && \text{Simplify} \\ &= 16 - 30i \end{aligned}$$

Therefore, $(5-3i)^2 = 16 - 30i$.

8. (A)

Use the quadratic formula to find the solutions to the quadratic equation $x^2 + 2x + 5 = 0$.

$$\begin{aligned} x &= \frac{-b \pm \sqrt{b^2 - 4ac}}{2a} = \frac{-2 \pm \sqrt{2^2 - 4(1)(5)}}{2(1)} \\ &= \frac{-2 \pm \sqrt{-16}}{2} = \frac{-2 \pm 4i}{2} \\ &= -1 \pm 2i \end{aligned}$$

Therefore, the solutions to $x^2 + 2x + 5 = 0$ are $-1 + 2i$ and $-1 - 2i$.

9. (B)

Use the quadratic formula to find the solutions to the quadratic equation $x^2 - 6x + 25 = 0$.

$$\begin{aligned} x &= \frac{-b \pm \sqrt{b^2 - 4ac}}{2a} = \frac{-(-6) \pm \sqrt{(-6)^2 - 4(1)(25)}}{2(1)} \\ &= \frac{6 \pm \sqrt{-64}}{2} = \frac{6 \pm 8i}{2} \\ &= 3 \pm 4i \end{aligned}$$

Since z is the solution of $x^2 - 6x + 25 = 0$, z can be either $3 + 4i$ or $3 - 4i$. Thus,

$$\text{If } z = 3 + 4i,\ |z| = \sqrt{3^2 + 4^2} = 5$$
$$\text{If } z = 3 - 4i,\ |z| = \sqrt{3^2 + (-4)^2} = 5$$

Therefore, $|z| = 5$.

10. (C)

Tip: Two complex numbers are equal if and only if their real parts are equal and their imaginary parts are equal.

Since $z_1 = (a-8) + bi$ and $z_2 = b + (4-a)i$ are equal, $(a-8) = b$, and $b = 4-a$. Since $(a-8) = b$ is equivalent to $a - b = 8$, and $b = 4 - a$ is equivalent to $a + b = 4$, use the linear combination method to find the values of a and b.

$$\begin{aligned} a - b &= 8 \\ a + b &= 4 \qquad \text{Add the two equations} \\ \hline 2a \quad &= 12 \\ a &= 6 \end{aligned}$$

Since $a = 6$ and $a - b = 8$, $b = -2$. Therefore, the sum of a and b is $6 + -2 = 4$.

LESSON 5 Polynomial Functions

Finding the Real Zeros of a Polynomial Function

A polynomial function $f(x)$ crosses the x-axis at three points as shown below.

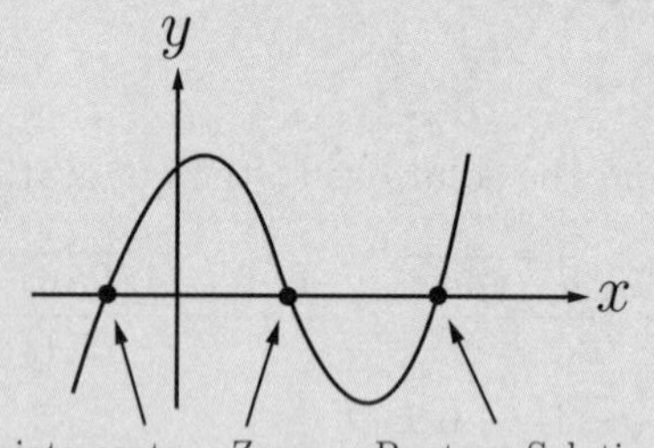

x-intercepts = Zeros = Roots = Solutions

These three points are called either the x-intercepts, zeros, roots, or solutions. That is, they are the values of x that make $f(x) = 0$.

Finding the real zeros of a polynomial function of degree 3 or higher is very complicated because there is no general formula like the quadratic formula available.

However, the following three theorems (remainder theorem, factor theorem, and rational zeros theorem) are very useful to find the real zeros of a polynomial function.

1. Remainder Theorem

If a polynomial function $f(x)$ is divided by $x - k$, the remainder is $f(k)$.

Tip

1. In order to evaluate the remainder, use either polynomial long division, synthetic division, or the remainder theorem. However, the remainder theorem is the easiest to use.

2. In order to determine the quotient polynomial when a polynomial function $f(x)$ is divided by $x - c$, use polynomial division or synthetic division. The remainder theorem doesn't tell you about the quotient polynomial.

	Quotient	Remainder
Polynomial long division	✓	✓
Synthetic division	✓	✓
Remainder theorem	Not available	✓

Example 1 Finding the quotient and remainder

If $f(x) = x^3 - 2x^2 + 3x + 2$ is divided by $x - 1$, find the quotient and remainder.

Solution Since we are trying to find the quotient and the remainder, we can use either polynomial long division as shown in Figure 1, or synthetic division as shown in Figure 2. However, synthetic division is easier and faster to find the quotient and remainder.

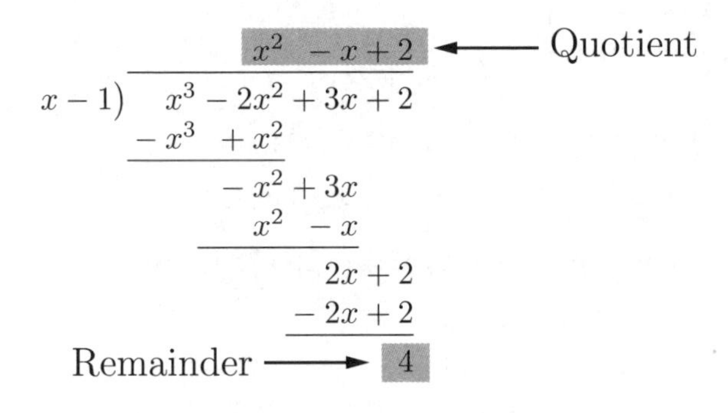

Fig. 1: Polynomial long division

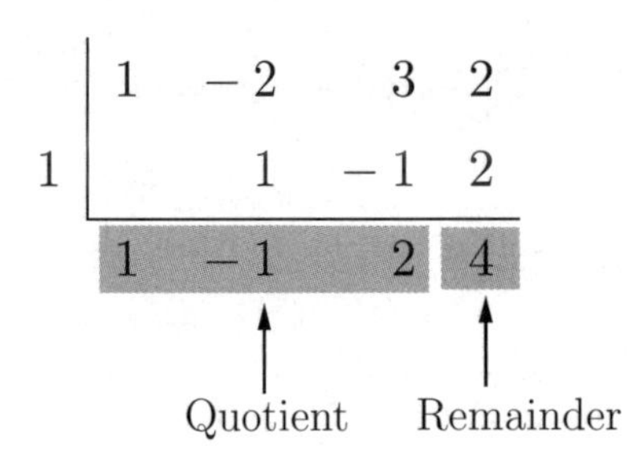

Fig. 2: Synthetic division

Therefore, the quotient polynomial is $x^2 - x + 2$ and the remainder is 4.

2. Factor Theorem

- If $x - k$ is a factor of $f(x)$, then the remainder $r = f(k) = 0$.
- If the remainder $r = f(k) = 0$, then k is a zero of $f(x)$ and $x - k$ is a factor of $f(x)$.

Tip The factor theorem states a relationship between a zero and a factor such that if a function has a zero k, then the function has a factor of $x - k$.

Example 2 Finding a factor of a polynomial function

Determine whether the function $f(x) = x^3 - 7x + 6$ has the factor $x - 2$.

Solution Use the factor theorem to determine whether $x - 2$ is the factor of $f(x)$. Substitute 2 for x in $f(x)$ to see if the remainder $r = f(2)$ is equal to 0.

$$f(x) = x^3 - 7x + 6 \qquad \text{Substitute 2 for } x$$
$$f(2) = 2^3 - 7(2) + 6 = 0$$

Since the remainder $r = f(2) = 0$, $x - 2$ is a factor of $f(x) = x^3 - 7x + 6$.

3. Rational Zeros Theorem

The rational zeros theorem provides a list of all possible rational zeros of a polynomial function with integer coefficients. Out of all possible rational zeros, use the factor theorem or synthetic division to find the real zeros of the polynomial function.

The rational zeros theorem states that for the polynomial function with integer coefficients, $f(x) = a_n x^n + a_{n-1}x^{n-1} + \cdots + a_1 x + a_0$, the possible rational zeros are as follows:

$$\text{Possible rational zeros} = \pm \frac{\text{factors of } a_o}{\text{factors of } a_n}$$

where a_n is the leading coefficient and a_0 is the constant.

For instance, if $f(x) = 2x^3 - 11x^2 + 13x - 4$, the possible rational zeros are shown below.

$$\begin{aligned}\text{Possible rational zeros} &= \pm \frac{\text{factors of } 4}{\text{factors of } 2}\\ &= \pm \frac{\{1, 2, 4\}}{\{1, 2\}}\\ &= \pm 1,\ \pm 2,\ \pm 4,\ \pm \frac{1}{2}\end{aligned}$$

There are 8 possible rational zeros. Use the factor theorem to find one real zero of $f(x)$.

$$f(-1) = -30 \neq 0, \qquad f(1) = 0$$

$f(1) = 0$ means that 1 is a zero of $f(x)$. Let's use synthetic division to factor $f(x)$.

$$\begin{array}{r|rrrr} & 2 & -11 & 13 & -4 \\ 1 & & 2 & -9 & 4 \\ \hline & 2 & -9 & 4 & 0 \end{array}$$

Since the quotient polynomial is $2x^2 - 9x + 4$, the function $f(x)$ can be factored as follows:

$$\begin{aligned}2x^3 - 11x^2 + 13x - 4 &= (x-1)(2x^2 - 9x + 4) && \text{Factor } 2x^2 - 9x + 4\\ &= (x-1)(x-4)(2x-1)\end{aligned}$$

Therefore, the real zeros of $f(x)$ are 1, 4, and $\frac{1}{2}$.

Tip — If a polynomial function has a rational zero, the zero is one of the possible rational zeros suggested by the rational zeros theorem. However, if a polynomial function does not have any rational zeros, the rational zeros theorem does not help you find complex zeros.

Conjugate Pairs Theorem

The conjugate pairs theorem states that complex zeros and irrational zeros always occur in conjugate pairs.

If $a + bi$ is a zero of f, $\Longrightarrow$ $a - bi$ is also a zero of f.

If $a + \sqrt{b}$ is a zero of f, $\Longrightarrow$ $a - \sqrt{b}$ is also a zero of f.

Writing a Polynomial Function with the Given Zeros

By the factor theorem, if 2 is a zero of f, then $x - 2$ is a factor of f. In order to write a polynomial function with the given zeros, convert each zero to a factor and expand. For instance,

$$\text{Given zeros: 2, and 3} \implies y = (x-2)(x-3) = x^2 - 5x + 6$$

$$\begin{aligned}\text{Given zeros: } 1+\sqrt{2}, \text{ and } 1-\sqrt{2} \implies y &= (x-(1+\sqrt{2}))(x-(1-\sqrt{2})) \\ &= ((x-1)-\sqrt{2})((x-1)+\sqrt{2}) \\ &= (x-1)^2 - 2 \\ &= x^2 - 2x - 1\end{aligned}$$

Vieta's Formulas

When irrational zeros or complex zeros are given, use Vieta's formula to easily write a quadratic function with leading coefficient 1. Vieta's formulas relate the coefficients of a polynomial to the sum and product of its zeros and are described below. For a quadratic function $f(x) = x^2 + bx + c$, let z_1 and z_2 be the zeros of f.

$z_1 + z_2 = -b$ Sum of zeros equals the opposite of the coefficient of x

$z_1 z_2 = c$ Product of zeros equals the constant term

For instance, when irrational zeros, $1 + \sqrt{2}$ and $1 - \sqrt{2}$, are given,

Sum of zeros: $(1+\sqrt{2}) + (1-\sqrt{2}) = 2 \xrightarrow{\text{Opposite}} -2$ (Coefficient of x)

Product of zeros: $(1+\sqrt{2})(1-\sqrt{2}) = -1 \xrightarrow{\text{Same}} -1$ (Constant term)

Thus, the quadratic function whose zeros are $1 + \sqrt{2}$ and $1 - \sqrt{2}$ is $x^2 - 2x - 1$.

1. In order to expand the expression $((x-1)-\sqrt{2})((x-1)+\sqrt{2})$ shown above, use $(a-b)(a+b) = a^2 - b^2$ formula.
2. In order to use Vieta's formulas, two zeros must be the same type: rational zeros, irrational zeros, or complex zeros.

Example 3 Writing a polynomial function with given zeros

If zeros of the polynomial function f are 1 and $2+i$, write a cubic function with leading coefficient 1.

Solution According to the conjugate pairs theorem, $2-i$, is also a zero of f. Thus, 1, $2+i$, and $2-i$ are zeros of f. Since complex zeros are given, use Vieta's formulas to write a quadratic function with leading coefficient 1.

$$\text{Sum of zeros:}\quad (2+i)+(2-i)=4 \xrightarrow{\text{Opposite}} -4 \text{ (Coefficient of } x)$$

$$\text{Product of zeros:}\quad (2+i)(2-i)=5 \xrightarrow{\text{Same}} 5 \text{ (Constant term)}$$

The quadratic function whose zeros are $2+i$ and $2-i$ is x^2-4x+5. Since 1 is a zero of f, $x-1$ is a factor of f. Thus,

$$\begin{aligned} f(x) &= (x-1)(x^2-4x+5) \\ &= x^3-5x^2+9x-5 \end{aligned}$$

Therefore, the cubic function whose zeros are 1, $2+i$, and $2-i$ is $f(x)=x^3-5x^2+9x-5$.

Graphing Cubic Functions

A polynomial function of degree 3, $f(x)=ax^3+bx^2+cx+d$, is a cubic function, where a is the leading coefficient and is nonzero. Depending on the values of a, either positive or negative, the graph of the cubic function varies.

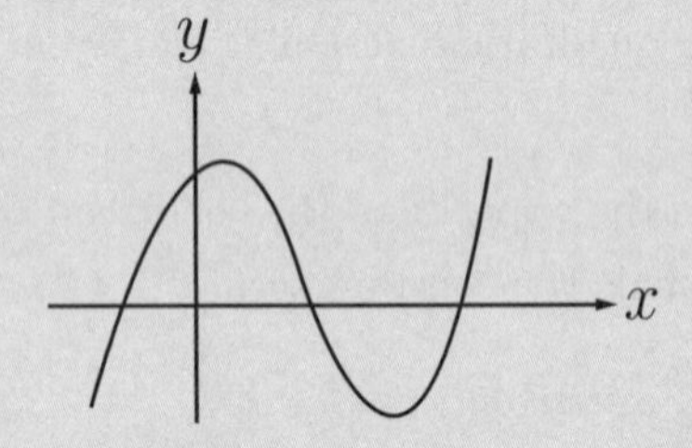

Figure 1: If $a>0$

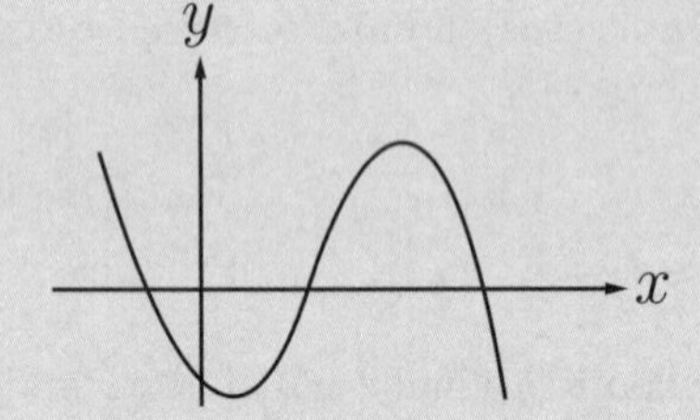

Figure 2: If $a<0$

If $a>0$, the graph of a cubic function shown in Figure 1 goes up as x increases and goes down as x decreases. Whereas, if $a<0$, the graph of a cubic function shown in Figure 2 goes down as x increases and goes up as x decreases.

Zeros of odd or even multiplicity

If $(x-c)^m$ is a factor of a polynomial function f, c is called a zero of multiplicity of m. Depending on the value of m, either odd or even, the graph of f either crosses or touches the x-axis at $x = c$.

$$\text{If } m = \text{odd} \implies \text{graph of } f \text{ crosses the } x\text{-axis at } x = c.$$
$$\text{If } m = \text{even} \implies \text{graph of } f \text{ touches the } x\text{-axis at } x = c.$$

For instance, let $f(x) = -5(x-1)^2(x-3)$. Since the leading coefficient is -5, the shape of the graph of f is similar to the graph in Figure 2.

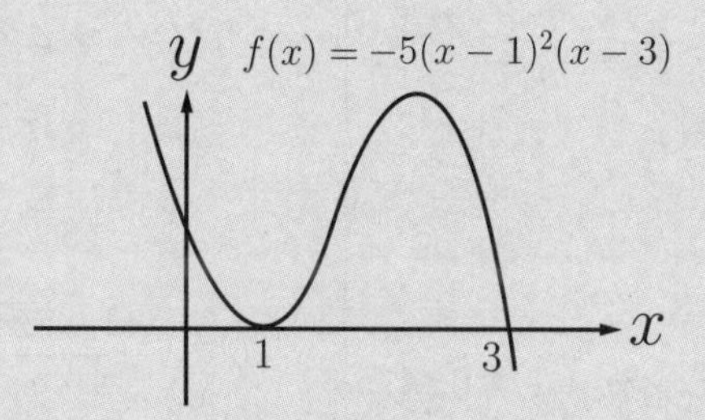

Figure 3

Since $(x-1)^2$ is a factor of f, 1 is a zero of multiplicity 2. Thus, the graph of f touches the x-axis at $x = 1$. Additionally, $(x-3)$ is a factor of f. Thus, 3 is a zero of multiplicity 1 and the graph of f crosses the x-axis at $x = 3$. The graph of f is shown in Figure 3 above.

Solving Polynomial Inequalities

Solving a polynomial inequality means finding the x-values for which the graph of the polynomial function f lies above or below the x-axis. A polynomial inequality can be solved algebraically. However, Solving a polynomial inequality **graphically** is highly recommended.

Let the polynomial inequality be $(x-1)(x-2)(x-3) > 0$. First, graph the polynomial function $f(x) = (x-1)(x-2)(x-3)$ with the leading coefficient 1. Since 1, 2, and 3 are zeros of multiplicity 1, the graph of f crosses the x-axis at $x = 1$, $x = 2$, and $x = 3$ as shown in Figure 4.

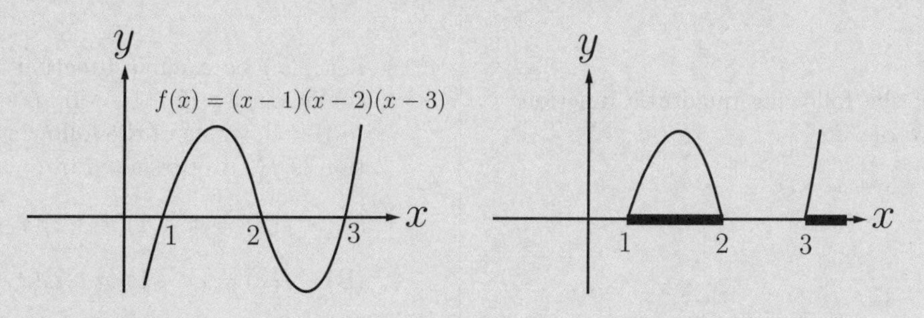

Figure 4 Figure 5

As shown in Figure 5, the graph of f lies above the x-axis when $1 < x < 2$ or $x > 3$. Therefore, the solution to $(x-1)(x-2)(x-3) > 0$ is $1 < x < 2$ or $x > 3$.

EXERCISES

1. Which of the following expression is a factor of $f(x) = x^3 - 2x - 4$?

   (A) $x + 2$

   (B) $x + 1$

   (C) $x - 2$

   (D) $x - 3$

   (E) $x - 4$

2. If α and β are solutions of $x^2 - 5x + 1 = 0$, what is the value of $\alpha + \beta$?

   (A) 1

   (B) 2

   (C) 3

   (D) 4

   (E) 5

3. If $f(x) = x^3 + 3x^2 + 4x - 1$ is divided by $x + 1$, what is the remainder?

   (A) -5

   (B) -4

   (C) -3

   (D) 1

   (E) 2

4. Which of the following quadratic function has a zero of i ?

   (A) $x^2 + x + 1$

   (B) $x^2 + 1$

   (C) $x^2 - 1$

   (D) $x^2 - x - 1$

   (E) $x^2 - 2x - 1$

5. What are the zeros of $f(x) = x^3 + x^2 - 12x$?

   (A) $x = -4,\ 1,\ 2$

   (B) $x = -4,\ 0,\ 3$

   (C) $x = -3,\ 0,\ 2$

   (D) $x = -3,\ 1,\ 4$

   (E) $x = -2,\ 2,\ 3$

6. $\dfrac{x^3 - 2x^2 - x + 5}{x - 2} =$

   (A) $x^2 + 1$

   (B) $x^2 + x + 1$

   (C) $x^2 + 1 + \frac{1}{x-2}$

   (D) $x^2 - 1 + \frac{3}{x-2}$

   (E) $x^2 + x + 1 + \frac{2}{x-2}$

7. If $f(x) = 2x^3 + 3x^2 + kx + 3$ is divisible by $x + 1$, what is the value of k ?

   (A) 4

   (B) 3

   (C) 1

   (D) -2

   (E) -3

8. Let $f(x)$ be a cubic function with leading coefficient 1. If $f(1) = 0$, $f(-2) = 0$, and $f(3) = 0$, which of the following cubic function is $f(x)$ represented by?

   (A) $f(x) = (x + 1)(x - 2)(x + 3)$

   (B) $f(x) = (x - 1)(x + 2)(x + 3)$

   (C) $f(x) = (x - 1)(x - 2)(x - 3)$

   (D) $f(x) = (x + 1)(x + 2)(x - 3)$

   (E) $f(x) = (x - 1)(x + 2)(x - 3)$

9. If $f(x)$ is a cubic function with leading coefficient 1 whose zeros are -1 and $3-2i$, which of the following is $f(x)$ represented by?

(A) $f(x) = x^3 - 7x^2 + 19x + 13$

(B) $f(x) = x^3 - 7x^2 + 11x - 5$

(C) $f(x) = x^3 - 5x^2 - x + 5$

(D) $f(x) = x^3 - 5x^2 + 7x + 13$

(E) $f(x) = x^3 - 3x^2 + 11x + 13$

10. Solve: $-(x+1)(x-2)(x-4) \le 0$

(A) $-1 \le x \le 2$ or $x \ge 4$

(B) $x \le -1$ or $2 \le x \le 4$

(C) $-1 \le x \le 4$

(D) $2 \le x \le 4$

(E) $x \le -1$ or $x \ge 4$

ANSWERS AND SOLUTIONS

1. (C)

Tip Factor Theorem: If $x-k$ is a factor of $f(x)$, then the remainder $r = f(k) = 0$.

Since $f(2) = 2^3 - 2(2) - 4 = 0$, $x-2$ is a factor of $x^3 - 2x - 4$.

2. (E)

Tip According to Vieta's formulas, for a quadratic equation $x^2 + bx + c = 0$, the sum of the zeros of α and β is the opposite of the coefficient of x. That is, $\alpha + \beta = -b$.

For the quadratic equation $x^2 - 5x + 1 = 0$, the zeros are α, and β. Since the coefficient of x is -5, $\alpha + \beta = -(-5) = 5$.

3. (C)

Tip Remainder Theorem: If a polynomial function $f(x)$ is divided by $x-k$, the remainder is $f(k)$.

In order to find the remainder when $f(x) = x^3 + 3x^2 + 4x - 1$ is divided by $x+1$, substitute -1 for x in $f(x)$. Therefore, $f(-1) = (-1)^3 + 3(-1)^2 + 4(-1) - 1 = -3$.

4. (B)

Tip Conjugate Pairs Theorem: Complex zeros and irrational zeros always occur in conjugate pairs.

According to the conjugate pairs theorem, $-i$ is also a zero of the function. Thus, i and $-i$ are the zeros of the quadratic function. In order to find the quadratic function, convert each zero into a factor and expand.

$$\begin{aligned}(x-i)(x-(-i)) &= (x-i)(x+i)\\ &= x^2 - i^2\\ &= x^2 + 1\end{aligned}$$

Therefore, (B) is the correct answer.

5. (B)

Substitute 0 for y and solve for x.

$$\begin{aligned} x^3 + x^2 - 12x &= 0 && \text{Factor out } x \\ x(x^2 + x - 12) &= 0 && \text{Factor } x^2 + x - 12 \\ x(x+4)(x-3) &= 0 && \text{Solve} \\ x &= -4,\ 0,\ 3 \end{aligned}$$

6. (D)

When $x^3 - 2x^2 - x + 5$ is divided by $x - 2$, the quotient polynomial is $x^2 - 1$ and the remainder is 3 as shown below.

$$\begin{array}{r} x^2 \qquad -1 \\ x-2 \overline{)\ x^3 - 2x^2 - x + 5} \\ \underline{-x^3 + 2x^2 \qquad\quad} \\ -x + 5 \\ \underline{x - 2} \\ 3 \end{array}$$

Therefore, $\dfrac{x^3 - 2x^2 - x + 5}{x - 2} = x^2 - 1 + \dfrac{3}{x - 2}$.

7. (A)

Since $2x^3 + 3x^2 + kx + 3$ is divisible by $x + 1$, the remainder, $f(-1)$, is 0 when $2x^3 + 3x^2 + kx + 3$ is divided by $x + 1$. Use the factor theorem to find the value of k.

$$\begin{aligned} f(-1) &= 2(-1)^3 + 3(-1)^2 + k(-1) + 3 && \text{Since } f(-1) = 0 \\ 0 &= 4 - k && \text{Solve for } k \\ k &= 4 \end{aligned}$$

Therefore, the value of k is 4.

8. (E)

Tip: Factor Theorem: If the remainder $r = f(k) = 0$, then k is a zero of $f(x)$ and $x - k$ is a factor of $f(x)$.

$f(1) = 0$, $f(-2) = 0$, and $f(3) = 0$. Thus, $x - 1$, $x + 2$, and $x - 3$ are factors of f. Therefore, $f(x) = (x-1)(x+2)(x-3)$.

9. (D)

According to the conjugate pairs theorem, $3+2i$ is also a zero of f. Thus, -1, $3-2i$, and $3+2i$ are zeros of f. Since complex zeros are given, use Vieta's formulas to write a quadratic function with a leading coefficient of 1.

$$\text{Sum of zeros:} \quad (3+2i)+(3-2i)=6 \quad \xrightarrow{\text{Opposite}} \quad -6 \text{ (Coefficient of } x\text{)}$$

$$\text{Product of zeros:} \quad (3+2i)(3-2i)=13 \quad \xrightarrow{\text{Same}} \quad 13 \text{ (Constant term)}$$

The quadratic function whose zeros are $3+2i$ and $3-2i$ is $x^2-6x+13$. Since -1 is a zero of f, $x+1$ is a factor of f. Thus,

$$\begin{aligned} f(x) &= (x+1)(x^2-6x+13) \\ &= x^3-5x^2+7x+13 \end{aligned}$$

Therefore, a cubic function whose zeros are -1, $3-2i$, and $3+2i$ is $f(x)=x^3-5x^2+7x+13$.

10. (A)

First, graph the polynomial function $f(x)=-(x+1)(x-2)(x-4)$ with a leading coefficient of -1. Since -1, 2, and 4 are zeros of multiplicity 1, the graph of f crosses the x-axis at $x=-1$, $x=2$, and $x=4$ as shown in Figure 1.

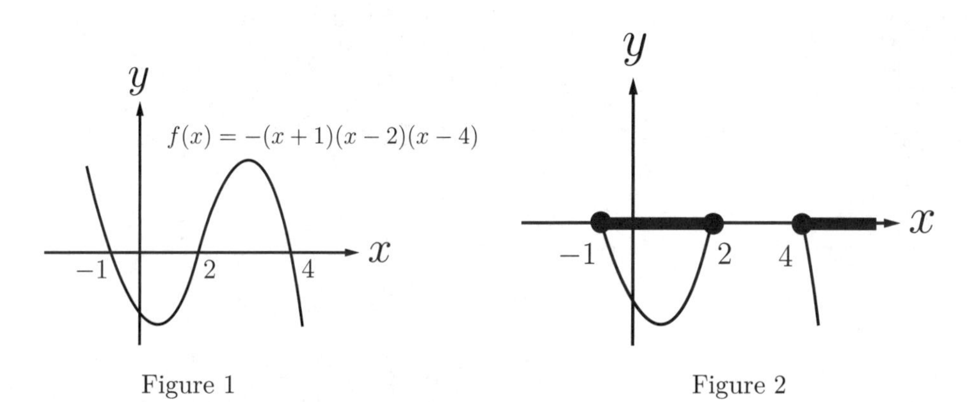

Figure 1 Figure 2

As shown in Figure 2, the graph of f lies below or on the x-axis when $-1 \le x \le 2$ or $x \ge 4$. Therefore, the solution to $-(x+1)(x-2)(x-4) \le 0$ is $-1 \le x \le 2$ or $x \ge 4$.

LESSON 6 Rational Functions and Limits

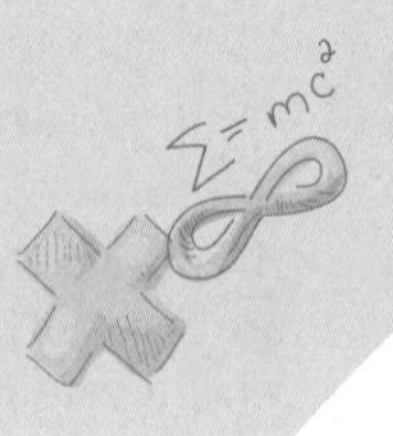

Rational Functions

A rational function $f(x)$ is a function that is expressed in the form

$$f(x) = \frac{p(x)}{q(x)}$$

where p and q are polynomial functions and q is not the zero polynomial function. The domain of a rational function f is a set of x-values for which the denominator q is not zero.

An asymptote is a line that the graph of a function approaches but never touches. There are three kinds of asymptotes: horizontal, vertical, and slant (or oblique) asymptotes. In general, a rational function has either a vertical asymptote or a horizontal asymptote or both.

A horizontal asymptote is a horizontal line that the graph of a function approaches as $x \to \infty$ (read as x approaches ∞) or as $x \to -\infty$. A vertical asymptote is a vertical line near which the values of f approach ∞ (increases without bound) or $-\infty$ (decreases without bound).

For instance, $f(x) = \dfrac{1}{x-2} + 1$ is a rational function whose graph is shown in Figure 1.

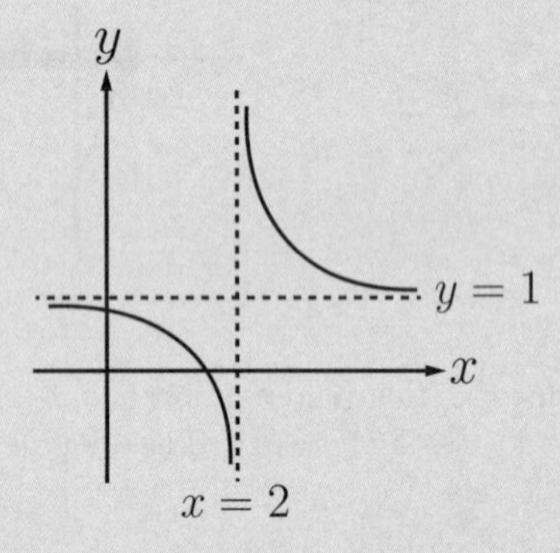

Figure 1: Graph of $f(x) = \frac{1}{x-2} + 1$

The horizontal asymptote of f is $y = 1$ because the values of f approach 1 as $x \to \infty$ or $x \to -\infty$. Whereas, the vertical asymptote is $x = 2$ because the values of f approach ∞ or $-\infty$ as $x \to 2$.

Tip: Since the values of f approach 1 as $x \to \infty$, this can be denoted by $\lim_{x\to\infty} f(x) = 1$. In general, $\lim_{x\to\infty} f(x)$, the limit of a function f at infinity, means to find the horizontal asymptote of f.

Finding Vertical Asymptotes

A rational function $f(x) = \frac{p(x)}{q(x)}$ has a vertical asymptote at $x = c$ if c is a real zero of the denominator $q(x)$.

For instance, let $f(x) = \frac{2x}{x^2-9}$. The zeros of the denominator of $x^2 - 9$ are 3 or -3. Thus, the rational function f has the vertical asymptotes at $x = 3$ and $x = -3$.

Tip If the denominator of a rational function has no real zeros, the rational function does not have a vertical asymptote. For instance, let $f(x) = \frac{1}{x^2+1}$. Since the zeros of $x^2 + 1$ are i and $-i$, the rational function f does not have a vertical asymptote.

Finding Horizontal Asymptotes $\left(\lim_{x\to\infty} f(x)\right)$

For the rational function

$$f(x) = \frac{p(x)}{q(x)} = \frac{ax^m + \cdots}{bx^n + \cdots}$$

where m is the degree of the numerator and n is the degree of the denominator, a horizontal asymptote can be determined by the following three cases.

- Case 1: If $n < m$, there is no horizontal asymptote. Whereas, there is a slant (or oblique) asymptote.
- Case 2: If $n = m$, f has a horizontal asymptote of $y = \frac{a}{b}$, where a and b are the leading coefficients of the numerator and denominator.
- Case 3: If $n > m$, f has a horizontal asymptote of $y = 0$.

For instance, for the rational function $f(x) = \frac{1}{x} = \frac{1\cdot x^0}{x^1}$ whose graph is shown below,

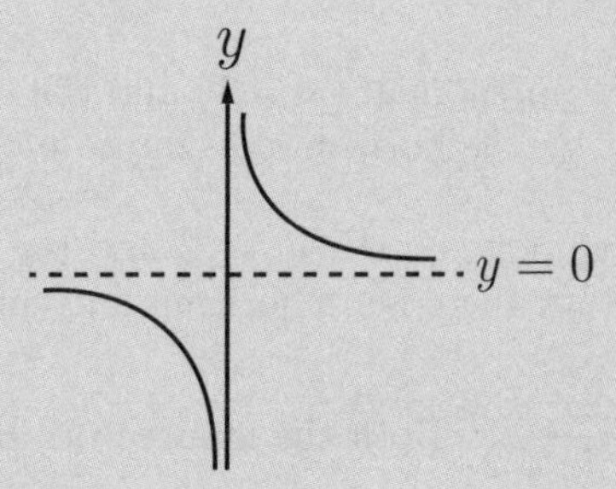

the degree of numerator is 0 and the degree of the denominator is 1. Thus, the rational function has the horizontal asymptote of $y = 0$, which can be denoted by $\lim_{x\to\infty} \frac{1}{x} = 0$ and $\lim_{x\to-\infty} \frac{1}{x} = 0$.

Example 1 Finding vertical asymptotes

Find the domain and vertical asymptotes of the following rational function.

$$y = \frac{2x+1}{x^2-5x+6}$$

Solution In order to find the domain and vertical asymptotes, set the denominator equal to zero and solve for x. The solutions to $x^2 - 5x + 6 = 0$ will be the values that are excluded in the domain, and will be the vertical asymptotes.

$$\begin{aligned} x^2 - 5x + 6 &= 0 && \text{Factor} \\ (x-2)(x-3) &= 0 && \text{Solve} \\ x = 2 \quad &\text{or} \quad x = 3 \end{aligned}$$

Therefore, the domain of the rational function is $x \neq 2$ and $x \neq 3$, and the vertical asymptotes are $x = 2$ and $x = 3$.

Example 2 Finding horizontal asymptotes

Find the horizontal asymptotes of each rational function.

(a) $f(x) = \dfrac{1-3x}{2x^2+3x}$

(b) $f(x) = \dfrac{-x^3+1}{x^2-2x+2}$

(c) $f(x) = \dfrac{2x^2-2x+3}{2-4x+3x^2}$

Solution

(a) The numerator is a first degree polynomial ($m = 1$) and the denominator is a second degree polynomial ($n = 2$). Since $n > m$, the horizontal asymptote of f is $y = 0$.

(b) The numerator is a third degree polynomial ($m = 3$) and the denominator is a second degree polynomial ($n = 2$). Since $n < m$, there is no horizontal asymptote.

(c) $f(x) = \dfrac{2x^2-2x+3}{2-4x+3x^2} = \dfrac{2x^2-2x+3}{3x^2-4x+2}$. Both the numerator and the denominator are 2nd degree polynomials. Thus, the horizontal asymptote of f is the ratio of the leading coefficients, or $y = \frac{2}{3}$.

Graphing Rational Functions

For a rational function $f(x) = \dfrac{a}{x-h} + k$, the vertical asymptote of f is $x = h$ (set the denominator $x - h$ equal to zero and solve for x) and the horizontal asymptote of f is $y = k$. Depending on the value of a, either positive or negative, the graph of f varies as shown in Figures 2 and 3.

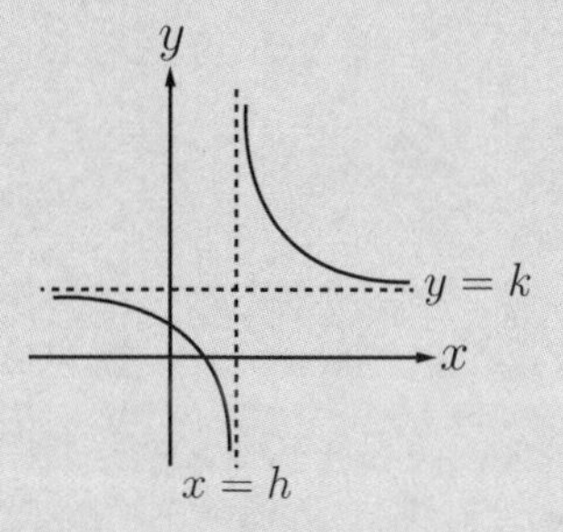

Figure 2: If $a > 0$

Figure 3: If $a < 0$

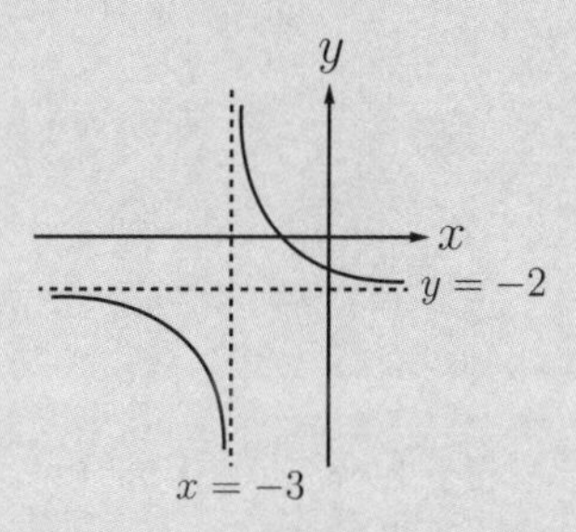

Figure 4: Graph of $\frac{1}{x+3} - 2$

For instance, for the rational function $f(x) = \dfrac{1}{x+3} - 2$, the vertical asymptote of f is $x = -3$ (set the denominator $x + 3$ equal to zero and solve for x), and the horizontal asymptote of y is $y = -2$. The graph of f is shown in Figure 4.

Graphing Rational Functions with a Hole

A hole in the graph of a rational function is produced by cancelling out the common factor $x - c$ from both the numerator and the denominator. A hole is a point at which f has no value and is plotted using an open circle.

For instance, let $f(x) = \frac{x-4}{(x-2)(x-4)}$. f has the common factor $x - 4$ from both the numerator and the denominator. Thus, cancelling out $x - 4$ produces a hole in the graph of f at $x = 4$. Since f simplifies to $\frac{1}{x-2}$, the vertical asymptote of f is $x = 2$ and the horizontal asymptote of f is $y = 0$. The graph of f is shown below.

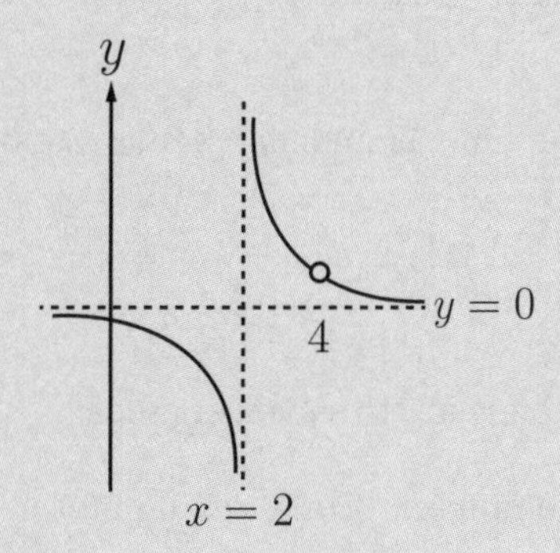

Finding Limits at a Constant $\left(\lim_{x\to c} f(x)\right)$

A limit of a function f, L, is a value that the function f approaches as $x \to c$.

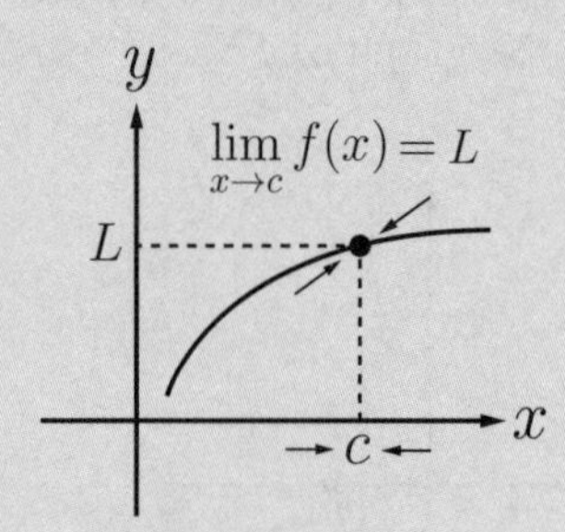

This can be denoted by $\lim_{x\to c} f(x) = L$ and is shown above.

Limit of a rational function at $x = c$

In order to find the limit of a rational function $f(x) = \frac{p(x)}{q(x)}$ at $x = c$, plug in $x = c$ to both the numerator and denominator. That is,

$$\lim_{x\to c} \frac{p(x)}{q(x)} = \frac{p(c)}{q(c)}$$

The value of $\frac{p(c)}{q(c)}$ determines the limit of f. Consider the following four cases.

- Case 1: If $\frac{p(c)}{q(c)} = \frac{\text{constant}}{\text{constant}}$, the limit is $\frac{\text{constant}}{\text{constant}}$. For instance, let $f(x) = \frac{x-1}{x+1}$.

$$\lim_{x\to 2} \frac{x-1}{x+1} = \frac{2-1}{2+1} = \frac{1}{3} \quad \Longrightarrow \quad \text{Limit of } f \text{ is } \frac{1}{3}$$

- Case 2: If $\frac{p(c)}{q(c)} = \frac{0}{\text{constant}}$, the limit is 0. For instance, let $f(x) = \frac{x-3}{x^2-1}$.

$$\lim_{x\to 3} = \frac{x-3}{x^2-1} = \frac{3-3}{3^2-1} = 0 \quad \Longrightarrow \quad \text{Limit of } f \text{ is } 0.$$

- Case 3: If $\frac{p(c)}{q(c)} = \frac{\text{constant}}{0}$, the limit is undefined. For instance, let $f(x) = \frac{1}{(x+1)^2}$.

$$\lim_{x\to -1} = \frac{1}{(x+1)^2} = \frac{1}{(-1+1)^2} = \frac{1}{0} \quad \Longrightarrow \quad \text{Limit of } f \text{ is undefined.}$$

- Case 4: If $\frac{p(c)}{q(c)} = \frac{0}{0}$, do the following three extra steps.
  - Step 1: Factor both the numerator and denominator.
  - Step 2: Cancel out a common factor.

– Step 3: Plug-in $x = c$.

For instance, let $f(x) = \frac{x^2-1}{x-1}$. Since $\lim_{x \to 1} \frac{x^2-1}{x-1} = \frac{0}{0}$, do the following steps.

$$\begin{aligned}
\lim_{x \to 1} \frac{x^2-1}{x-1} &= \lim_{x \to 1} \frac{(x+1)(x-1)}{x-1} && \text{Factor the numerator} \\
&= \lim_{x \to 1} \frac{(x+1)\cancel{(x-1)}}{\cancel{x-1}} && \text{Cancel out } x-1 \\
&= \lim_{x \to 1} (x+1) && \text{Plug-in } x = 1 \\
&= 2
\end{aligned}$$

Example 3 Finding limits at a constant

Determine the limit of the following functions.

(a) $f(x) = \lim_{x \to 2} \frac{2}{x^2 - x - 2}$

(b) $f(x) = \lim_{x \to 1} \frac{x-1}{x^2+1}$

(c) $f(x) = \lim_{x \to -1} \frac{x^3+1}{x+1}$

Solution

(a) Plug-in $x = 2$ to both the numerator and the denominator.

$$\lim_{x \to 2} \frac{2}{x^2 - x - 2} = \frac{2}{2^2 - 2 - 2} = \frac{2}{0}$$

Therefore, the limit of the function is undefined.

(b) Plug-in $x = 1$ to both the numerator and the denominator.

$$\lim_{x \to 1} \frac{x-1}{x^2+1} = \frac{1-1}{1^2+1} = \frac{0}{2}$$

Therefore, the limit of the function is 0.

(c) Plug-in $x = -1$ to both the numerator and the denominator.

$$\lim_{x \to -1} \frac{x^3+1}{x+1} = \frac{(-1)^3+1}{-1+1} = \frac{0}{0}$$

Since $\lim_{x\to-1}\frac{x^3+1}{x+1}=\frac{0}{0}$, do the following steps.

$$\lim_{x\to-1}\frac{x^3+1}{x+1}=\frac{(x+1)(x^2-x+1)}{x+1} \qquad \text{Factor the numerator}$$

$$=\lim_{x\to-1}\frac{\cancel{(x+1)}(x^2-x+1)}{\cancel{x+1}} \qquad \text{Cancel out } x+1$$

$$=\lim_{x\to-1}(x^2-x+1) \qquad \text{Plug-in } x=-1$$

$$=3$$

Therefore, the limit of the function is 3.

EXERCISES

1. What is the vertical asymptote of $\frac{x-2}{x-1}$?

   (A) $x=1$

   (B) $x=2$

   (C) $x=\frac{1}{2}$

   (D) $y=1$

   (E) $y=2$

2. What is the horizontal asymptote of $\frac{1-2x}{3x}$?

   (A) $x=0$

   (B) $y=0$

   (C) $x=\frac{2}{3}$

   (D) $y=-\frac{3}{2}$

   (E) $y=-\frac{2}{3}$

3. $\lim_{x\to2}\frac{3x-2}{2x-3}=$

   (A) Undefined

   (B) 1

   (C) 2

   (D) 3

   (E) 4

4. If $f(x)=\frac{x^2-4}{x+2}$, which of the following statement is true about f ?

   (A) f has a hole at $x=-2$

   (B) f has the vertical asymptote at $x=-2$

   (C) f has the horizontal asymptote at $y=0$

   (D) f has the horizontal asymptote at $y=1$

   (E) f is undefined at $x=2$

5. $\lim_{x\to\infty} \dfrac{x-5}{x^2-x-2} =$

(A) $x=-1$

(B) $x=2$

(C) $y=0$

(D) $y=1$

(E) $y=5$

6. If $f(x)=\frac{2}{x^2+1}$, which of the following statement is NOT true about f ?

(A) There is no x-intercept.

(B) The y-intercept is 2.

(C) There is no vertical asymptote.

(D) As x increases without bound, f increases without bound.

(E) The horizontal asymptote is $y=0$.

7. If $f(x)=\frac{x+3}{2x+1}$ and $g(x)=2f(x)$, what is the horizontal asymptote of g ?

(A) $y=3$

(B) $y=2$

(C) $y=1$

(D) $y=\frac{1}{2}$

(E) $y=0$

8. $\lim_{x\to 2} \dfrac{x^3-8}{x-2} =$

(A) 16

(B) 12

(C) 8

(D) 4

(E) 1

9. The vertical asymptote and the horizontal asymptote of f are $x=\frac{3}{2}$ and $y=\frac{1}{2}$, respectively. If the x-intercept and y-intercept of f are 4 and $\frac{4}{3}$, respectively, which of the following function can be f ?

(A) $f(x)=\frac{x-4}{x-3}$

(B) $f(x)=\frac{x-4}{2x-3}$

(C) $f(x)=\frac{2x-3}{x-4}$

(D) $f(x)=\frac{x-3}{x-4}$

(E) $f(x)=\frac{x+4}{2x-3}$

10. $\lim_{x\to 0} \dfrac{\sin x}{x} =$

(A) 0

(B) 0.017

(C) 0.5

(D) 1

(E) Undefined

ANSWERS AND SOLUTIONS

1. (A)

In order to find the vertical asymptote, set the denominator equal to zero and solve for x: $x-1=0$. Thus, $x=1$. Therefore, the vertical asymptote of $\frac{x-2}{x-1}$ is $x=1$.

2. (E)

Since $\frac{1-2x}{3x}=\frac{-2x+1}{3x}$, both the numerator and the denominator are first degree polynomials. Thus, the horizontal asymptote of $\frac{-2x+1}{3x}$ is the ratio of the leading coefficients, or $y=-\frac{2}{3}$.

3. (E)

In order to find the limit of the rational function $\frac{3x-2}{2x-3}$ at $x = 2$, plug-in $x = 2$ to both the numerator and denominator.

$$\lim_{x \to 2} \frac{3x-2}{2x-3} = \frac{3(2)-2}{2(2)-3} = 4$$

Therefore, (E) is the correct answer.

4. (A)

f has a common factor of $x + 2$ in both the numerator and the denominator.

$$f(x) = \frac{x^2-4}{x+2} = \frac{\cancel{(x+2)}(x-2)}{\cancel{x+2}} = x - 2$$

Thus, cancelling out $x + 2$ produces a hole in the graph of f at $x = -2$. Since f simplifies to $x - 2$, f does not have vertical and horizontal asymptotes.

5. (C)

$\lim_{x \to \infty} \frac{x-5}{x^2-x-2}$ means finding the horizontal asymptote of $\frac{x-5}{x^2-x-2}$. Since the numerator is a first degree polynomial ($m = 1$) and the denominator is a second degree polynomial ($n = 2$), the horizontal asymptote is $y = 0$.

6. (D)

Since there is no x-value that makes $\frac{2}{x^2+1} = 0$, f does not have a x-intercept. In order to find the y-intercept of f, substitute $x = 0$. Thus, the y-intercept of f is 2. Since the denominator of f, $x^2 + 1$, cannot be zero, there is no vertical asymptote. Since the numerator of f is a constant ($m = 0$) and the denominator of f is a second degree polynomial ($n = 2$), the horizontal asymptote is $y = 0$. Therefore, (D) is the correct answer.

7. (C)

$f(x) = \frac{x+3}{2x+1}$. Since $g(x) = 2f(x)$,

$$g(x) = \frac{2(x+3)}{2x+1} = \frac{2x+6}{2x+1}$$

Since both the numerator and the denominator of g are first degree polynomials, the horizontal asymptote of g is the ratio of the leading coefficients, or $y = \frac{2}{2} = 1$.

8. (B)

Plug-in $x = 2$ to both the numerator and the denominator.

$$\lim_{x\to 2} \frac{x^3-8}{x-2} = \frac{(2)^3-8}{2-2} = \frac{0}{0}$$

Since $\lim_{x\to 2} \frac{x^3-8}{x-2} = \frac{0}{0}$, do the following steps.

$$\begin{aligned}
\lim_{x\to 2} \frac{x^3-8}{x-2} &= \frac{(x-2)(x^2+2x+4)}{x-2} && \text{Factor the numerator} \\
&= \lim_{x\to 2} \frac{\cancel{(x-2)}(x^2+2x+4)}{\cancel{x-2}} && \text{Cancel out } x-2 \\
&= \lim_{x\to 2} (x^2+2x+4) && \text{Plug-in } x=2 \\
&= 12
\end{aligned}$$

Therefore, the limit of the function at $x = 2$ is 12.

9. (B)

The functions in answer choices (B) and (E) both have vertical asymptotes at $x = \frac{3}{2}$. Since the function in answer choice (B) has an x-intercept at $x = 4$, (B) is the correct answer.

10. (D)

Tip: $\lim_{x\to 0}$ means that the value of x is getting closer to 0. It does not mean that $x = 0$.

Set the angle mode to Radians on your calculator. Then, substitute $x = 0.01$ into $\frac{\sin x}{x}$ to find the limit.

$$\lim_{x\to 0} \frac{\sin x}{x} = \frac{\sin(0.01)}{0.01} \approx 1$$

Therefore, (D) is the correct answer.

LESSON 7 Exponential and Logarithmic Functions

Graphs of Exponential and Logarithmic Functions

The exponential functions, $y = a^x$, and logarithmic functions, $y = \log_a x$, are inverses of each other. For instance, the inverse function of 3^x is $\log_3 x$. The following two equations relate exponents to logarithms or vice versa and are used to find the inverse function of an exponential function or a logarithmic function.

$$a^x = y \quad \Longleftrightarrow \quad x = \log_a y$$

Note that the two equations above are equivalent.

Since exponential functions and logarithmic functions are inverse functions, their graphs are symmetric with respect to the line $y = x$ as shown in Figures 1 and 2.

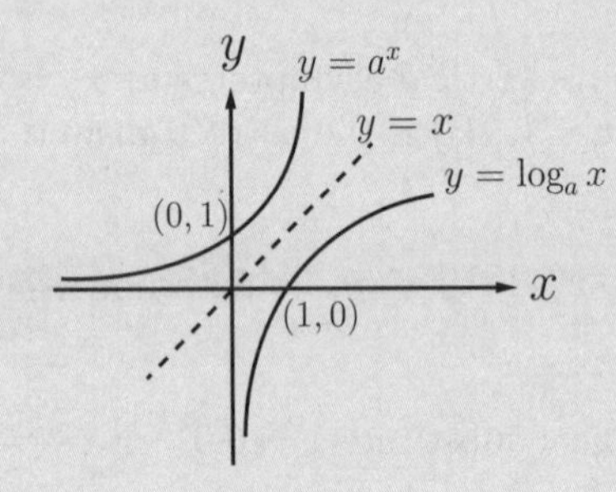

Figure 1: when $a > 1$

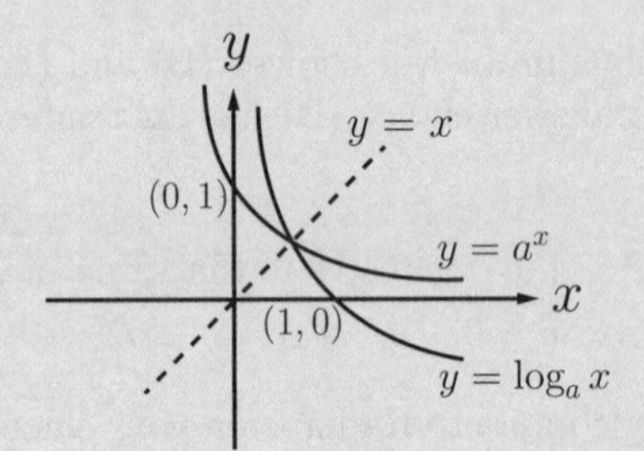

Figure 2: when $0 < a < 1$

Tip: The graphs of all exponential functions pass through the point $(0, 1)$. Whereas, the graphs of all logarithmic functions pass through the point $(1, 0)$.

Domain of a Logarithmic function

For a logarithmic function $y = \log_a h(x)$, where $h(x)$ is an algebraic expression or the argument of a logarithmic function, the domain of a logarithmic function is a set of all x-values for which $h(x) > 0$. In other words, solve the inequality $h(x) > 0$ to find the domain of a logarithmic function. For instance,

$$\log_2(x-2) \implies \text{Solve } x-2>0 \implies \text{Domain: } x>2$$
$$\log_3(6-2x) \implies \text{Solve } 6-2x>0 \implies \text{Domain: } x<3$$
$$\log_{10}(x^2-2x) \implies \text{Solve } x^2-2x>0 \implies \text{Domain: } x<0 \text{ or } x>2$$

Properties of Exponents

The table below summarizes the properties of exponents.

Properties of Exponents	Example
1. $a^m \cdot a^n = a^{m+n}$	1. $2^4 \cdot 2^6 = 2^{10}$
2. $\frac{a^m}{a^n} = a^{m-n}$	2. $\frac{2^{10}}{2^3} = 2^{10-3} = 2^7$
3. $(a^m)^n = a^{mn} = (a^n)^m$	3. $(2^3)^4 = 2^{12} = (2^4)^3$
4. $a^0 = 1$	4. $(-2)^0 = 1,\ (3)^0 = 1,\ (100)^0 = 1$
5. $a^{-1} = \frac{1}{a}$	5. $2^{-1} = \frac{1}{2}$
6. $a^{\frac{1}{n}} = \sqrt[n]{a}$	6. $2^{\frac{1}{2}} = \sqrt{2}, \quad x^{\frac{1}{3}} = \sqrt[3]{x}$
7. $a^{\frac{m}{n}} = (a^m)^{\frac{1}{n}} = \sqrt[n]{a^m}$	7. $2^{\frac{3}{2}} = \sqrt[2]{2^3}, \quad x^{\frac{3}{4}} = \sqrt[4]{x^3}$
8. $a^{-\frac{m}{n}} = (a^{\frac{m}{n}})^{-1} = \frac{1}{\sqrt[n]{a^m}}$	8. $2^{-\frac{3}{4}} = (2^{\frac{3}{4}})^{-1} = \frac{1}{\sqrt[4]{2^3}}$
9. $(ab)^n = a^n b^n$	9. $(2 \cdot 3)^6 = 2^6 \cdot 3^6, \quad (2x)^2 = 2^2 x^2$
10. $\left(\frac{a}{b}\right)^n = \frac{a^n}{b^n}$	10. $(\frac{2}{x})^3 = \frac{2^3}{x^3}$
11. $\frac{b^{-n}}{a^{-m}} = \frac{a^m}{b^n}$	11. $\frac{y^{-3}}{x^{-2}} = \frac{x^2}{y^3}$

Properties of Logarithms

The table below summarizes the properties of logarithms.

Properties of Logarithms	Example
1. $\log_a 0 = \text{undefined}$	1. $\log_a 0 = \text{undefined}$
2. $\log_a 1 = 0$	2. $\log_2 1 = 0$
3. $\log_a a = 1$	3. $\log_2 2 = 1$
4. $\log_a x^n = n \log_a x$	4. $\log_2 125 = \log_2 5^3 = 3 \log_2 5$
5. $\log_{a^n} x = \frac{1}{n} \log_a x$	5. $\log_8 5 = \log_{2^3} 5 = \frac{1}{3} \log_2 5$
6. $\log_a x = \frac{\log_c x}{\log_c a}$	6. $\log_2 3 = \frac{\log_{10} 3}{\log_{10} 2} = \frac{\ln 3}{\ln 2}$
7. $\log_a xy = \log_a x + \log_a y$	7. $\log_2 15 = \log_2 (3 \cdot 5) = \log_2 3 + \log_2 5$
8. $\log_a \frac{x}{y} = \log_a x - \log_a y$	8. $\log_2 \frac{5}{3} = \log_2 5 - \log_2 3$
9. $a^{\log_a x} = x^{\log_a a} = x$	9. $2^{\log_2 3} = 3^{\log_2 2} = 3^1 = 3$

1. The common logarithm is the logarithm with base 10, which is denoted by either $\log_{10} x$ or $\log x$.
2. e is an irrational number and is approximately $2.718\cdots$. The natural logarithm of x can be expressed as either $\log_e x$ or $\ln x$.

Example 1 Finding the domain of a logarithmic function

Find the domain of $y = \log_3(4 - x^2)$.

Solution The domain of the logarithmic function $y = \log_3(4 - x^2)$ is a set of all x-values for which $4 - x^2 > 0$. Thus,

$4 - x^2 > 0$	Subtract 4 from each side
$-x^2 > -4$	Multiply each side by -1 and reverse the inequality symbol
$x^2 < 4$	Solve the inequality
$-2 < x < 2$	

Therefore, the domain of $y = \log_3(4 - x^2)$ is $-2 < x < 2$ or $(-2, 2)$.

Example 2 Finding the inverse function of an exponential function

Find the inverse function of $y = 3^{x-1} + 2$.

Solution In order to find the inverse function, switch the x and y variables and solve for y.

$y = 3^{x-1} + 2$	Switch the x and y variables
$x = 3^{y-1} + 2$	Subtract 2 from each side
$3^{y-1} = x - 2$	Convert the equation to a logarithmic equation
$y - 1 = \log_3(x - 2)$	Add 1 to each side
$y = \log_3(x - 2) + 1$	

Therefore, the inverse function of $y = 3^{x-1} + 2$ is $f^{-1}(x) = \log_3(x - 2) + 1$.

Example 3 Writing a logarithmic expression

If $\log 2 = x$ and $\log 3 = y$, write $\log 72$ in terms of x and y.

Solution Since the prime factorization of $72 = 2^3 \cdot 3^2$,

$\log 72 = \log(2^3 \cdot 3^2)$	Use $\log_a xy = \log_a x + \log_a y$
$= \log 2^3 + \log 3^2$	Use $\log_a x^n = n \log_a x$
$= 3\log 2 + 2\log 3$	Since $\log 2 = x$ and $\log 3 = y$
$= 3x + 2y$	

Therefore, $\log 72$ can be written as $3x + 2y$.

Solving Exponential and Logarithmic Equations

Below shows how to solve an exponential equation and a logarithmic equation.

- When each side of an equation (either exponential or logarithmic) has the same base.

$$\text{If } a^x = a^y \implies x = y \qquad \text{e.g. If } 2^x = 2^3, \text{ then } x = 3$$
$$\text{If } \log_a x = \log_a y \implies x = y \qquad \text{e.g. If } \log_2 x = \log_2 5, \text{ then } x = 5$$

- When each side of an equation (either exponential or logarithmic) has a different base, convert the exponential equation to a logarithmic equation or vice versa.

$$\text{If } a^x = b \implies x = \log_a b \qquad \text{e.g. If } 2^x = 7, \text{ then } x = \log_2 7$$
$$\text{If } \log_a x = b \implies x = a^b \qquad \text{e.g. If } \log_2 x = 3, \text{ then } x = 2^3$$

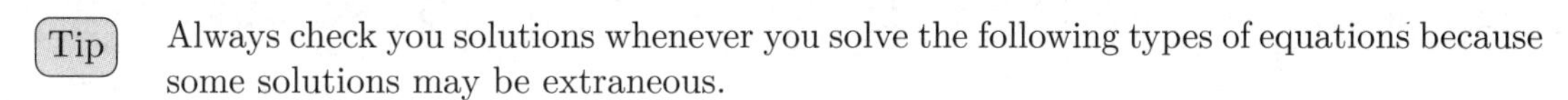

Tip Always check you solutions whenever you solve the following types of equations because some solutions may be extraneous.

- Radical equations: e.g. $x - 1 = \sqrt{x+5}$
- Rational equations: e.g. $\dfrac{1}{x-1} = \dfrac{2}{x(x-1)}$
- Logarithmic equations: e.g. $\ln(x+1) + \ln(x-2) = \ln 4$

Example 4 Solving an exponential equation

If $2^x = 3^y$, find the value of $\frac{x}{y}$.

Solution Since each side of equation has a different base (left side has the base 2 and right side has the base 3), convert the exponential equation to a logarithmic equation.

$2^x = 3^y$	Convert the equation to a logarithmic equation
$x = \log_2 3^y$	Use $\log_a x^n = n \log_a x$
$x = y \log_2 3$	Divide each side by y
$\dfrac{x}{y} = \log_2 3$	

Therefore, the value of $\frac{x}{y}$ is $\log_2 3 \approx 1.585$.

Example 5 Solving a logarithmic equation

Solve: $\ln(x+1)+\ln(x-2)=\ln 4$

Solution Express the left side as a single logarithm and solve for x.

$\ln(x+1)+\ln(x-2)=\ln 4$	Express the left side as a single logarithm
$\ln\left[(x+1)(x-2)\right]=\ln 4$	Since each side has the same base, e
$(x+1)(x-2)=4$	Subtract 4 from each side
$x^2-x-6=0$	Use the factoring method
$(x+2)(x-3)=0$	Use the zero product property
$x=-2$ or $x=3$	

Substitute -2 and 3 for x in the original equation to check the solutions.

$\ln(-2+1)+\ln(-2-2)=\ln 4$
undefined $\neq \ln 4$ (Not a solution)

$\ln(3+1)+\ln(3-2)=\ln 4$
$\ln 4=\ln 4$ ✓ (Solution)

Therefore, the only solution to $\ln(x+1)+\ln(x-2)=\ln 4$ is $x=3$.

Compound Interest

If an initial amount P is invested at an annual interest rate (expressed as a decimal) r compounded n times per year, the amount of money A accumulated in t years is as follows:

$$A=P\left(1+\frac{r}{n}\right)^{nt}$$

For instance, the initial amount of $1000 is deposited into a savings account which yields an annual interest of 4%. The amounts after 3 years if the interest is compounded annually, quarterly, and monthly are shown below.

- Annually compounded:

$$A=\$1000\left(1+\frac{0.04}{1}\right)^{1\cdot 3}=\$1124.86$$

- Quarterly compounded:

$$A=\$1000\left(1+\frac{0.04}{4}\right)^{4\cdot 3}=\$1126.83$$

- Monthly compounded:

$$A=\$1000\left(1+\frac{0.04}{12}\right)^{12\cdot 3}=\$1127.27$$

Tip If the interest rate is compounded continuously, the compound interest formula above becomes $A=Pe^{rt}$. Thus, the amount after 3 years is $A=\$1000e^{(0.04)(3)}=\$1127.5$

EXERCISES

1. Which of the following expression is equal to $\sqrt[4]{4}$?

   (A) $\sqrt{2}$

   (B) $\frac{1}{2}$

   (C) 1

   (D) $\frac{3}{2}$

   (E) 2

2. What is the domain of $y = \ln(4 - 2x)$?

   (A) $x \leq 2$

   (B) $x < 2$

   (C) $x > 0$

   (D) $x > 2$

   (E) $x \geq 2$

3. Solve: $9^{5+x} = 27^{x+2}$

   (A) 1

   (B) 2

   (C) 3

   (D) 4

   (E) 5

4. $\log_{\frac{1}{9}} 3 + \log_9 27 =$

   (A) 1

   (B) $\frac{3}{2}$

   (C) 2

   (D) $\frac{5}{2}$

   (E) 3

5. If $f(x) = \log_2(x+1) - 3$, what is the inverse function of f ?

   (A) $3^{x+1} - 2$

   (B) $3^{x+2} + 1$

   (C) $3^{x-2} - 1$

   (D) $2^{x+1} + 3$

   (E) $2^{x+3} - 1$

6. If $3^m = 4^n$, what is the value of $\frac{n^2}{m^2}$?

   (A) 1.59

   (B) 1.26

   (C) 0.63

   (D) 0.79

   (E) 0.87

7. Solve: $\log_{10} 5x + \log_{10}(x+1) = 2$

   (A) $x = 2$ only

   (B) $x = 2$ or $x = -3$

   (C) $x = -5$ only

   (D) $x = 4$ only

   (E) $x = 4$ or $x = -5$

8. If $\log 2 = n$, what is $\log 5$ in terms of n ?

   (A) $1 - 2n$

   (B) $1 - n$

   (C) $n - 1$

   (D) $n + 1$

   (E) $2n - 1$

9. An initial amount of \$2000 is deposited into savings account which yields an annual interest of 2%. What is the amount after 5 years if the interest is compounded semiannually?

   (A) \$2158.13

   (B) \$2187.48

   (C) \$2209.24

   (D) \$2437.99

   (E) \$10201.11

10. Which of the following expression is equal to $3^{2\log_3 x}$?

   (A) $\frac{1}{x^2}$

   (B) $\frac{1}{x}$

   (C) x

   (D) x^2

   (E) x^3

ANSWERS AND SOLUTIONS

1. (A)

   Tip $\sqrt[n]{a} = a^{\frac{1}{n}}$

$$\sqrt[4]{4} = 4^{\frac{1}{4}} = (2^2)^{\frac{1}{4}} = 2^{\frac{1}{2}} = \sqrt{2}$$

   Therefore, $\sqrt[4]{4} = \sqrt{2}$.

2. (B)

   The domain of the logarithmic function $y = \ln(4-2x)$ is the set of all x-values for which $4-2x > 0$.

$4 - 2x > 0$	Subtract 4 from each side
$-2x > -4$	Divide each side by -2 and reverse the inequality symbol
$x < 2$	

   Therefore, the domain of $y = \ln(4 - 2x)$ is $x < 2$.

3. (D)

   Change 9 to 3^2, and 27 to 3^3.

$9^{5+x} = 27^{x+2}$	Since $9 = 3^2$ and $27 = 3^3$
$(3^2)^{5+x} = (3^3)^{x+2}$	Use the exponent property: $(a^m)^n = a^{mn}$
$3^{10+2x} = 3^{3x+6}$	Since each side of an equation has the same base
$10 + 2x = 3x + 6$	Solve for x
$x = 4$	

   Therefore, the solution to $9^{5+x} = 27^{x+2}$ is $x = 4$.

4. (A)

> **Tip**
>
> Use the following properties of logarithms.
>
> 1. $\log_a a = 1$
> 2. $\log_a x^n = n \log_a x$
> 3. $\log_{a^n} x = \frac{1}{n} \log_a x$

Since $\log_{\frac{1}{9}} 3 = -\frac{1}{2}$ and $\log_9 27 = \frac{3}{2}$ as shown below,

$$\log_{\frac{1}{9}} 3 = \log_{3^{-2}} 3 = -\frac{1}{2}\log_3 3 = -\frac{1}{2}$$

$$\log_9 27 = \log_{3^2} 3^3 = 3\log_{3^2} 3 = \frac{3}{2}\log_3 3 = \frac{3}{2}$$

$\log_{\frac{1}{9}} 3 + \log_9 27 = -\frac{1}{2} + \frac{3}{2} = 1$. Therefore, (A) is the correct answer.

5. (E)

In order to find the inverse function, switch the x and y variables and solve for y.

$y = \log_2(x+1) - 3$	Switch the x and y variables
$x = \log_2(y+1) - 3$	Add 3 to each side
$x + 3 = \log_2(y+1)$	Convert the equation to an exponential equation
$y + 1 = 2^{x+3}$	Subtract 1 from each side
$y = 2^{x+3} - 1$	

Therefore, the inverse function of $y = \log_2(x+1) - 3$ is $2^{x+3} - 1$.

6. (C)

Since each side of the equation has a different base (the left side has a base of 3 and the right side has a base of 4), convert the exponential equation to a logarithmic equation.

$3^m = 4^n$	Convert the equation to a logarithmic equation
$m = \log_3 4^n$	Use $\log_a x^n = n\log_a x$
$m = n\log_3 4$	Divide each side by n
$\frac{m}{n} = \log_3 4 = 1.26$	Take the reciprocal of each side
$\frac{n}{m} = 0.79$	Square each side
$\frac{n^2}{m^2} = 0.63$	

Therefore, (C) is the correct answer.

7. (D)

Express the left side as a single logarithm and solve for x.

$\log_{10} 5x + \log_{10}(x+1) = 2$	Express the left side as a single logarithm
$\log_{10}\left[5x(x+1)\right] = 2$	Convert the equation to an exponential equation
$5x(x+1) = 10^2$	Divide each side by 5
$x^2 + x - 20 = 0$	Use the factoring method
$(x+5)(x-4) = 0$	Use the zero product property
$x = -5 \quad \text{or} \quad x = 4$	

Substitute -5 and 4 for x in the original equation to check the solutions.

$$\log_{10}(5(-5)) + \log_{10}(-5+1) = 2$$
$$\text{undefined} \neq 2$$

$$\log_{10}(5(4)) + \log_{10}(4+1) = 2$$
$$\log_{10} 100 = 2 \qquad \checkmark \text{ (Solution)}$$

Therefore, the only solution to $\log_{10} 5x + \log_{10}(x+1) = 2$ is $x = 4$.

8. (B)

$\log_{10} 5 = \log_{10} \frac{10}{2}$	Use $\log_a \frac{x}{y} = \log_a x - \log_a y$
$= \log_{10} 10 - \log_{10} 2$	Since $\log_{10} 10 = 1$ and $\log_{10} 2 = n$
$= 1 - n$	

Therefore, $\log 5$ in terms of n is $1 - n$.

9. (C)

$$A = P\left(1 + \frac{r}{n}\right)^{nt} \qquad \text{where } P = \$2000,\ r = 0.02,\ n = 2, \text{ and } t = 5$$
$$= \$2000\left(1 + \frac{0.02}{2}\right)^{2\cdot 5}$$
$$= \$2209.24$$

Therefore, (C) is the correct answer.

10. (D)

Tip

Use the following properties of logarithms.

1. $\log_a a = 1$
2. $n \log_a x = \log_a x^n$
3. $a^{\log_a x} = x^{\log_a a} = x$

Since $2\log_3 x = \log_3 x^2$,

$$3^{2\log_3 x} = 3^{\log_3 x^2} = (x^2)^{\log_3 3} = x^2$$

Therefore, $3^{2\log_3 x} = x^2$.

LESSON 8 Trigonometric Functions

Angles

Two rays form an angle. One ray is called the **initial side** and the other ray is called the **terminal side**. An angle θ is in **standard position** if its vertex is at the origin and the initial side is on the positive x-axis shown in Figure 1.

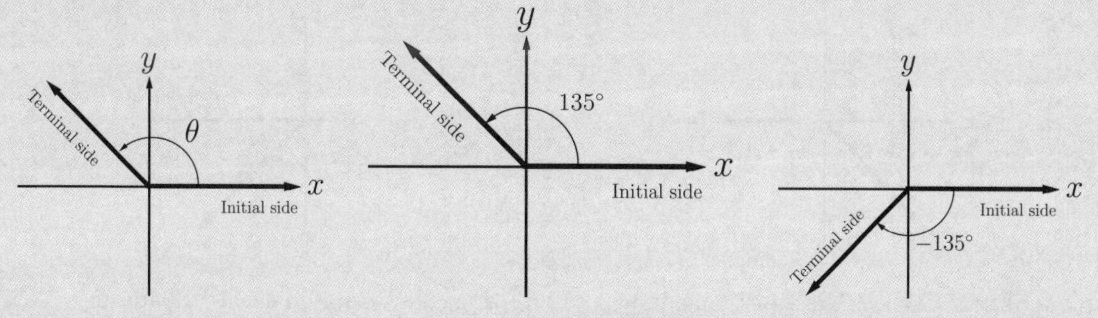

Fig. 1: Standard position　　Fig. 2: Positive angle　　Fig. 3: Negative angle

The angle shows the direction and amount of rotation from the initial side to the terminal side. If the rotation is in a counterclockwise direction, the angle is **positive** shown in Figure 2. Whereas, if the rotation is in a clockwise direction, the angle is **negative** shown in Figure 3.

Coterminal angle are angles in standard position that have a common terminal side. For instance, angles $135°$ and $-225°$, shown in Figure 4, are coterminal angles.

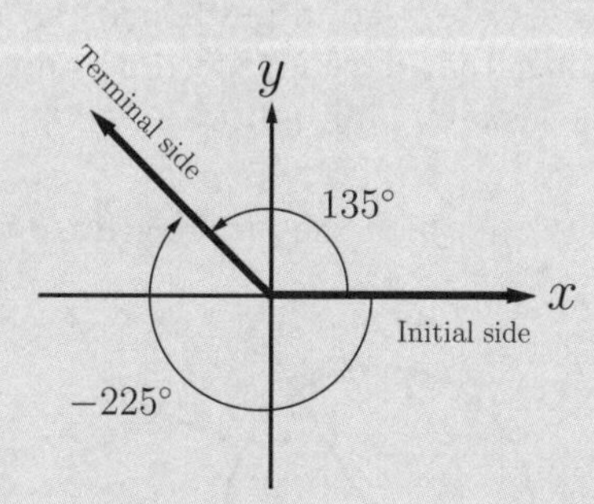

Fig. 4: Coterminal angles

Reference angle, β, is a positive acute angle formed by the terminal side and the closest x-axis, not the y-axis. If the terminal side lies in the first quadrant, reference angle β is the same as angle θ as shown in Figure 5. However, if the terminal side lies in other quadrants, reference angle β is different from the angle θ as shown in Figures 6, 7, and 8.

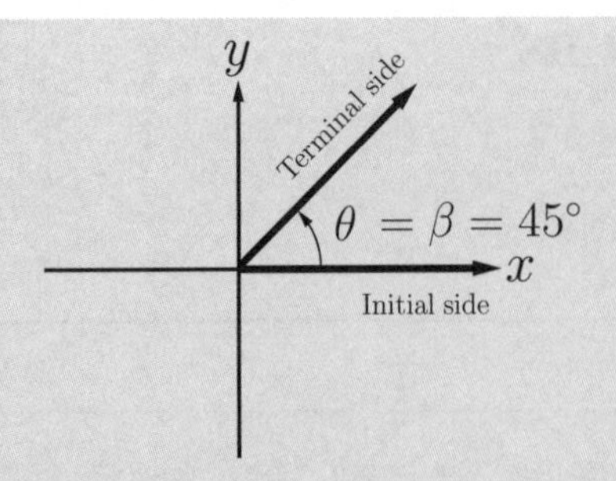

Fig. 5: When θ in the 1st quadrant

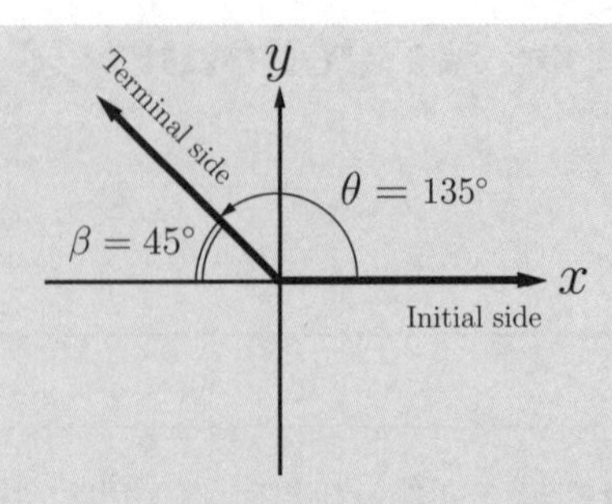

Fig. 6: When θ in the 2nd quadrant

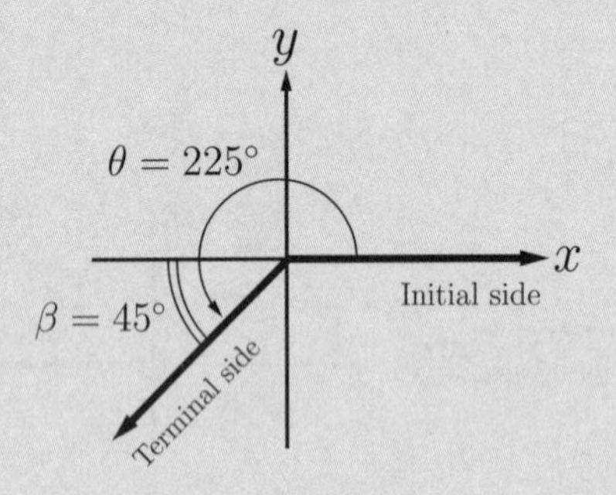

Fig. 7: When θ in the 3rd quadrant

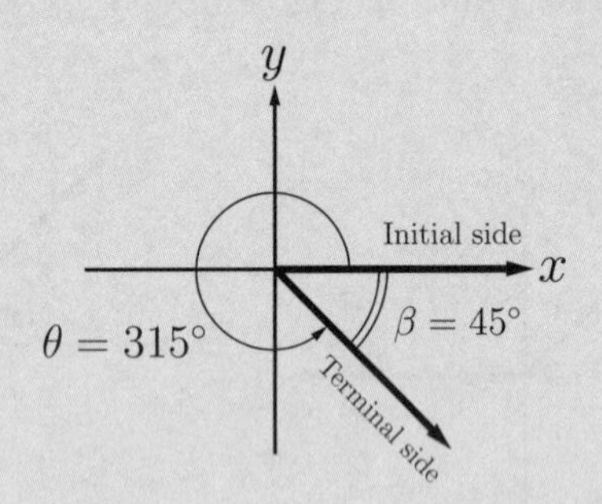

Fig. 8: When θ in the 4th quadrant

Radians

The radian is a very useful angular measure used in mathematics. Mathematicians prefer the radian to the degree (°) because it is a number that does not need an unit symbol. Although the radian can be denoted by the symbol "rad", it is usually omitted.

A **radian** is the measure of an angle θ at which the arc length is equal to the radius of the circle as shown in Figure 9.

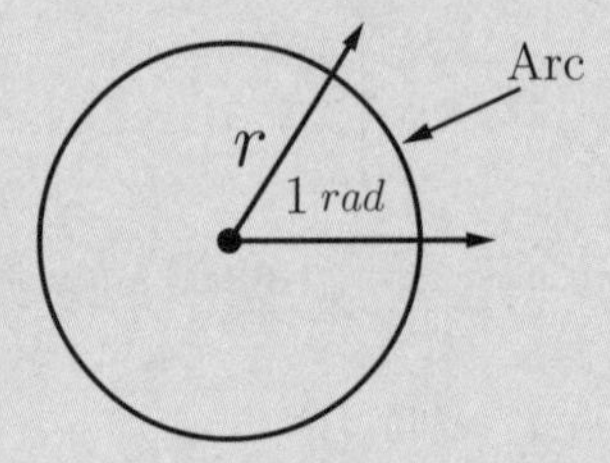

Fig. 9: At 1 radian, arc length = radius

Measuring one radian with a protractor, it is approximately $57.3°$. Thus,

$$1\,rad \approx 57.3° \qquad \text{Multiply each side by } \pi = 3.141592\cdots$$
$$\pi\,rad = 180°$$

Since $\pi\,rad = 180°$, $2\pi\,rad = 360°$.

Arc length and Area of a sector

An arc, shown in Figure 10, is a part of the circumference of a circle. A part can be expressed as the ratio of the central angle to $360°$ or 2π.

When θ (°) is given: $\quad$ Arc length $= 2\pi r \times \dfrac{\theta}{360°}$

When θ (rad) is given: $\quad$ Arc length $= \cancel{2\pi} r \times \dfrac{\theta}{\cancel{2\pi}} = r\theta\ (rad)$

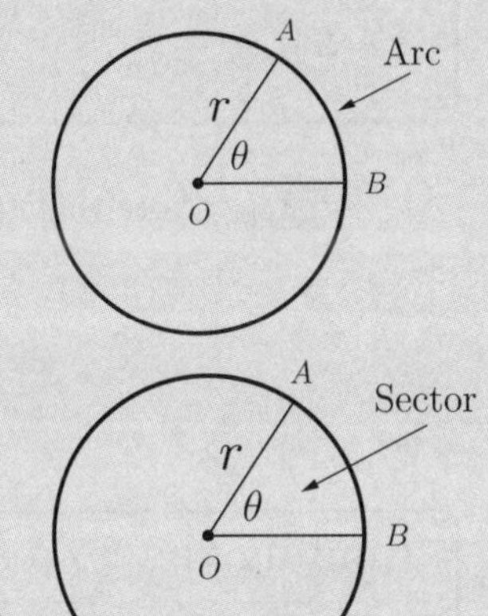

Figure 10

A sector, shown in Figure 10, is a part of the area of a circle. A part can be expressed as the ratio of the central angle to $360°$ or 2π.

When θ (°) is given: $\quad$ Area of a sector $= \pi r^2 \times \dfrac{\theta}{360°}$

When θ (rad) is given: $\quad$ Area of a sector $= \cancel{\pi} r^2 \times \dfrac{\theta}{2\cancel{\pi}} = \dfrac{1}{2}r^2\theta\ (rad)$

Conversion of Angles Between Degrees and Radians

The table below shows the conversion of angles between degrees and radians.

Degrees	30°	45°	60°	90°	120°	135°	150°	180°
Radians	$\frac{\pi}{6}$	$\frac{\pi}{4}$	$\frac{\pi}{3}$	$\frac{\pi}{2}$	$\frac{2\pi}{3}$	$\frac{3\pi}{4}$	$\frac{5\pi}{6}$	π

Degrees	210°	225°	240°	270°	300°	315°	330°	360°
Radians	$\frac{7\pi}{6}$	$\frac{5\pi}{4}$	$\frac{4\pi}{3}$	$\frac{3\pi}{2}$	$\frac{5\pi}{3}$	$\frac{7\pi}{4}$	$\frac{11\pi}{6}$	2π

1. To convert degrees to radians, multiply degrees by $\frac{\pi}{180°}$.
2. To convert radians to degrees, multiply radians by $\frac{180°}{\pi}$.

Definitions of Six Trigonometric Functions

The six trigonometric functions are defined as ratios of sides in a right triangle. In the right triangle, shown to the right, the definitions of the six trigonometric functions are as follows:

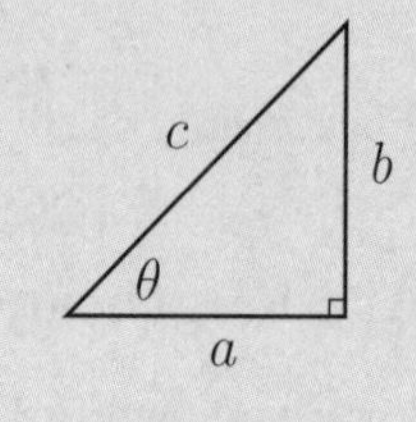

$$\sin\theta = \frac{\text{opposite}}{\text{hypotenuse}} = \frac{b}{c} \qquad \cos\theta = \frac{\text{adjacent}}{\text{hypotenuse}} = \frac{a}{c} \qquad \tan\theta = \frac{\text{opposite}}{\text{adjacent}} = \frac{b}{a}$$

$$\csc\theta = \frac{1}{\sin\theta} = \frac{c}{b} \qquad \sec\theta = \frac{1}{\cos\theta} = \frac{c}{a} \qquad \cot\theta = \frac{1}{\tan\theta} = \frac{a}{b}$$

Tip Note that $\tan\theta = \frac{\sin\theta}{\cos\theta}$ and $\cot\theta = \frac{\cos\theta}{\sin\theta}$.

Finding the Exact Value of Trigonometric Functions

The table below shows the exact value of $\sin\theta$, $\cos\theta$, and $\tan\theta$ at reference angle, β.

	Reference angle, β		
	$30°\ (\frac{\pi}{6})$	$45°\ (\frac{\pi}{4})$	$60°\ (\frac{\pi}{3})$
$\sin\theta$	$\frac{1}{2}$	$\frac{\sqrt{2}}{2}$	$\frac{\sqrt{3}}{2}$
$\cos\theta$	$\frac{\sqrt{3}}{2}$	$\frac{\sqrt{2}}{2}$	$\frac{\sqrt{1}}{2}$
$\tan\theta$	$\frac{\sqrt{3}}{3}$	1	$\sqrt{3}$

II (y)	I
$\sin\theta > 0$ $\csc\theta > 0$	All positive
$\tan\theta > 0$ $\cot\theta > 0$	$\cos\theta > 0$ $\sec\theta > 0$ (x)
III	IV

The sign of a trigonometric function is determined by the quadrant the terminal side of the angle (or simply the angle) lies in. The chart above shows which quadrants the six trigonometric functions are positive. For instance, $\sin\theta$ and $\csc\theta$ are positive in the 1st and 2nd quadrants, $\cos\theta$ and $\sec\theta$ are positive in the 1st and 4th quadrants, and $\tan\theta$ and $\cot\theta$ are positive in the 1st and 3rd quadrants.

Evaluating trigonometric functions using the reference angle

When the angle θ lies in the either 2nd, or 3rd, or 4th quadrant, use the following formulas to evaluate the trigonometric functions.

$$\sin\theta = \pm\sin\beta \qquad \cos\theta = \pm\cos\beta \qquad \tan\theta = \pm\tan\beta$$

where β is the reference angle. Note that the sign of a trigonometric function, either $+$ or $-$, is determined by the quadrant the terminal side of the angle lies in.

Let's evaluate $\cos 225°$. As shown in the figure below, the angle $\theta = 225°$ lies in the third quadrant.

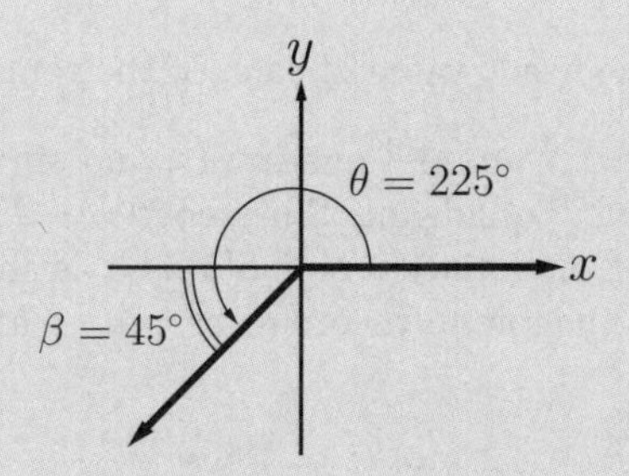

The reference angle β, an angle formed by the terminal side and the closest x-axis, is 45°. Since the angle θ lies in the 3rd quadrant, the sign of the cosine function is negative. Thus,

$$\begin{aligned} \cos 225° &= \pm \cos 45° && \text{Since cosine is negative in the 3rd quadrant} \\ &= -\cos 45° && \text{Since } \cos 45° = \frac{\sqrt{2}}{2} \\ &= -\frac{\sqrt{2}}{2} \end{aligned}$$

Therefore, the exact value of $\cos 225°$ is $-\frac{\sqrt{2}}{2}$.

Example 1 Finding the quadrant where an angle θ lies

Find the quadrant where each of the following angles θ lies.

(a) If $\sin\theta < 0$ and $\cos\theta > 0$

(b) If $\tan\theta < 0$ and $\cos\theta < 0$

(c) If $\cot\theta > 0$ and $\csc\theta < 0$

Solution

(a) $\sin\theta < 0$ indicates that θ lies in the 3rd quadrant or 4th quadrant. Furthermore, $\cos\theta > 0$ indicates that θ lies in the 1st quadrant or 4th quadrant. Thus, θ must be in the 4th quadrant.

(b) $\tan\theta < 0$ indicates that θ lies in the 2nd quadrant or 4th quadrant. Furthermore, $\cos\theta < 0$ indicates that θ lies in the 2nd quadrant or 3rd quadrant. Thus, θ must be in the 2nd quadrant.

(c) $\cot\theta > 0$ indicates that θ lies in the 1st quadrant or 3rd quadrant. Furthermore, $\csc\theta < 0$ indicates that θ lies in the 3rd quadrant or 4th quadrant. Thus, θ must be in the 3rd quadrant.

Example 2 Finding exact values of trigonometric functions

If $\cos\theta = -\frac{3}{5}$ and $\sin\theta > 0$, find the exact value of each of the remaining trigonometric functions.

Solution $\cos\theta$ is negative in the 2nd and 3rd quadrants, and $\sin\theta$ is positive in the 1st and 2nd quadrants. Thus, θ must be in the 2nd quadrant. Since $\cos\theta = -\frac{3}{5}$, the length of the hypotenuse of a right triangle is 5 and the length of the adjacent side of θ is 3 as shown below. Thus, the length of the opposite side of θ is 4 using the Pythagorean theorem: $c^2 = a^2 + b^2$.

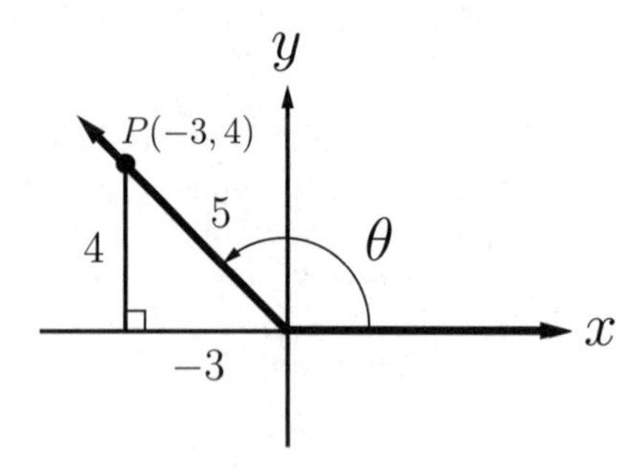

Suppose point $P(x, y)$ is on the terminal side of θ. Since θ lies in the 2nd quadrant, the x and y coordinates of point P is $(-3, 4)$. Use the definition of the trigonometric functions to find the exact values of the remaining trigonometric functions.

$$\sin\theta = \frac{\text{opposite}}{\text{hypotenuse}} = \frac{4}{5} \qquad \cos\theta = \frac{\text{adjacent}}{\text{hypotenuse}} = -\frac{3}{5} \qquad \tan\theta = \frac{\text{opposite}}{\text{adjacent}} = -\frac{4}{3}$$

$$\csc\theta = \frac{1}{\sin\theta} = \frac{5}{4} \qquad \sec\theta = \frac{1}{\cos\theta} = -\frac{5}{3} \qquad \cot\theta = \frac{1}{\tan\theta} = -\frac{3}{4}$$

Pythagorean Identities

$$\sin^2\theta + \cos^2\theta = 1 \qquad 1 + \tan^2\theta = \sec^2\theta \qquad 1 + \cot^2\theta = \csc^2\theta$$

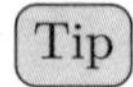

1. Note that $\sin^2\theta = (\sin\theta)^2$ and $\cos^2\theta = (\cos\theta)^2$.
2. The following variations of the Pythagorean identities are often used.

$$\sin^2\theta = 1 - \cos^2\theta, \qquad \cos^2\theta = 1 - \sin^2\theta$$

Even-Odd Properties

Knowing whether a trigonometric function is odd or even is useful when evaluating the trigonometric function of a negative angle. $\sin\theta$, $\csc\theta$, $\tan\theta$, and $\cot\theta$ are odd functions that satisfy $f(-\theta) = -f(\theta)$ for all θ. Whereas, $\cos\theta$ and $\sec\theta$ are even functions that satisfy $f(-\theta) = f(\theta)$ for all θ. Below is a summary of the even-odd properties for the six trigonometric functions.

$$\sin(-\theta) = -\sin\theta \qquad \csc(-\theta) = -\csc\theta$$
$$\cos(-\theta) = \cos\theta \qquad \sec(-\theta) = \sec\theta$$
$$\tan(-\theta) = -\tan\theta \qquad \cot(-\theta) = -\cot\theta$$

Example 3 Finding exact values using even-odd properties

Find the exact value of the following expressions.

(a) $\sin(-30°)$

(b) $\cos(-\frac{\pi}{6})$

Solution

(a) Since $\sin(-\theta) = -\sin\theta$, $\sin(-30°) = -\sin(30°) = -\frac{1}{2}$.

(b) Since $\cos(-\theta) = \cos\theta$, $\cos(-\frac{\pi}{6}) = \cos(\frac{\pi}{6}) = \frac{\sqrt{3}}{2}$.

Sum and Difference Formulas and Double-Angle Formulas

- Sum and Difference Formulas

$$\sin(\alpha+\beta) = \sin\alpha\cos\beta + \cos\alpha\sin\beta \qquad \sin(\alpha-\beta) = \sin\alpha\cos\beta - \cos\alpha\sin\beta$$
$$\cos(\alpha+\beta) = \cos\alpha\cos\beta - \sin\alpha\sin\beta \qquad \cos(\alpha-\beta) = \cos\alpha\cos\beta + \sin\alpha\sin\beta$$
$$\tan(\alpha+\beta) = \frac{\tan\alpha + \tan\beta}{1 - \tan\alpha\tan\beta} \qquad \tan(\alpha-\beta) = \frac{\tan\alpha - \tan\beta}{1 + \tan\alpha\tan\beta}$$

- Double-Angle Formulas

$$\sin 2\theta = 2\sin\theta\cos\theta$$
$$\begin{aligned}\cos 2\theta &= \cos^2\theta - \sin^2\theta \\ &= 2\cos^2\theta - 1 \\ &= 1 - 2\sin^2\theta\end{aligned}$$

Tip Using the sum formula for cosine, the double-angle formula for cosine can be derived as follows:

$$\begin{aligned}\cos 2\theta &= \cos(\theta+\theta) \\ &= \cos\theta\cdot\cos\theta - \sin\theta\cdot\sin\theta \\ &= \cos^2\theta - \sin^2\theta\end{aligned}$$

Cofunctions Identities

Two angles are **complementary** if their sum is equal to 90°. The following cofunction identities show relationships between sine and cosine, tangent and cotangent, and secant and cosecant. The value of a trigonometric function of an angle is equal to the value of the cofunction of the complementary of the angle.

$$\sin(90^\circ - \theta) = \cos\theta \qquad \tan(90^\circ - \theta) = \cot\theta \qquad \sec(90^\circ - \theta) = \csc\theta$$

For instance, 50° and 40° are complementary angles. Thus,

$$\sin 50^\circ = \cos 40^\circ \qquad \tan 50^\circ = \cot 40^\circ \qquad \sec 50^\circ = \csc 40^\circ$$

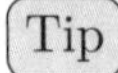

Since $180^\circ = \pi$ radians, $90^\circ = \frac{\pi}{2}$ radians. Thus, the cofunction identities can be expressed in radians.

$$\sin\left(\frac{\pi}{2} - \theta\right) = \cos\theta \qquad \tan\left(\frac{\pi}{2} - \theta\right) = \cot\theta \qquad \sec\left(\frac{\pi}{2} - \theta\right) = \csc\theta$$

Solving Trigonometric Equations

A trigonometric equation is an equation that involves trigonometric functions. The solutions of the trigonometric equation are the angles that satisfy the equation. In most cases, the Pythagorean Identities, factoring, the distributive property, or other algebraic skills are used to find the solutions of the equation.

Let's solve the trigonometric equation: $\sin\theta = -\frac{\sqrt{2}}{2}$, $0 \le \theta < 2\pi$.

First, find the reference angle β for which $\sin\beta = \frac{\sqrt{2}}{2}$. The reference angle is $\beta = 45^\circ$. Since the sine function is negative, the angle θ lies in either 3rd quadrant or 4th quadrant as shown below.

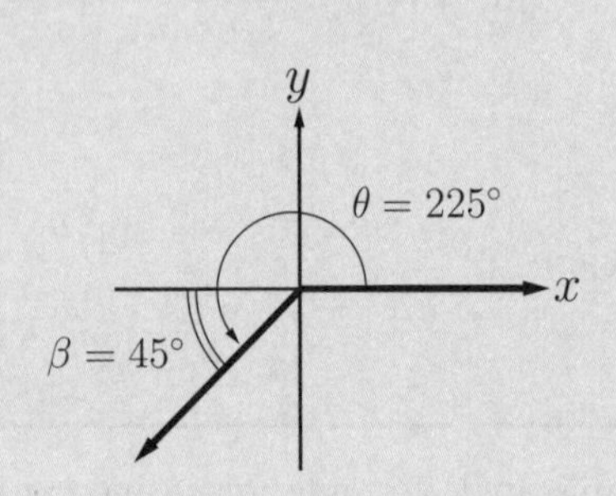

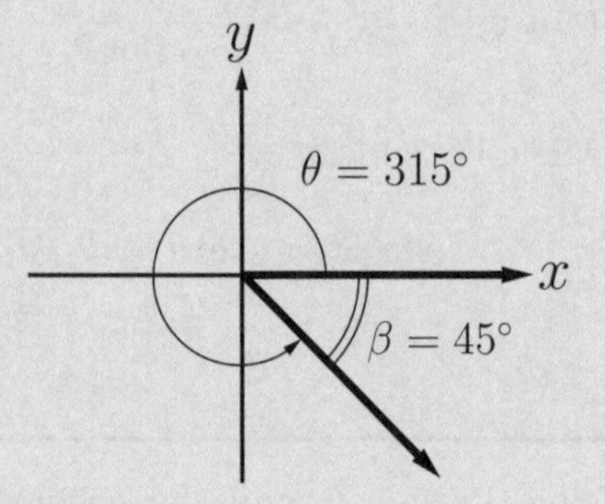

Therefore, the solutions of the equation are 225° (or $\frac{5\pi}{4}$) or 315° (or $\frac{7\pi}{4}$).

Example 4 Finding exact values using double-angle formula

If $\sin\theta = -\frac{12}{13}$, $\pi < \theta < \frac{3\pi}{2}$, find the exact values of $\sin 2\theta$ and $\cos 2\theta$.

Solution $\pi < \theta < \frac{3\pi}{2}$ indicates that θ lies in the 3rd quadrant. Since $\sin\theta = -\frac{12}{13}$, the length of the hypotenuse of a right triangle is 13 and the length of the opposite side of θ is 12 as shown below. Thus, the length of the adjacent side of θ is 5 using the Pythagorean theorem.

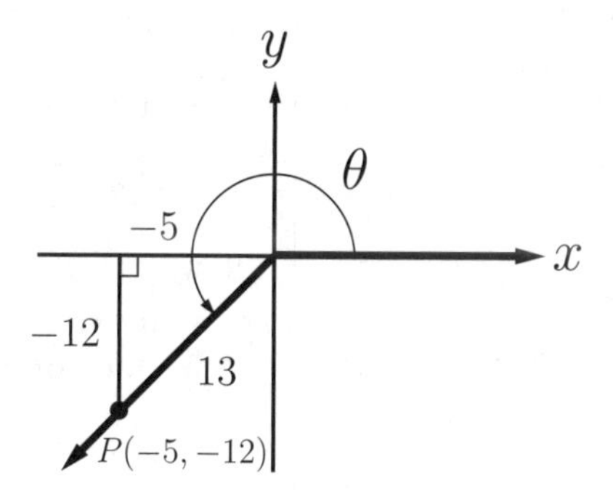

Suppose point $P(x, y)$ is on the terminal side of θ. Since θ lies in the 3rd quadrant, the x and y coordinates of point P is $(-5, -12)$ and $\cos\theta = -\frac{5}{13}$. Thus,

$$\sin 2\theta = 2\sin\theta\cos\theta = 2\left(-\frac{12}{13}\right)\left(-\frac{5}{13}\right) = \frac{120}{169}$$

$$\cos 2\theta = 1 - 2\sin^2\theta = 1 - 2\left(-\frac{12}{13}\right)^2 = 1 - 2\left(\frac{144}{169}\right) = -\frac{119}{169}$$

Example 5 Solving a trigonometric equation

Solve the equation: $2\sin^2\theta - 5\sin\theta + 2 = 0,\ 0 \le \theta < 2\pi$.

Solution Factor the expression $2\sin^2\theta - 5\sin\theta + 2$ and solve for θ.

$$2\sin^2\theta - 5\sin\theta + 2 = 0 \qquad \text{Factor}$$

$$(\sin\theta - 2)(2\sin\theta - 1) = 0 \qquad \text{Solve}$$

$$\sin\theta = 2 \quad \text{or} \quad \sin\theta = \frac{1}{2}$$

Since $-1 \le \sin\theta \le 1$ for any angle θ, the equation $\sin\theta = 2$ has no solution. The solutions to $\sin\theta = \frac{1}{2}$, where $0 \le \theta < 2\pi$ are $\frac{\pi}{6}$ and $\frac{5\pi}{6}$ or 30° and 150°.

EXERCISES

1. Which of the following degree measure is equal to $\frac{7\pi}{12}$?

(A) $75°$

(B) $105°$

(C) $135°$

(D) $150°$

(E) $210°$

2. Which of the following expression is equal to $\cos 150°$?

(A) $-\cos 30°$

(B) $-\sin 30°$

(C) $\cos 30°$

(D) $\sin 30°$

(E) $\tan 30°$

3. $\csc x \cdot \cos x =$

(A) $\sin x$

(B) $\tan x$

(C) $\cot x$

(D) $\sec x$

(E) 1

4. If $\sin\theta > 0$ and $\cos\theta < 0$, which quadrant does θ lie in?

(A) I

(B) II

(C) III

(D) IV

(E) Cannot be determined

5. If $\cos\theta = -\frac{3}{4}$, what is the value of $\cos 2\theta$?

(A) $\frac{1}{16}$

(B) $\frac{1}{8}$

(C) $\frac{1}{4}$

(D) $\frac{1}{3}$

(E) $\frac{1}{2}$

6. If $\tan\theta = -\frac{5}{12}$ and $\sin\theta < 0$, what is the value of $\cos\theta$?

(A) $-\frac{12}{13}$

(B) $-\frac{12}{5}$

(C) $-\frac{5}{13}$

(D) $\frac{12}{13}$

(E) $\frac{5}{13}$

7. If $\sin(3x - 10) = \cos(2x + 20)$, what is the value of x ?

(A) 45

(B) 30

(C) 24

(D) 20

(E) 16

8. $(\sin\theta + \cos\theta)^2 =$

(A) 1

(B) $\sin^2\theta + \cos^2\theta$

(C) $2\sin\theta\cos\theta$

(D) $1 + \sin 2\theta$

(E) $1 + \cos 2\theta$

9. What is the exact value of $\sec\frac{5\pi}{3}$?

(A) 2

(B) $\frac{2\sqrt{3}}{3}$

(C) $\frac{2\sqrt{2}}{2}$

(D) 1

(E) $\frac{1}{2}$

10. Solve the equation: $2\cos^2\theta+3\cos\theta+1=0$, where $0<\theta<2\pi$.

(A) $\theta=0, \frac{2\pi}{3}, \frac{4\pi}{3}$

(B) $\theta=\frac{2\pi}{3}, \pi, \frac{4\pi}{3}$

(C) $\theta=0, \frac{\pi}{3}, \frac{5\pi}{3}$

(D) $\theta=\frac{\pi}{2}, \frac{2\pi}{3}, \pi$

(E) $\theta=0, \frac{\pi}{2}, \pi$

ANSWERS AND SOLUTIONS

1. (B)

Since π rad is $180°$,

$$\frac{7\pi}{12}=\frac{7\times 180°}{12}=105°$$

Therefore, (B) is the correct answer.

2. (A)

As shown in the figure below, the angle $\theta=150°$ lies in the second quadrant.

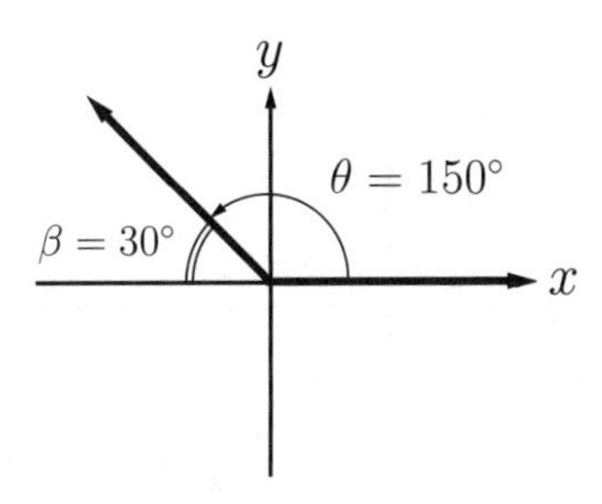

The reference angle β, an angle formed by the terminal side and the closest x-axis, is $30°$. Since the angle θ lies in the 2nd quadrant, the sign of the cosine function is negative. Thus,

$$\begin{aligned}\cos 150° &= \pm\cos\beta && \text{Since } \beta=30° \\ &= \pm\cos 30° && \text{Since cosine is negative in the } 2^{\text{nd}} \text{ quadrant} \\ &= -\cos 30°\end{aligned}$$

Therefore, $\cos 150°$ is equal to $-\cos 30°$.

3. (C)

> Tip
> 1. $\csc x = \frac{1}{\sin x}$
> 2. $\cot x = \frac{\cos x}{\sin x}$

$$\begin{aligned}\csc x \cdot \cos x &= \frac{1}{\sin x} \cdot \cos x \\ &= \frac{\cos x}{\sin x} \\ &= \cot x\end{aligned}$$

Therefore, $\csc x \cdot \cos x = \cot x$.

4. (B)

$\sin\theta > 0$ indicates that θ lies in the 1st quadrant or 2nd quadrant. Furthermore, $\cos\theta < 0$ indicates that θ lies in the 2nd quadrant or 3rd quadrant. Thus, θ must be in the 2nd quadrant.

5. (B)

$$\begin{aligned}\cos 2\theta &= 2\cos^2\theta - 1 && \text{Since } \cos\theta = -\frac{3}{4} \\ &= 2\left(-\frac{3}{4}\right)^2 - 1 && \text{Simplify} \\ &= \frac{1}{8}\end{aligned}$$

Therefore, the value of $\cos 2\theta = \frac{1}{8}$.

6. (D)

$\tan\theta$ is negative in the 2nd and 4th quadrants, and $\sin\theta$ is negative in the 3rd and 4th quadrants. Thus, θ must lie in the 4th quadrant. Since $\tan\theta = -\frac{5}{12}$, the length of the side adjacent to θ is 12, and the length of side opposite to θ is 5. Using the Pythagorean theorem: $c^2 = 5^2 + 12^2$, the length of the hypotenuse of the right triangle is 13 as shown in the figure below.

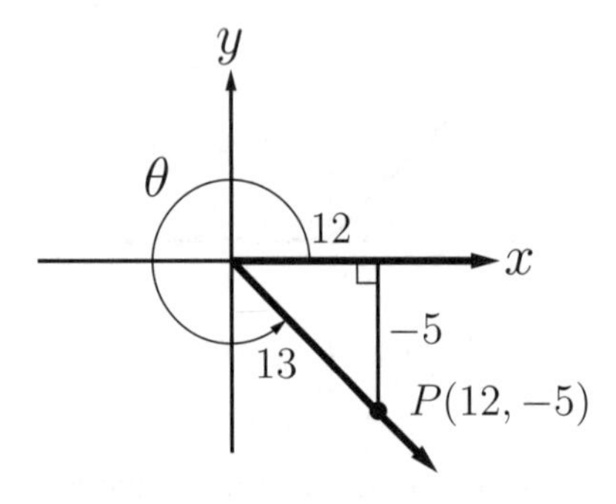

Suppose point $P(x, y)$ is on the terminal side of θ. Since θ lies in the 4th quadrant, the x and y coordinates of point P is $(12, -5)$. Thus,

$$\cos\theta = \frac{\text{adjacent side}}{\text{hypotenuse}} = \frac{12}{13}$$

Therefore, the value of $\cos\theta = \frac{12}{13}$.

7. (E)

This problem is about cofunction identities. The value of the sine function is equal to the value of the cosine function when two angles are complementary (their sum is 90°).

$$\begin{aligned} 3x - 10 + 2x + 20 &= 90 \\ 5x + 10 &= 90 \\ x &= 16 \end{aligned}$$

Therefore, the value of x is 16.

8. (D)

Tip	
	1. $(x+y)^2 = x^2 + 2xy + y^2$
	2. $\sin^2\theta + \cos^2\theta = 1$
	3. $\sin 2\theta = 2\sin\theta\cos\theta$

$$\begin{aligned} (\sin\theta + \cos\theta)^2 &= \sin^2\theta + 2\sin\theta\cos\theta + \cos^2\theta && \text{Since } \sin^2\theta + \cos^2\theta = 1 \\ &= 1 + 2\sin\theta\cos\theta && \text{Since } \sin 2\theta = 2\sin\theta\cos\theta \\ &= 1 + \sin 2\theta \end{aligned}$$

Therefore, $(\sin\theta + \cos\theta)^2 = 1 + \sin 2\theta$.

9. (A)

As shown in the figure below, the angle $\theta = \frac{5\pi}{3}$ lies in the fourth quadrant.

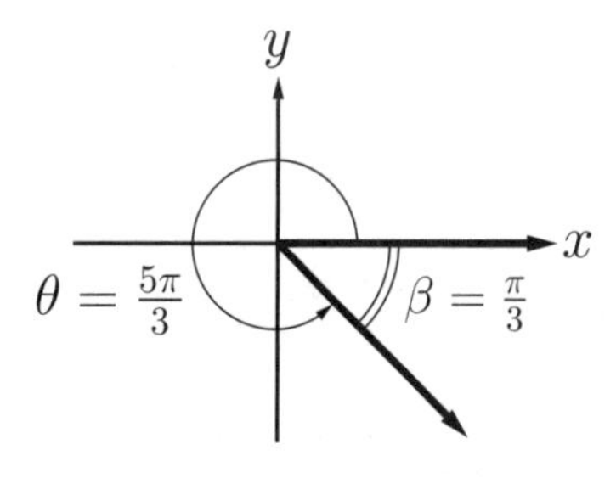

The reference angle β, an angle formed by the terminal side and the closest x-axis, is $\frac{\pi}{3}$. Since the angle θ lies in the 4th quadrant, the sign of the cosine function is positive. Thus,

$$\begin{aligned} \sec\frac{5\pi}{3} &= \frac{1}{\cos\frac{5\pi}{3}} && \text{Since } \sec\theta = \frac{1}{\cos\theta} \text{ and } \beta = \frac{\pi}{3} \\ &= \frac{1}{\pm\cos\frac{\pi}{3}} && \text{Since cosine is positive in the } 4^{\text{th}} \text{ quadrant} \\ &= \frac{1}{\frac{1}{2}} \\ &= 2 \end{aligned}$$

Therefore, $\sec\frac{5\pi}{3} = 2$.

10. (B)

Factor the expression $2\cos^2\theta + 3\cos\theta + 1$ and solve for θ.

$$2\cos^2\theta + 3\cos\theta + 1 = 0 \qquad \text{Factor}$$
$$(\cos\theta + 1)(2\cos\theta + 1) = 0 \qquad \text{Solve}$$
$$\cos\theta = -1 \quad \text{or} \quad \cos\theta = -\frac{1}{2}$$

The solution to $\cos\theta = -1$, where $0 \le \theta < 2\pi$, is π. Furthermore, the solutions to $\cos\theta = -\frac{1}{2}$, where $0 \le \theta < 2\pi$, are $\frac{2\pi}{3}$ and $\frac{4\pi}{3}$. Therefore, the solutions to $2\cos^2\theta + 3\cos\theta + 1 = 0$ are $\frac{2\pi}{3}$, π, and $\frac{4\pi}{3}$.

Graphs of Trigonometric Functions

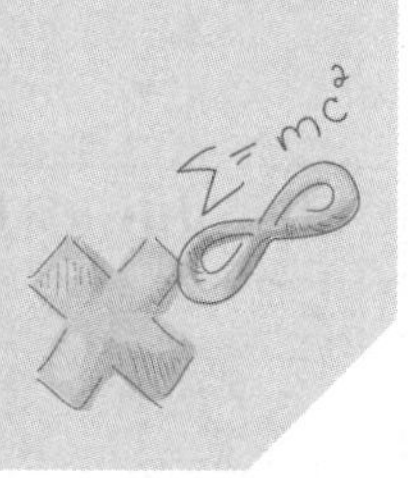

Graphs of Six Trigonometric Functions

Below are the graphs of the six trigonometric functions: sine, cosine, tangent, cosecant, secant, and cotangent.

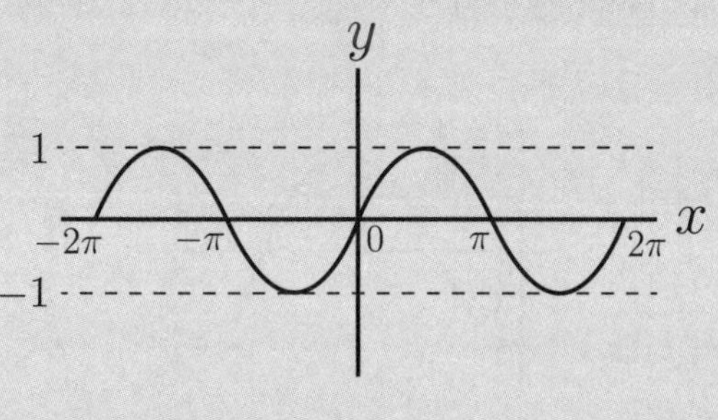

$y = \sin x$

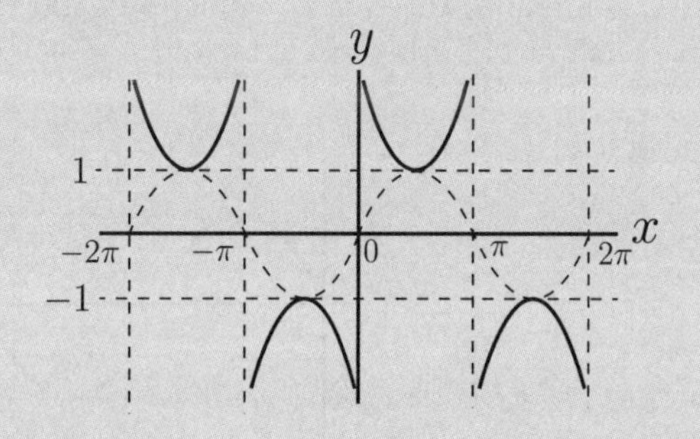

$y = \csc x$

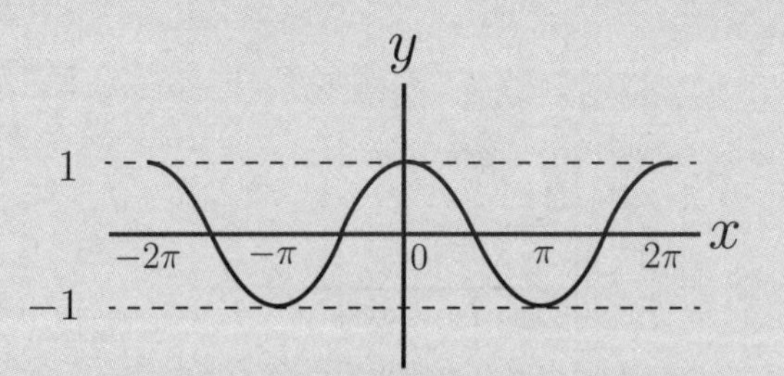

$y = \cos x$

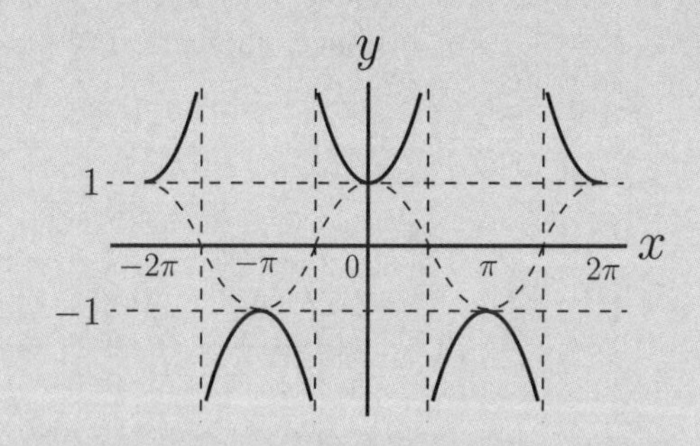

$y = \sec x$

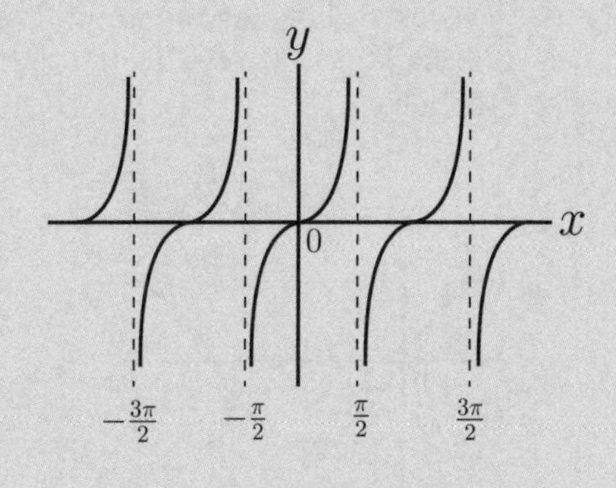

$y = \tan x$

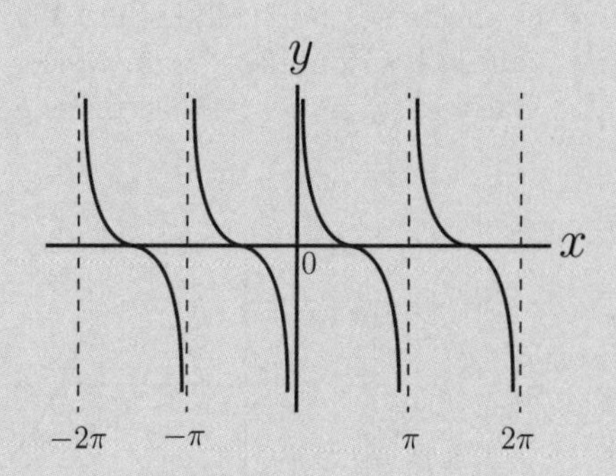

$y = \cot x$

Graphing the Sine and Cosine Functions

The general forms of the sine function and cosine function are as follows:

$$y = A\sin\left(B(x-C)\right)+D \qquad \text{or} \qquad y = A\cos\left(B(x-C)\right)+D$$

where A, B, C and D affect the amplitude, period, horizontal shift, and vertical shift of the graphs of the sine and cosine functions.

- A affects the amplitude. The amplitude is half the distance between the maximum and minimum values of the function. Since distance is always positive, the amplitude is $|A|$. For instance, both $y = 2\sin x$ and $y = -2\cos x$ have an amplitude of 2.

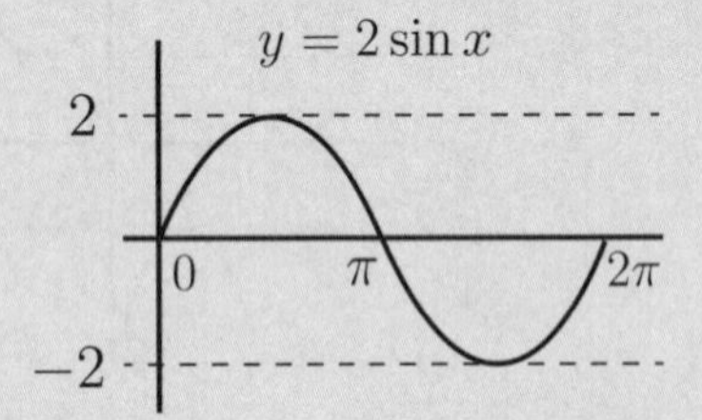

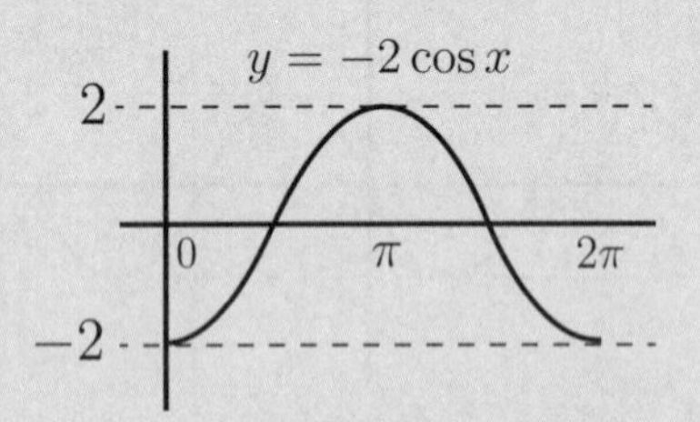

- B affects the period. The period, P, is the horizontal length of one complete cycle obtained by $P = \frac{2\pi}{B}$. For instance, the period of $y = \sin(2x)$ is $\frac{2\pi}{2} = \pi$, and the period of $y = \cos(\frac{1}{3}x)$ is $\frac{2\pi}{\frac{1}{3}} = 6\pi$.

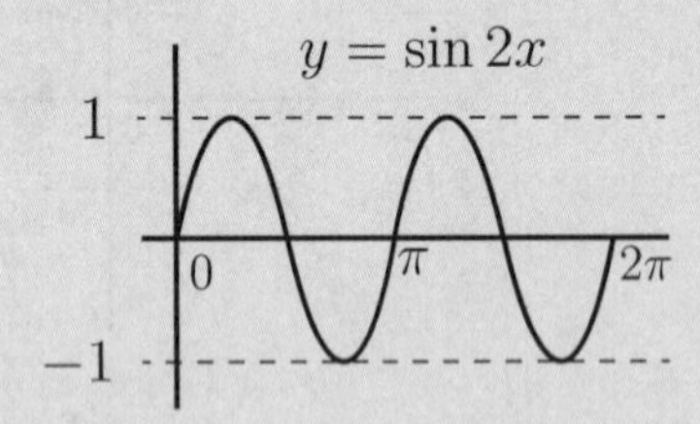

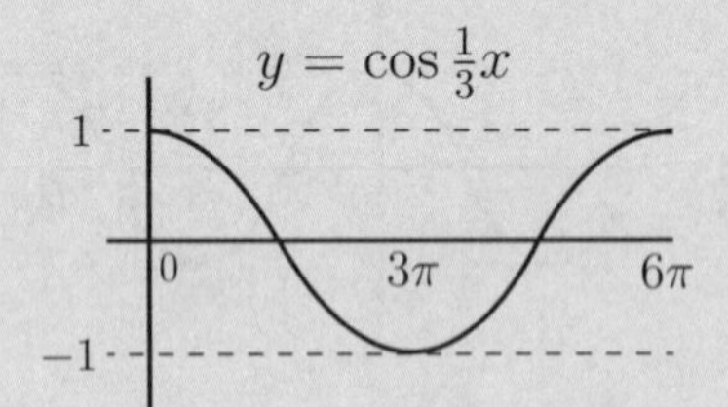

- C affects the horizontal shift. The horizontal shift is the measure of how far the graph has shifted horizontally. For instance, $y = \sin(x - \frac{\pi}{4})$ means a horizontal shift of $\frac{\pi}{4}$ to the right. Whereas, $y = \cos(x + \frac{\pi}{2})$ means a horizontal shift of $\frac{\pi}{2}$ to the left.

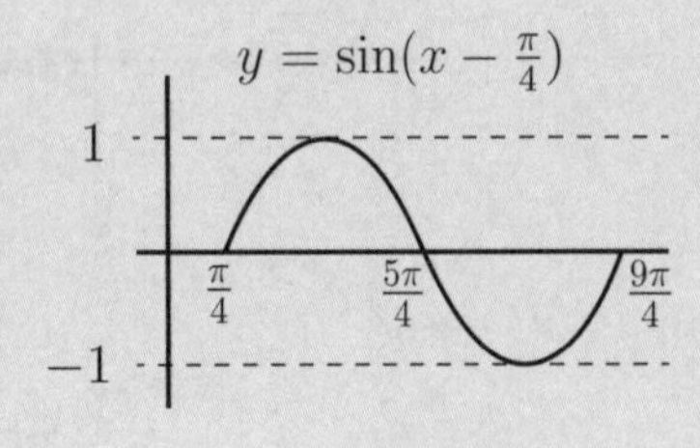

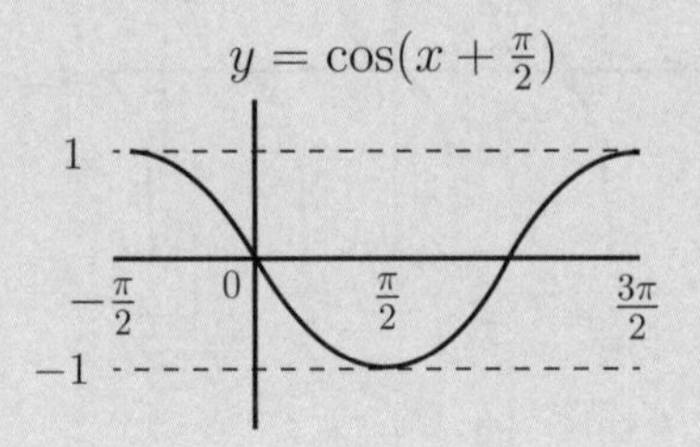

- D affects the vertical shift. The vertical shift is the measure of how far the graph has shifted vertically. For instance, $y = \sin x + 1$ means a vertical shift of 1 up. Whereas, $y = \cos x - 2$ means a vertical shift of 2 down.

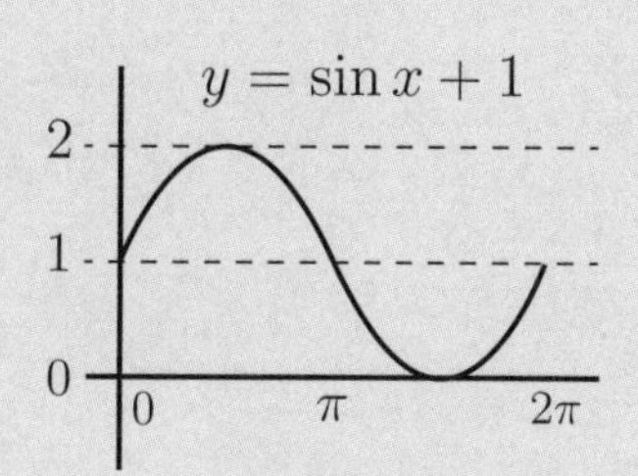

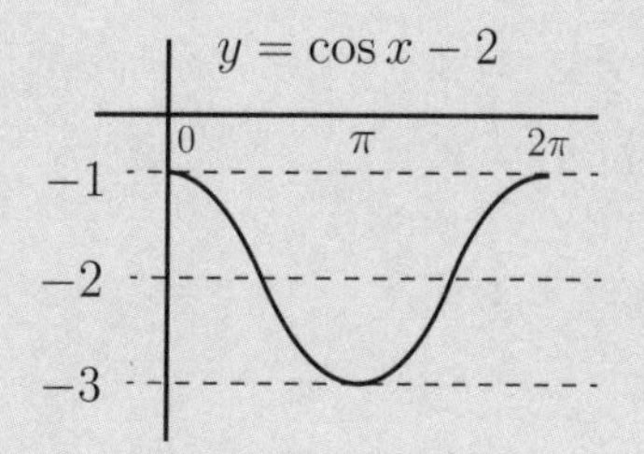

Graphing the Tangent Functions

The general form of the tangent functions is as follows:

$$y = A \tan\big(B(x - C)\big) + D$$

where B, C and D affect the period, horizontal shift, and vertical shift of the graphs of the sine and cosine functions.

- The tangent function does not have an amplitude because it has no maximum or minimum value. If $|A| > 1$, the graph of the tangent function is vertically stretched. Whereas, if $0 < |A| < 1$, the graph of the tangent function is vertically compressed.
- B affects the period. The tangent function has the period of $\frac{\pi}{B}$ as opposed to the sine and cosine functions which have the period of $\frac{2\pi}{B}$. For instance, the period of $y = \tan(2x)$ is $\frac{\pi}{2}$. The graph of $y = \tan(2x)$ is shown in Figure 1.
- C affects the horizontal shift. For instance, $y = \tan(x - \frac{\pi}{2})$ means a horizontal shift of $\frac{\pi}{2}$ to the right. The graph of $y = \tan(x - \frac{\pi}{2})$ is shown in Figure 2.
- D affects the vertical shift. For instance, $y = \tan x + 3$ means a vertical shift of 3 up. The graph of $y = \tan x + 3$ is shown in Figure 3.

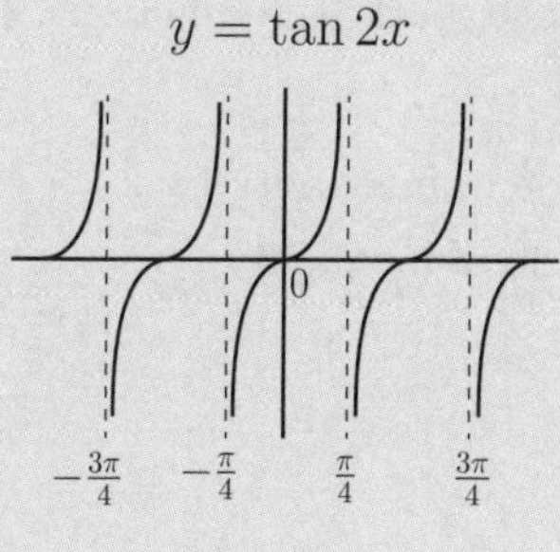

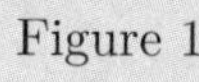
Figure 1

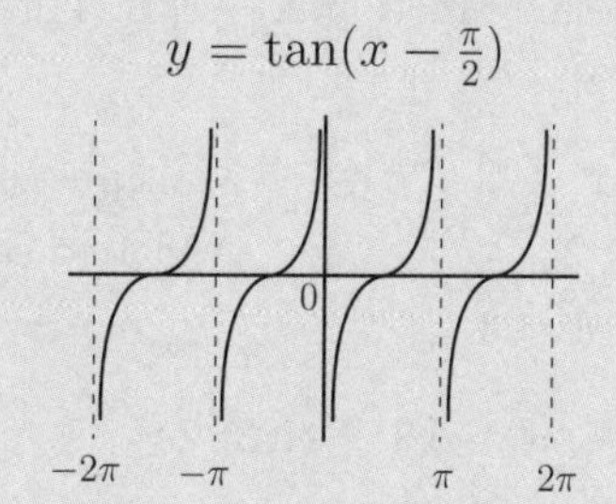

Figure 2

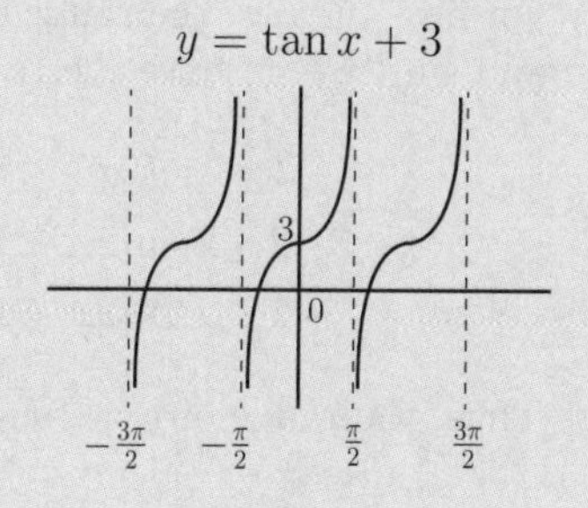

Figure 3

Example 1 Finding amplitude, period, horizontal shift, and vertical shift

Find the amplitude, period, horizontal shift, and vertical shifts of the following trigonometric functions.

(a) $y = 2\cos(x - \frac{\pi}{6}) + 1$

(b) $y = -3\sin(2x + \frac{\pi}{2}) - 2$

(c) $y = \tan(\frac{1}{2}x - \frac{\pi}{4}) + 3$

Solution

(a) Comparing $y = 2\cos(x - \frac{\pi}{6}) + 1$ to the general form of $y = A\cos\big(B(x - C)\big) + D$, we found that $A = 2$, $B = 1$, $C = \frac{\pi}{6}$, and $D = 1$. Thus, the amplitude is 2. The period, P, is $P = \frac{2\pi}{B} = \frac{2\pi}{1} = 2\pi$. The horizontal shift is $\frac{\pi}{6}$ to the right, and the vertical shift is 1 up.

(b) Change $y = -3\sin(2x + \frac{\pi}{2}) - 2$ to $y = -3\sin\big(2(x - (-\frac{\pi}{4}))\big) - 2$. Then, compare $y = -3\sin\big(2(x - (-\frac{\pi}{4}))\big) - 2$ to $y = A\sin\big(B(x - C)\big) + D$. We found that $A = -3$, $B = 2$, $C = -\frac{\pi}{4}$, and $D = -2$. Thus, the amplitude is $|-3| = 3$. The period, P, is $P = \frac{2\pi}{B} = \frac{2\pi}{2} = \pi$. The horizontal shift is $\frac{\pi}{4}$ to the left, and the vertical shift is 2 down.

(c) Change $y = \tan(\frac{1}{2}x - \frac{\pi}{4}) + 3$ to $y = \tan(\frac{1}{2}(x - \frac{\pi}{2})) + 3$. Then compare $y = \tan(\frac{1}{2}(x - \frac{\pi}{2})) + 3$ to $y = A\tan\big(B(x - C)\big) + D$. We found that $A = 1$, $B = \frac{1}{2}$, $C = \frac{\pi}{2}$, and $D = 3$. The tangent function does not have an amplitude because it has no maximum or minimum value. The period, P, is $P = \frac{\pi}{B} = \frac{\pi}{\frac{1}{2}} = 2\pi$. The horizontal shift is $\frac{\pi}{2}$ to the right, and the vertical shift is 3 up.

Example 2 Finding range of a trigonometric function

Find the range of the following trigonometric functions.

(a) $y = 2\sin(2x - 1) + 1$

(b) $y = -3\cos(x + \frac{\pi}{2}) - 2$

Solution

(a) For any angle x, the range of a sine function is $[-1, 1]$. For instance, both sine functions, $\sin x$ and $\sin(2x - 1)$ have the range of $[-1, 1]$.

$-1 \leq \sin(2x - 1) \leq 1$	Multiply each side of inequality by 2
$-2 \leq 2\sin(2x - 1) \leq 2$	Add 1 to each side of inequality
$-1 \leq 2\sin(2x - 1) + 1 \leq 3$	

Thus, the range of $y = 2\sin(2x - 1) + 1$ is $[-1, 3]$.

(b) For any angle x, the range of a cosine function is $[-1, 1]$.

$$-1 \le \cos\left(x + \frac{\pi}{2}\right) \le 1 \qquad \text{Multiply each side of inequality by } -3$$

$$-3 \le -3\cos\left(x + \frac{\pi}{2}\right) \le 3 \qquad \text{Subtract 2 from each side of inequality}$$

$$-5 \le -3\cos\left(x + \frac{\pi}{2}\right) - 2 \le 1$$

Thus, the range of $y = -3\cos(x + \frac{\pi}{2}) - 2$ is $[-5, 1]$.

Example 3 Finding period from the graph of a trigonometric function

Graph $y = |\sin 2x|$, $0 \le x \le 2\pi$ and find the period of $|\sin 2x|$.

Solution The period, P, of $y = \sin 2x$ is $P = \frac{2\pi}{2} = \pi$, which indicates that there are two complete cycles in the interval $[0, 2\pi]$ as shown in Figure 4.

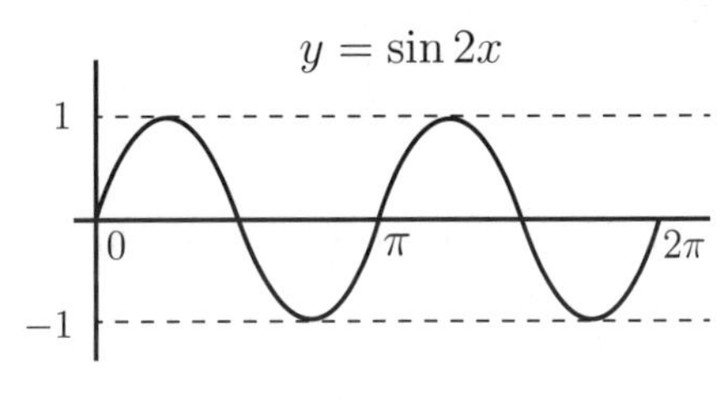

Figure 4

In order to graph $y = |\sin 2x|$, determine the part of the graph of $y = \sin 2x$ that lies below the x-axis as shown in Figure 5.

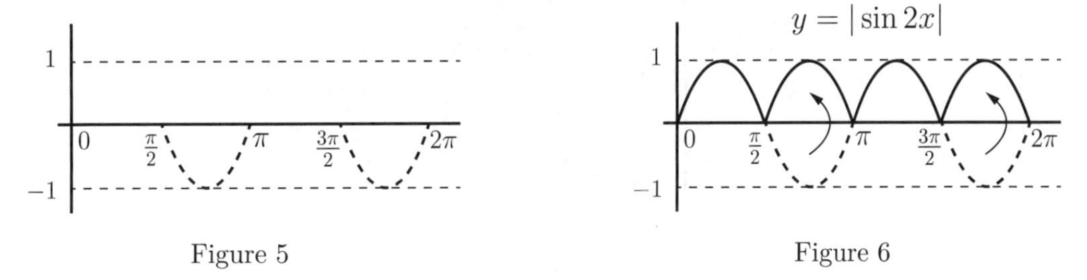

Figure 5 Figure 6

Lastly, reflect the part of the graph that lies below the x-axis about the x-axis as shown in Figure 6. Since the graph of $y = |\sin 2x|$ completes its cycle every $\frac{\pi}{2}$, the period $y = |\sin 2x|$ is $\frac{\pi}{2}$.

Example 4 Solving the trigonometric equation: $\cos 2\theta = \frac{1}{2}$

Solve the trigonometric equation: $\cos 2\theta = \frac{1}{2},\ 0 \le \theta < 2\pi$

Solution The period, P, of $y = \cos 2\theta$ is $P = \frac{2\pi}{2} = \pi$, which indicates that there are two complete cycles in the interval $[0, 2\pi)$ as shown in Figure 7.

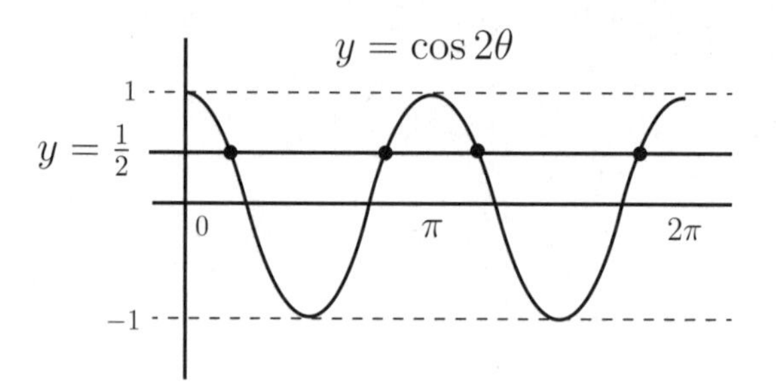

Figure 7

The graph of $y = \cos 2\theta$ intersects $y = \frac{1}{2}$ four times in the interval $[0, 2\pi)$ as shown in Figure 7. Thus, there are four solutions to the equation $\cos 2\theta = \frac{1}{2}$. Since $\cos\frac{\pi}{3} = \frac{1}{2}$ and $\cos\frac{5\pi}{3} = \frac{1}{2}$, the general solutions of $\cos 2\theta = \frac{1}{2},\ 0 \le \theta < 2\pi$ are as follows:

$$2\theta = \frac{\pi}{3} + 2n\pi \quad \Longrightarrow \quad \theta = \frac{\pi}{6} + n\pi,\ n \text{ is any integer}$$

$$2\theta = \frac{5\pi}{3} + 2n\pi \quad \Longrightarrow \quad \theta = \frac{5\pi}{6} + n\pi,\ n \text{ is any integer}$$

Substitute $n = 0$ and $n = 1$ to find all solutions in the interval $[0, 2\pi)$.

$$\text{Substitute } n = 0: \qquad \theta = \frac{\pi}{6}, \frac{5\pi}{6}$$

$$\text{Substitute } n = 1: \qquad \theta = \frac{7\pi}{6}, \frac{11\pi}{6}$$

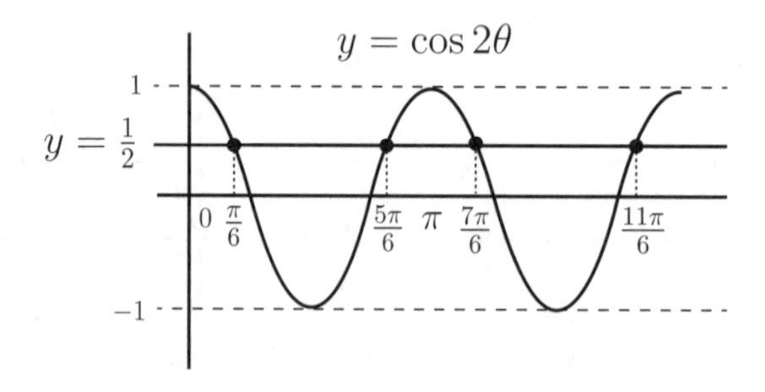

Figure 8

Therefore, solutions to $\cos 2\theta = \frac{1}{2},\ 0 \le \theta < 2\pi$ are $\theta = \frac{\pi}{6}, \frac{5\pi}{6}, \frac{7\pi}{6}$, and $\frac{11\pi}{6}$ as shown in Figure 8.

EXERCISES

1. What is the amplitude of $y = -3\sin 2x$?

   (A) -3

   (B) 1

   (C) 3

   (D) $\frac{\pi}{2}$

   (E) 2π

2. What is the period of $y = -\cos(\pi x + 4)$?

   (A) -4

   (B) 2

   (C) 4

   (D) π

   (E) 2π

3. What is the range of $y = 4\cos(x - \frac{\pi}{2}) + 3$?

   (A) $-1 \le y \le 7$

   (B) $0 \le y \le 7$

   (C) $0 \le y \le -1$

   (D) $-7 \le y \le 0$

   (E) $-7 \le y \le 7$

4. Which of the following function has a period of $\frac{\pi}{2}$?

   (A) $\sin\frac{x}{2}$

   (B) $\sin 2x$

   (C) $\tan\frac{x}{2}$

   (D) $\cos\frac{x}{4}$

   (E) $\cos 4x$

5. If $f(x) = 2\sin(3x - \pi)$, what is the horizontal shift of f ?

   (A) 2π to the right

   (B) π to the left

   (C) $\frac{2\pi}{3}$ to the right

   (D) $\frac{\pi}{2}$ to the left

   (E) $\frac{\pi}{3}$ to the right

6. Which of the following cosine function has an amplitude of 2, a period of 4π, and a vertical shift of 3 units down?

   (A) $2\cos(2x) - 3$

   (B) $2\cos(\frac{1}{2}x) + 3$

   (C) $-2\cos(\frac{1}{2}x) - 3$

   (D) $-2\cos(2x) - 3$

   (E) $-3\cos(2x) + 2$

7. What is the period of $y = \tan(\frac{x}{3}) + 1$?

   (A) 3π

   (B) $\frac{3\pi}{2}$

   (C) π

   (D) $\frac{\pi}{2}$

   (E) $\frac{\pi}{3}$

8. What is the period of $y = |\cos x|$?

   (A) 2π

   (B) π

   (C) $\frac{2\pi}{3}$

   (D) $\frac{\pi}{2}$

   (E) $\frac{\pi}{4}$

9. What is the amplitude of the trigonometric function shown below ?

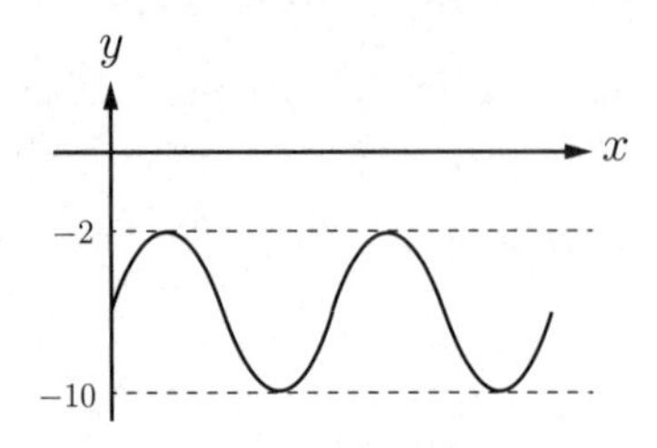

(A) 12

(B) 10

(C) 8

(D) 4

(E) 2

10. Solve: $\sin 2\theta = -\frac{\sqrt{3}}{2}$, where $0 \le \theta < 2\pi$.

(A) $\theta = \frac{4\pi}{3}, \frac{5\pi}{3}$

(B) $\theta = \frac{2\pi}{3}, \frac{5\pi}{6}$

(C) $\theta = \frac{4\pi}{3}, \frac{3\pi}{2}, \frac{5\pi}{3}$

(D) $\theta = \frac{2\pi}{3}, \frac{5\pi}{6}, \frac{5\pi}{3}, \frac{11\pi}{6}$

(E) $\theta = \frac{2\pi}{3}, \frac{5\pi}{6}, \frac{4\pi}{3}, \frac{5\pi}{3}$

ANSWERS AND SOLUTIONS

1. (C)

The amplitude of $y = -3\sin 2x$ is $|-3| = 3$. Therefore, (C) is the correct answer.

2. (B)

The period, P, is the horizontal length of one complete cycle and is obtained by $P = \frac{2\pi}{B}$ for the sine and cosine functions. Thus, the period of $y = -\cos(\pi x + 4)$ is $P = \frac{2\pi}{\pi} = 2$.

3. (A)

For any angle x, the range of a cosine function is $[-1, 1]$.

$$-1 \le \cos\left(x - \frac{\pi}{2}\right) \le 1$$ Multiply each side of the inequality by 4

$$-4 \le 4\cos\left(x - \frac{\pi}{2}\right) \le 4$$ Add 3 to each side of the inequality

$$-1 \le 4\cos\left(x - \frac{\pi}{2}\right) + 3 \le 7$$

Thus, the range of $y = 4\cos(x - \frac{\pi}{2}) + 3$ is $[-1, 7]$. Therefore, (A) is the correct answer.

4. (E)

The period of $\sin\frac{x}{2}$ is $\frac{2\pi}{\frac{1}{2}} = 4\pi$ and the period of $\sin 2x$ is $\frac{2\pi}{2} = \pi$. Furthermore, the period of $\tan\frac{x}{2} = \frac{\pi}{\frac{1}{2}} = 2\pi$ and the period of $\cos\frac{x}{4}$ is $\frac{2\pi}{\frac{1}{4}} = 8\pi$. Since the period of $\cos 4x$ is $\frac{2\pi}{4} = \frac{\pi}{2}$, (E) is the correct answer.

5. (E)

Since $2\sin(3x-\pi) = 2\sin\left(3(x-(\frac{\pi}{3}))\right)$, the horizontal shift is $\frac{\pi}{3}$ to the right. Therefore, (E) is the correct answer.

6. (C)

The cosine function $-2\cos(\frac{1}{2}x)-3$ has an amplitude of $|-2|=2$, a period of $\frac{2\pi}{\frac{1}{2}}=4\pi$, and a vertical shift of 3 units down. Therefore, (C) is the correct answer.

7. (A)

The period, P, of the tangent function $y=\tan\frac{x}{3}+1$ is $\frac{\pi}{\frac{1}{3}}=3\pi$. Therefore, (A) is the correct answer.

8. (B)

The graph of $y=\cos x$ in the interval $[-2\pi, 2\pi]$ is shown in Figure 9.

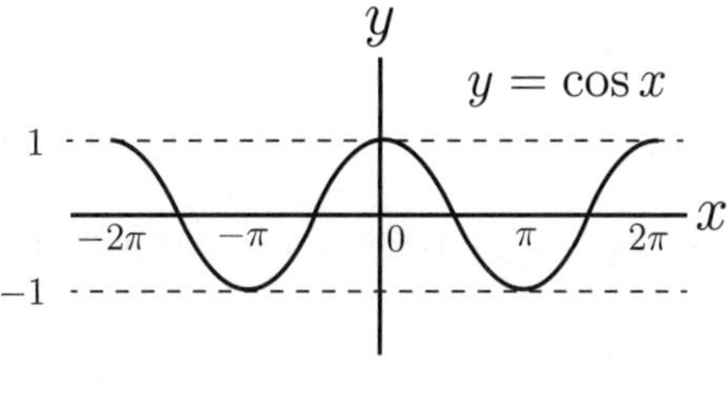

Figure 9

In order to graph $y=|\cos x|$, determine the part of the graph of $y=\cos x$ that lies below the x-axis as shown in Figure 10.

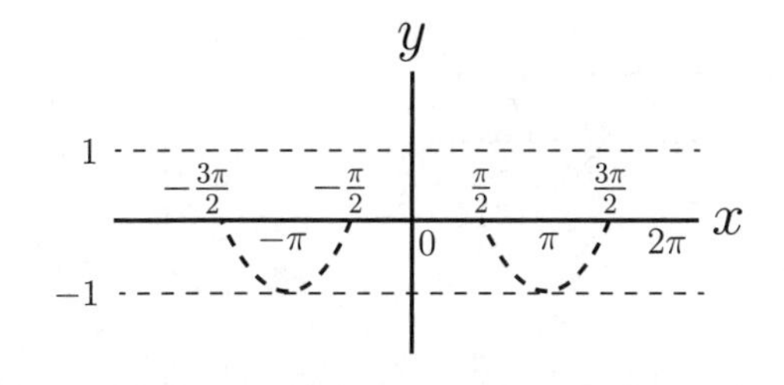

Figure 10

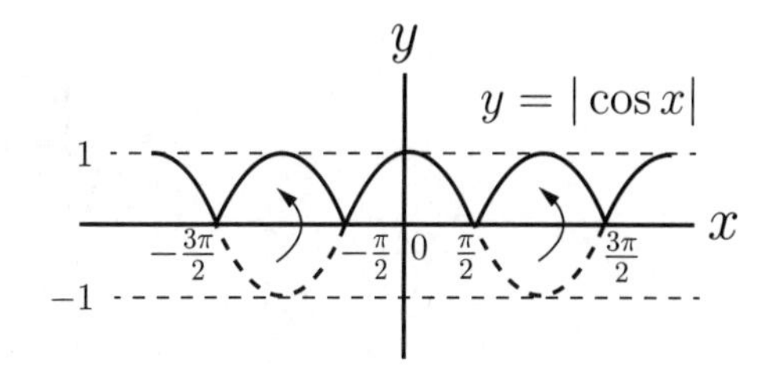

Figure 11

Then, reflect the part of the graph that lies below the x-axis about the x-axis as shown in Figure 11. Since the graph of $y=|\cos x|$ completes a cycle every π, the period $y=|\cos x|$ is π. Therefore, (B) is the correct answer.

9. (D)

The amplitude is half the distance between the maximum and the minimum value of the function. Since the maximum is -2 and the minimum is -10, the distance is $-2-(-10)=8$. Thus, the amplitude is $\frac{1}{2}(8)=4$. Therefore, (D) is the correct answer.

10. (D)

The period, P, of $y = \sin 2\theta$ is $P = \frac{2\pi}{2} = \pi$, which indicates that there are two complete cycles in the interval $[0, 2\pi)$ as shown in Figure 12.

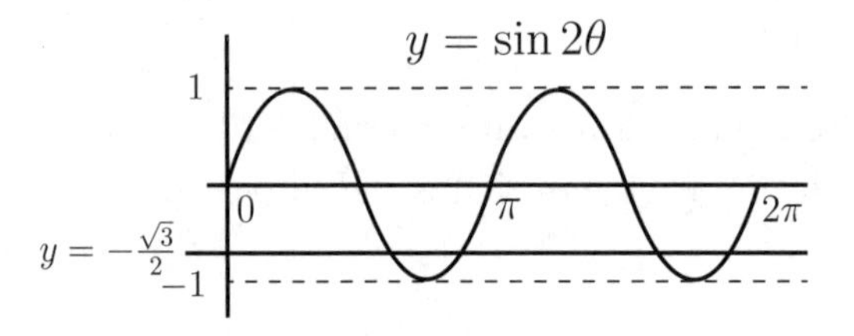

Figure 12

The graph of $y = \sin 2\theta$ intersects $y = -\frac{\sqrt{3}}{2}$ four times in the interval $[0, 2\pi)$ as shown in Figure 12. Thus, there are four solutions to the equation $\sin 2\theta = -\frac{\sqrt{3}}{2}$. Since $\sin \frac{4\pi}{3} = -\frac{\sqrt{3}}{2}$ and $\sin \frac{5\pi}{3} = -\frac{\sqrt{3}}{2}$, the general solutions of $\sin 2\theta = -\frac{\sqrt{3}}{2}$, $0 \le \theta < 2\pi$ are as follows:

$$2\theta = \frac{4\pi}{3} + 2n\pi \quad \Longrightarrow \quad \theta = \frac{2\pi}{3} + n\pi,\ n \text{ is any integer}$$

$$2\theta = \frac{5\pi}{3} + 2n\pi \quad \Longrightarrow \quad \theta = \frac{5\pi}{6} + n\pi,\ n \text{ is any integer}$$

Substitute $n = 0$ and $n = 1$ to find all solutions in the interval $[0, 2\pi)$.

$$\text{Substitute } n = 0: \qquad \theta = \frac{2\pi}{3}, \frac{5\pi}{6}$$

$$\text{Substitute } n = 1: \qquad \theta = \frac{5\pi}{3}, \frac{11\pi}{6}$$

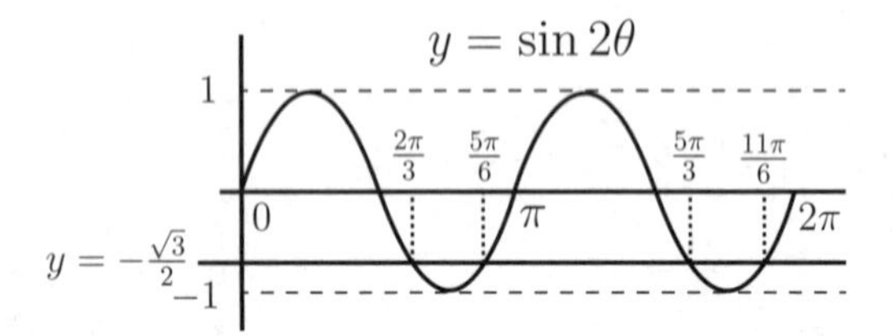

Figure 13

Therefore, the solutions to $\sin 2\theta = -\frac{\sqrt{3}}{2}$, $0 \le \theta < 2\pi$ are $\theta = \frac{2\pi}{3}, \frac{5\pi}{6}, \frac{5\pi}{3}$, and $\frac{11\pi}{6}$ as shown in Figure 13.

LESSON 10 Solving Triangles

Solving Triangles

Solving a triangle means finding the missing lengths of its sides and the measures of its angles. In general, solving a triangle can be classified as either solving a right triangle or solving a non-right triangle. The general rules for solving triangles are as follows:

- To solve a right triangle, use the definition of the six trigonometric functions or the Pythagorean theorem.
- To solve a non-right triangle, use the Law of Sines or Law of Cosines.

Solving Right Triangles

To solve a right triangle, the following information about a right triangle must be given: either the length of a side and an angle (other than $90°$) or the lengths of two sides. In order to find the missing lengths of its sides and the measures of its angles of a right triangle, use the definition of the six trigonometric functions or the Pythagorean theorem: $c^2 = a^2 + b^2$.

Let's solve a right triangle ABC shown at the right. If $c = 10$, and $m\angle A = 40°$, find a, b, and $m\angle B$.

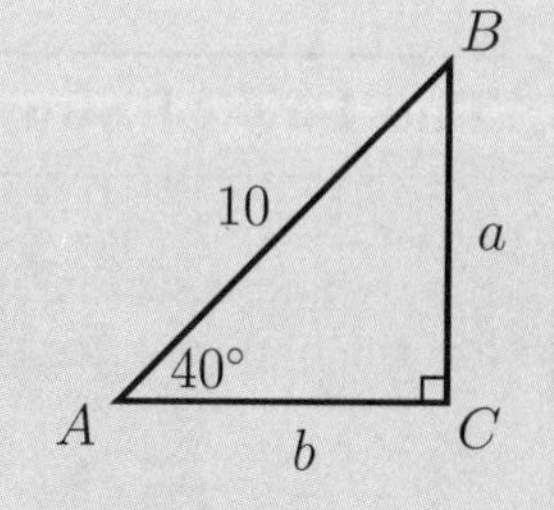

Since $m\angle A + m\angle B = 90°$, $m\angle B = 50°$. To find the side a, use the definition of $\sin 40°$.

$$\sin 40° = \frac{a}{10} \quad \Longrightarrow \quad a = 10\sin 40° \approx 6.43$$

To find the side b, use the definition of $\cos 40°$.

$$\cos 40° = \frac{b}{10} \quad \Longrightarrow \quad b = 10\cos 40° \approx 7.66$$

Tip: Once the side a is known in triangle ABC above, you can use the Pythagorean theorem to find the side b.

$$
\begin{aligned}
c^2 &= a^2 + b^2 && \text{Since } c = 10 \text{ and } a = 6.43 \\
10^2 &= 6.43^2 + b^2 && \text{Solve for } b \\
b &= \sqrt{10^2 - 6.43^2} && \text{Since } b > 0 \\
b &= 7.66
\end{aligned}
$$

Solving Non-Right Triangles

Classifying non-right triangles

Depending on the information about the sides and angles given, non-right triangles can be classified as follows:

- ASA triangle: Two angles and the included side are known.
- SAA triangle: One side and two angles are known.
- SAS triangle: Two sides and the included angle are known.
- SSS triangle: Three sides are known.

The general rules for solving non-right triangles are as follows:

- To solve ASA and SAA triangles, use the Law of Sines.
- To solve SAS and SSS triangles, use the Law of Cosines.

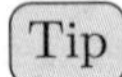

1. If none of the angles of a triangle is a right angle, the triangle is called either a non-right triangle or an oblique triangle.
2. SSA (Two sides and one angle opposite one of them) triangle is referred to as the **ambiguous case** because the given information may result in one triangle, two triangles, or no triangle. It is less likely that the ambiguous case will be on the SAT II Math level II test.

The Law of Sines (ASA and SAA)

If a, b, and c are the lengths of the sides of a triangle, and A, B, and C are the opposite angles, then

$$\frac{a}{\sin A} = \frac{b}{\sin B} = \frac{c}{\sin C}$$

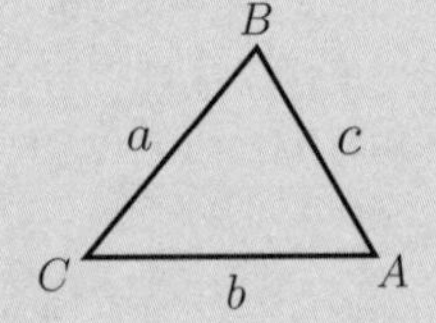

The Law of Sines implies that the largest angle is opposite the longest side and the smallest angle is opposite the shortest side. For instance, let a and c be the sides opposite the angles A and C of the triangle above, respectively. The Law of Sines satisfies the following:

$$\text{If } m\angle C < m\angle A \implies c < a$$
$$\text{If } c < a \implies m\angle C < m\angle A$$

Let's solve a non-right triangle ABC shown at the right. If $a = 8$, $m\angle A = 60°$ and $m\angle C = 40°$, find $m\angle B$, b, and c.

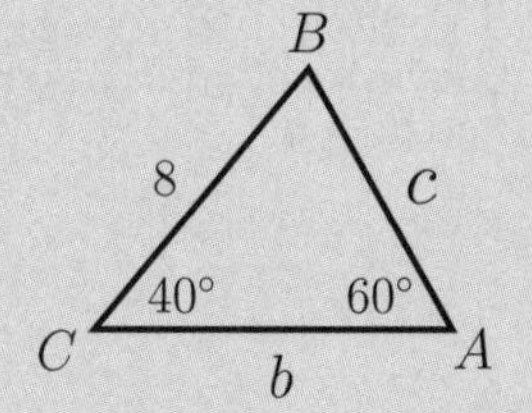

Since $m\angle A + m\angle B + m\angle C = 180°$, $m\angle B = 80$. Triangle ABC is a SAA triangle. Use the Law of Sines to find the sides b and c.

$$\frac{8}{\sin 60°} = \frac{b}{\sin 80°} \quad \Longrightarrow \quad b = \frac{8 \sin 80°}{\sin 60°} \approx 9.1$$

$$\frac{8}{\sin 60°} = \frac{c}{\sin 40°} \quad \Longrightarrow \quad c = \frac{8 \sin 40°}{\sin 60°} \approx 5.94$$

The Law of Cosines (SAS and SSS)

- If triangle ABC shown at the right is a SAS triangle (a, b, and $m\angle C$ are known), side c can be calculated by the Law of Cosines.

$$c^2 = a^2 + b^2 - 2ab \cos C$$

  Note that side c is opposite angle C.

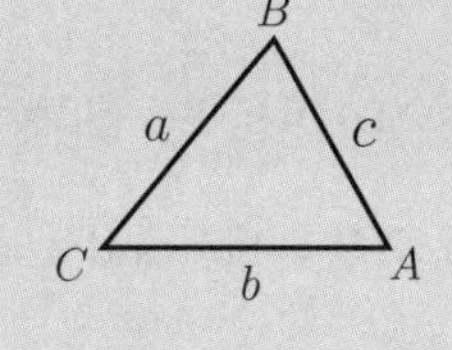

- If triangle ABC shown at the right is a SSS triangle (a, b, and c are known), the measure of angle A can be calculated by the Law of Cosines.

$$m\angle A = \cos^{-1}\left(\frac{a^2 - b^2 - c^2}{-2bc}\right)$$

  Note that side a is opposite angle A.

Let's find the side c and $m\angle B$ of a non-right triangle ABC shown at the right when $a = 4$, $b = 5$, and $m\angle C = 70°$.

Since triangle ABC is a SAS triangle, use the Law of Cosines to find the side c.

$$c^2 = 4^2 + 5^2 - 2(4)(5)\cos 70° \quad \Longrightarrow \quad c \approx 5.23$$

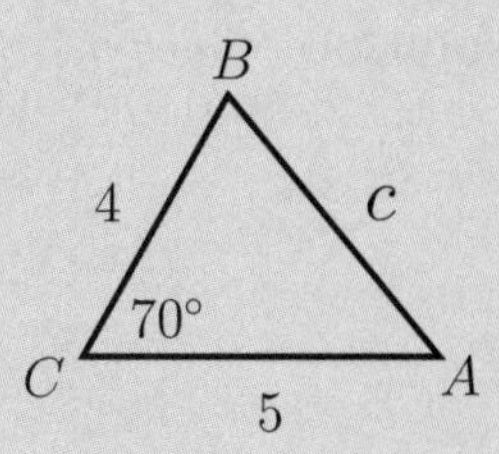

To find the measure of angle B, use the Law of Cosines. Since side b is opposite angle B,

$$m\angle B = \cos^{-1}\left(\frac{b^2 - a^2 - c^2}{-2ac}\right) = \cos^{-1}\left(\frac{5^2 - 4^2 - 5.23^2}{-2(4)(5.23)}\right) \approx 63.98°$$

The Area of a SAS Triangle

If triangle ABC shown at the right is a SAS triangle (a, b, and $m\angle C$ are known), the area of triangle ABC is as follows:

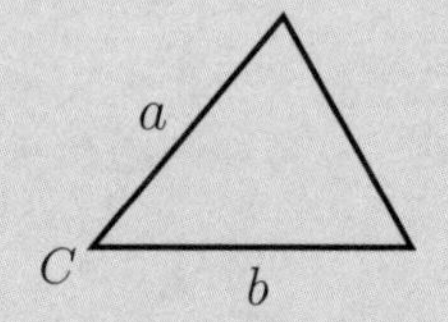

$$A = \frac{1}{2}ab\sin C$$

For instance, if triangle ABC shown at the right is a SAS triangle with $a = 5$, $b = 8$, and $m\angle C = 50°$, the area of triangle ABC is shown below.

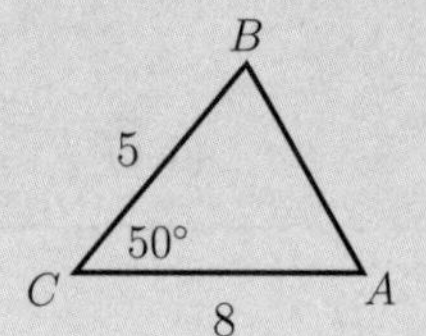

$$A = \frac{1}{2}(5)(8)\sin 50° \approx 15.32$$

Example 1 Solving a right triangle

If $m\angle B = 65°$ and $a = 5$, solve the triangle shown below.

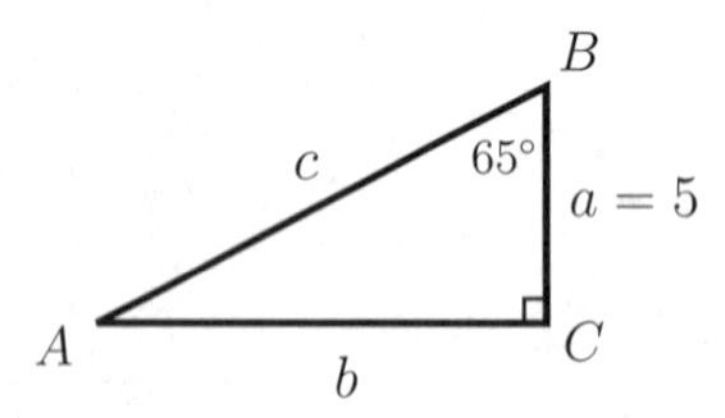

Solution Since $m\angle B = 65°$ and $m\angle A + m\angle B = 90°$, $m\angle A = 25°$. In order to find the sides b and c, use the definition of the tangent and cosine functions.

$$\tan 65° = \frac{b}{5} \implies b = 5\tan 65° \approx 10.72$$

$$\cos 65° = \frac{5}{c} \implies c = \frac{5}{\cos 65°} \approx 11.83$$

Therefore, $m\angle A = 25°$, $b = 10.72$, and $c = 11.83$.

Example 2 Solving a ASA triangle using the Law of Sines

If $m\angle A = 55°$, $m\angle C = 45°$, and $b = 8$, solve the triangle shown below.

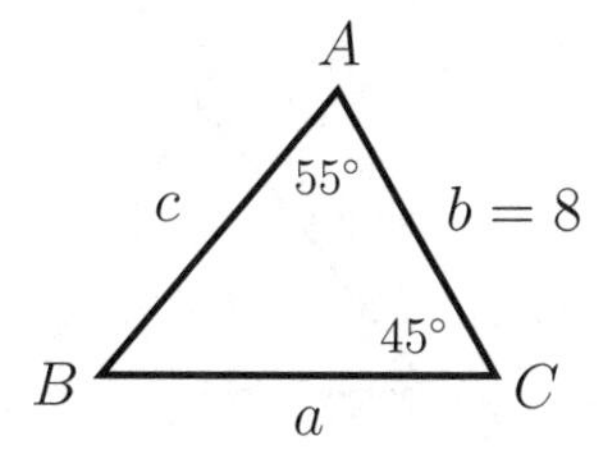

Solution The sum of the measures of interior angles of a triangle is 180°. Since $m\angle A = 55°$ and $m\angle C = 45°$, $m\angle B = 80°$. In order to find the sides a and c, use the Law of Sines.

$$\frac{8}{\sin 80°} = \frac{a}{\sin 55°} \implies a = \frac{8\sin 55°}{\sin 80°} \approx 6.65$$

$$\frac{8}{\sin 80°} = \frac{c}{\sin 45°} \implies c = \frac{8\sin 45°}{\sin 80°} \approx 5.74$$

Therefore, $m\angle B = 80°$, $a = 6.65$, and $c = 5.74$.

Example 3 Solving a SSS triangle using the Law of Cosines

Find the largest angle of a triangle under the given following conditions.

(a) if the ratio of the measures of three interior angles of the triangle is $2 : 3 : 4$.

(b) if the ratio of the three sides of the triangle is $2 : 3 : 4$.

Solution

(a) The ratio of the measures of three interior angles of the triangle is $2 : 3 : 4$. Let $2x$, $3x$, and $4x$ be the measures of the interior angles. Since the sum of the measures of interior angles of a triangle is 180°,

$$\begin{aligned} 2x + 3x + 4x &= 180 \\ 9x &= 180 \\ x &= 20 \end{aligned}$$

Therefore, the largest angle is $4x = 4(20) = 80°$.

(b) For simplicity, let the three sides a, b and c be 4, 3, and 2, respectively, since the ratio of the three sides is $2:3:4$. The Law of Sines implies that the largest angle is opposite the longest side. Thus, $\angle A$ is the largest angle since side a is the longest side.

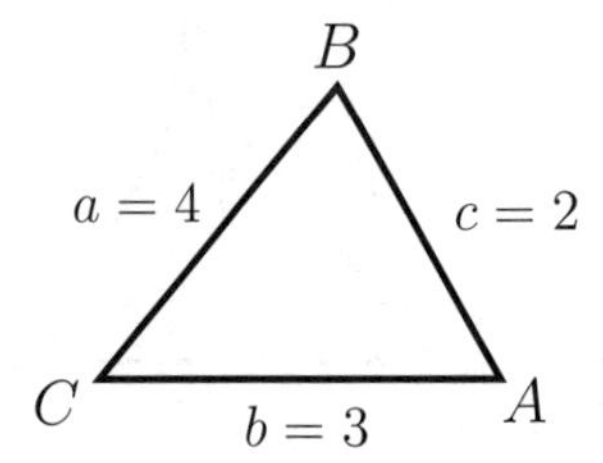

In order to find the measure of angle A, use the Law of Cosines. Since side a is opposite angle A,

$$m\angle A = \cos^{-1}\left(\frac{a^2 - b^2 - c^2}{-2bc}\right) = \cos^{-1}\left(\frac{4^2 - 3^2 - 2^2}{-2(3)(2)}\right) \approx 104.48^\circ$$

Therefore, the largest angle of the triangle is 104.48°.

EXERCISES

1. In $\triangle ABC$ shown below, $m\angle ACB = 55^\circ$ and $BC = 10$. what is AB ?

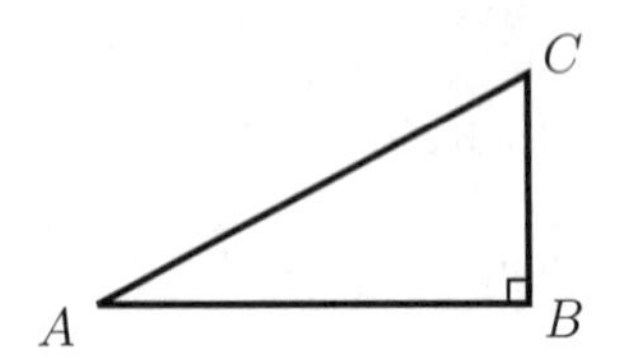

(A) 5.72

(B) 10

(C) 14.28

(D) 20

(E) 25.72

2. In $\triangle ABC$ shown below, $AB = BC$. If $AC = 25$, what is BC ?

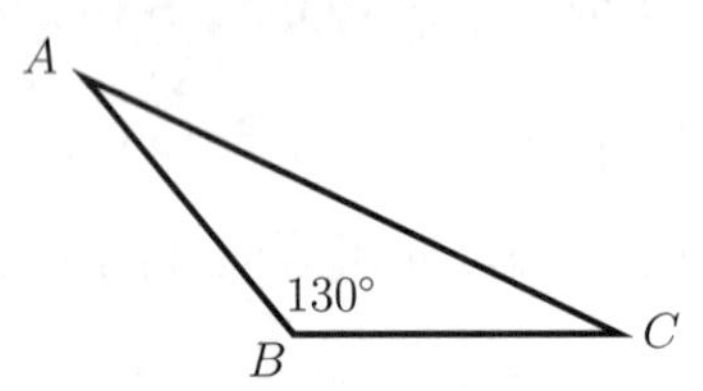

(A) 13.79

(B) 10.96

(C) 7.17

(D) 5.32

(E) 3.56

3. If $\tan\theta = 2$, what is the value of $\sec^2\theta$?

(A) 1

(B) $\sqrt{3}$

(C) $\sqrt{5}$

(D) 3

(E) 5

4. In the right triangle ABC shown below, what is BC in terms of θ ?

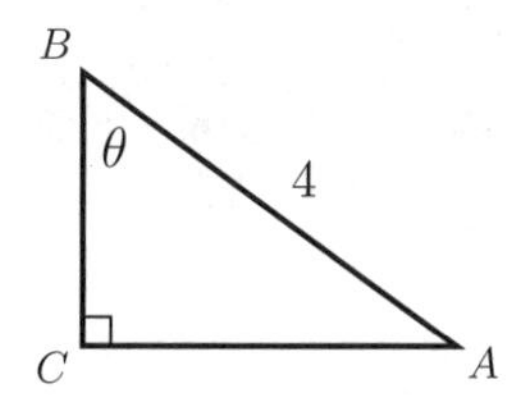

(A) $4\sin\theta$

(B) $\frac{4}{\sin\theta}$

(C) $4\cos\theta$

(D) $\frac{4}{\cos\theta}$

(E) $4\tan\theta$

5. In $\triangle ABC$ shown below, $AB = 7$, $BC = 5$, and $m\angle B = 75°$. What is AC ?

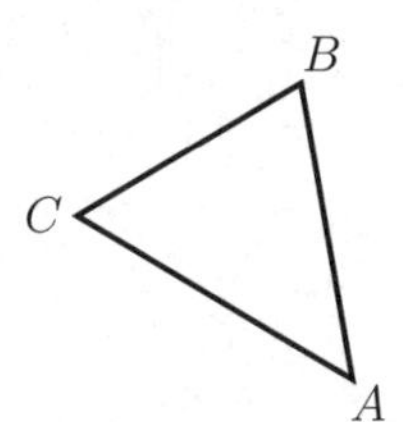

(A) 6.24

(B) 7.48

(C) 9.14

(D) 10.62

(E) 13.56

6. In $\triangle ABC$ shown below, $AB = BC$. If $m\angle C = 40°$, what is the area of the triangle?

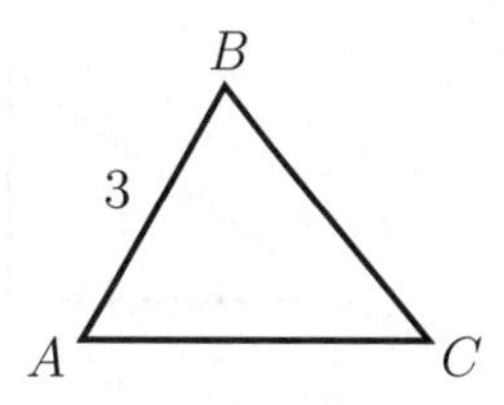

(A) 9.28

(B) 7.83

(C) 6.93

(D) 5.51

(E) 4.43

7. If the ratio of the three sides of a triangle is $4 : 6 : 7$, what is the smallest angle of the triangle?

(A) 23.45°

(B) 34.77°

(C) 42.36°

(D) 57.15°

(E) 61.94°

8. In the figure below, $ABCD$ is a square and triangle CDE is an isosceles right triangle. What is the measure of angle AEB ?

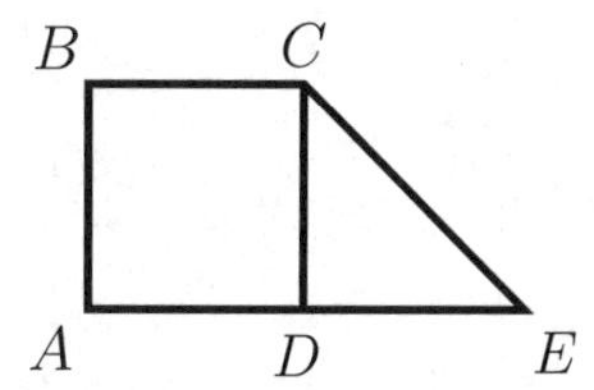

(A) 24.29°

(B) 26.57°

(C) 28.84°

(D) 31.35°

(E) 35.44°

9. In the right triangle ABC shown below, $AB = 4$. If $m\angle A = \theta$, what is the area of the triangle in terms of θ ?

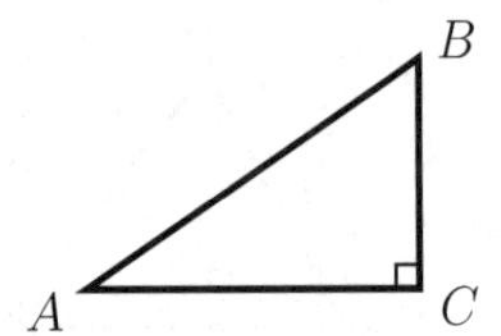

(A) $\sin\theta\cos\theta$

(B) $2\sin\theta\cos\theta$

(C) $4\sin\theta\cos\theta$

(D) $6\sin\theta\cos\theta$

(E) $8\sin\theta\cos\theta$

10. In $\triangle ABC$, $m\angle A : m\angle B : m\angle C = 2:3:4$, and $BC = 6$. What is AB ?

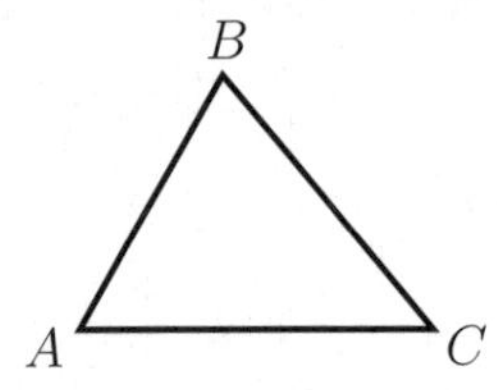

(A) 5.21

(B) 7.93

(C) 8.31

(D) 9.19

(E) 13.89

ANSWERS AND SOLUTIONS

1. (C)

$\triangle ABC$ is a right triangle as shown below.

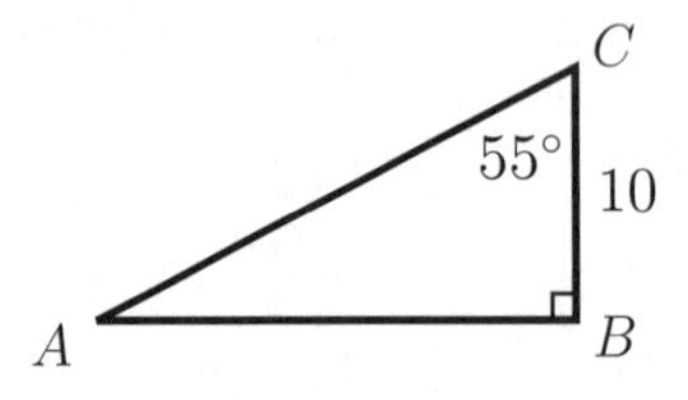

In order to find AB, use the definition of the tangent function: $\tan\theta = \frac{\text{Opposite side}}{\text{Adjacent side}}$.

$$\tan 55^\circ = \frac{AB}{10} \quad \Longrightarrow \quad AB = 10\tan 55^\circ = 14.28$$

Therefore, $AB = 14.28$.

2. (A)

Since $AB = BC$, $\triangle ABC$ is an isosceles triangle as shown below.

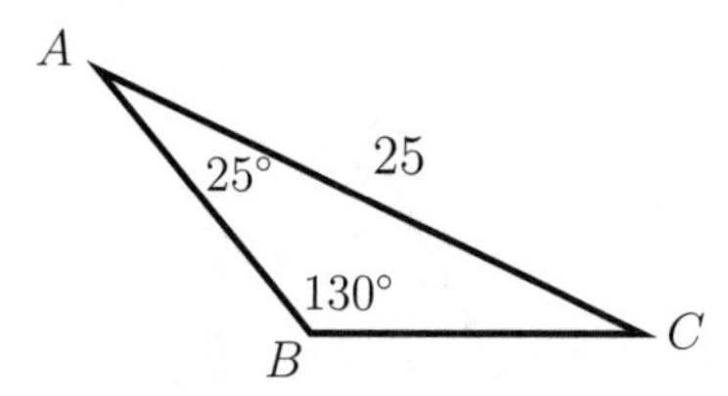

Thus, $m\angle A = m\angle C = \frac{180-130}{2}$ or $25°$. Since one side ($AC = 25$) and two angles are known ($130°$ and $25°$), $\triangle ABC$ is a SAA triangle. In order to find BC, use the Law of Sines.

$$\frac{BC}{\sin 25°} = \frac{25}{\sin 130°} \implies BC = \frac{25\sin 25°}{\sin 130°} = 13.79$$

Therefore, $BC = 13.79$.

3. (E)

Since $\tan\theta = 2$, $\theta = \tan^{-1} 2 = 63.43°$. Thus,

$$\sec^2\theta = \frac{1}{\cos^2\theta} = \frac{1}{(\cos 63.43°)^2} = 5$$

Therefore, the value of $\sec^2\theta = 5$.

4. (C)

Since $\triangle ABC$ is a right triangle, in order to find BC, use the definition of the cosine function: $\cos\theta = \frac{\text{Adjacent side}}{\text{Hypotenuse}}$

$$\cos\theta = \frac{\text{BC}}{4} \implies BC = 4\cos\theta$$

Therefore, BC in terms of θ is $4\cos\theta$.

5. (B)

Since the triangle ABC shown below is a SAS triangle (AB, BC, and $m\angle B$ are known),

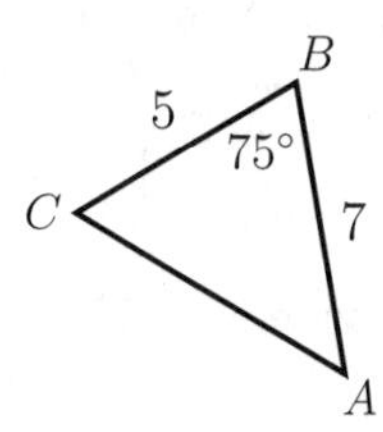

AC can be calculated by the Law of Cosines: $AC^2 = AB^2 + BC^2 - 2 \cdot AB \cdot BC \cdot \cos B$. Thus,

$$AC^2 = 7^2 + 5^2 - 2(7)(5)\cos 75° \implies AC = 7.48$$

Therefore, $AC = 7.48$.

6. (E)

$\triangle ABC$ shown below is an isosceles triangle where $AB = BC = 3$, and $m\angle A = m\angle C = 40°$. Thus, $m\angle B = 100°$.

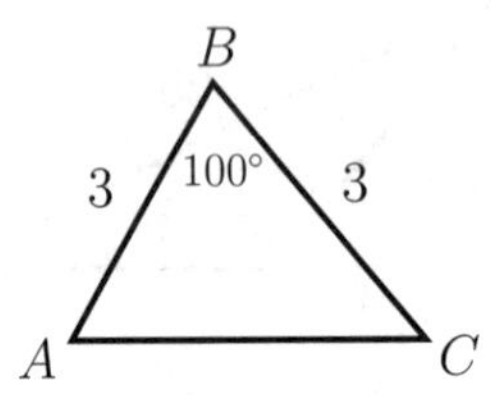

Since $\triangle ABC$ is a SAS triangle (AB, BC, and $m\angle B$ are known), the area of $\triangle ABC$ is as follows:

$$A = \frac{1}{2} \cdot AB \cdot BC \cdot \sin 100° = 4.43$$

Therefore, the area of $\triangle ABC$ is 4.43.

7. (B)

The ratio of the three sides of $\triangle ABC$ shown below is $4 : 6 : 7$. The Law of Sines implies that the smallest angle is opposite the shortest side. Thus, $\angle C$ is the smallest angle since $\overline{AB}$ is the shortest side.

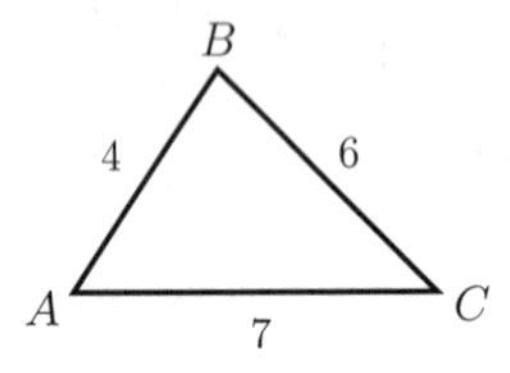

In order to find the measure of $\angle C$, use the Law of Cosines. Since $\overline{AB}$ is opposite $\angle C$,

$$m\angle C = \cos^{-1}\left(\frac{AB^2 - BC^2 - AC^2}{-2 \cdot BC \cdot AC}\right) = \cos^{-1}\left(\frac{4^2 - 6^2 - 7^2}{-2(6)(7)}\right) = 34.77°$$

Therefore, the smallest angle of the triangle is 34.77°.

8. (B)

For simplicity, let the length of the square equal 1. Thus, $AB = AD = 1$ as shown below. Since $\triangle CDE$ is an isosceles triangle, $DE = 1$.

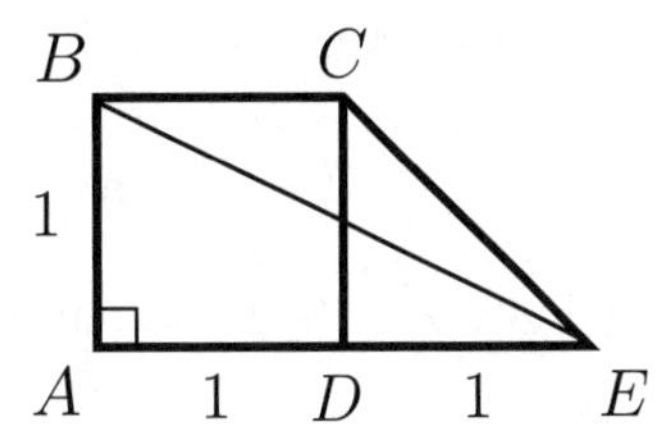

$\triangle AEB$ is a right triangle where $AE = 2$ and $AB = 1$. In order to find $m\angle AEB$, use the definition of the tangent function: $\tan\theta = \frac{\text{Opposite side}}{\text{Adjacent side}}$. Let $\theta = m\angle AEB$.

$$\tan\theta = \frac{1}{2} \quad \Longrightarrow \quad \theta = \tan^{-1}\left(\frac{1}{2}\right) = 26.57^\circ$$

Therefore, the measure of angle AEB is 26.57°.

9. (E)

Since $\triangle ABC$ is a right triangle as shown below,

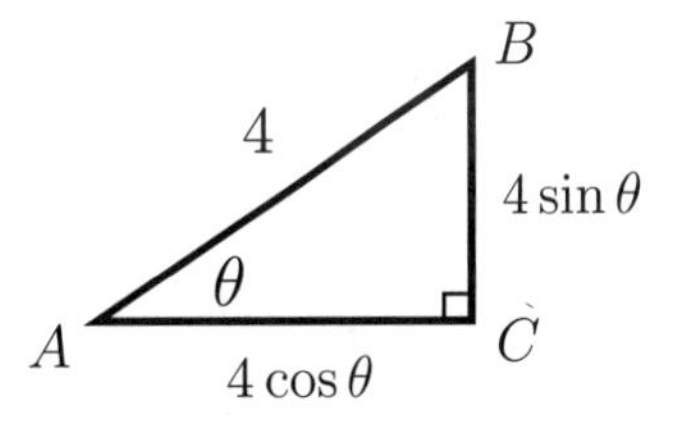

use the definition of the sine function and the cosine function to find AC and BC, respectively.

$$\cos\theta = \frac{AC}{4} \quad \Longrightarrow \quad AC = 4\cos\theta$$
$$\sin\theta = \frac{BC}{4} \quad \Longrightarrow \quad BC = 4\sin\theta$$

Thus, the area of $\triangle ABC$ in terms of θ is as follows:

$$A = \frac{1}{2}\cdot AC \cdot BC = \frac{1}{2}\cdot 4\cos\theta \cdot 4\sin\theta = 8\sin\theta\cos\theta$$

Therefore, the area of $\triangle ABC$ is $8\sin\theta\cos\theta$.

10. (D)

Let $2x$, $3x$, and $4x$ be $m\angle A$, $m\angle B$, and $m\angle C$, respectively. Since the sum of the measures of the interior angles of a triangle is $180°$,

$$\begin{aligned} 2x + 3x + 4x &= 180 \\ 9x &= 180 \\ x &= 20 \end{aligned}$$

Thus, $m\angle A = 40°$, $m\angle B = 60°$, and $m\angle C = 80°$ as shown below.

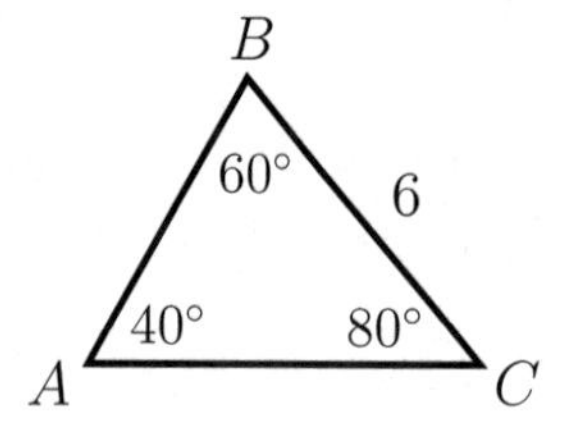

Since $\triangle ABC$ is a SAA triangle ($BC = 6$, $m\angle A$, and $m\angle C$ are known), use the Law of Sines to find AB.

$$\frac{6}{\sin 40°} = \frac{AB}{\sin 80°} \quad \Longrightarrow \quad AB = \frac{6\sin 80°}{\sin 40°} = 9.19$$

Therefore, $AB = 9.19$.

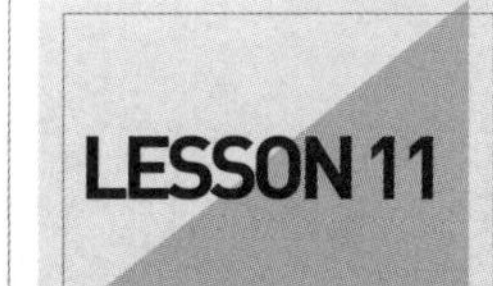

Polar Coordinates and Vectors

Polar Coordinates

In rectangular coordinates shown in Figure 1, a point is determined by (x, y), where x, and y represent where the point is x units horizontally and y units vertically from the origin, respectively.

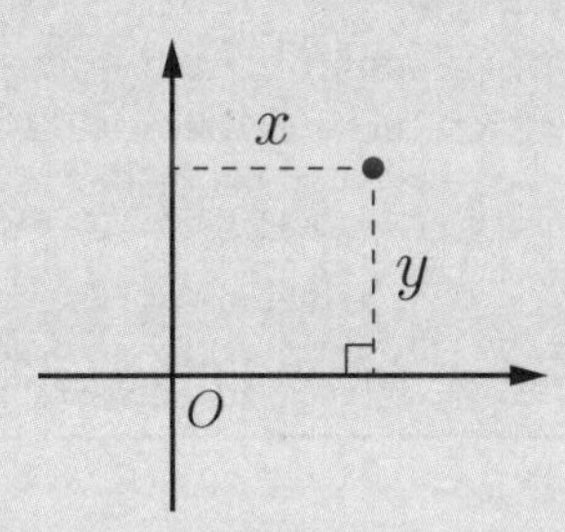

Fig. 1: Rectangular Coordinates

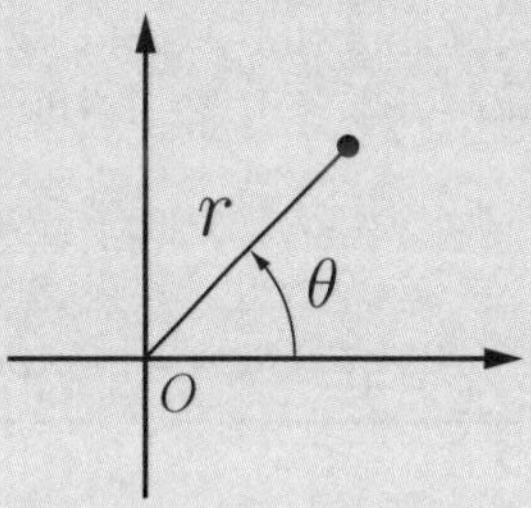

Fig. 2: Polar Coordinates

Whereas, in polar coordinates shown in Figure 2, a point is determined by (r, θ), where r represents the distance between the point and the origin, and θ represents the counterclockwise angle formed by the positive x-axis and the terminal side.

Identical Points in Polar Coordinates

In polar coordinates, two points are considered the same in the following two cases.

- Case 1: $(r, \theta) = (r, \theta + 2\pi k)$, where k is an integer.

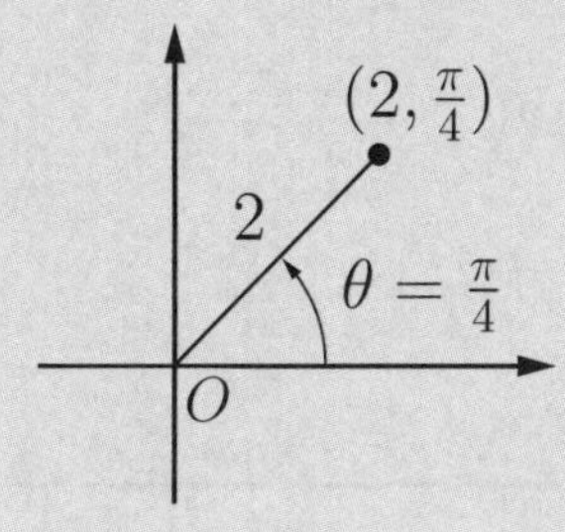

Figure 3: $(2, \frac{\pi}{4})$

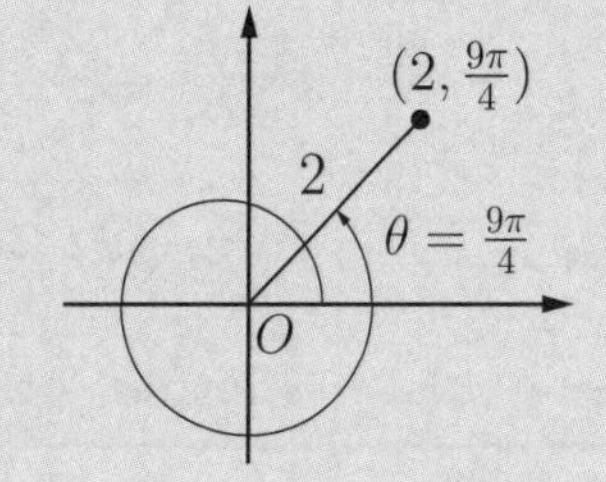

Figure 4: $(2, \frac{9\pi}{4})$

For instance, point $(2, \frac{\pi}{4})$ is plotted in the polar coordinates shown in Figure 3. Point $(2, \frac{9\pi}{4})$ is the same point as $(2, \frac{\pi}{4})$, and is plotted in Figure 4.

- Case 2: $(-r, \theta) = (r, \theta \pm \pi)$ Note: If $\theta + \pi > 2\pi$, use $\theta - \pi$ instead.

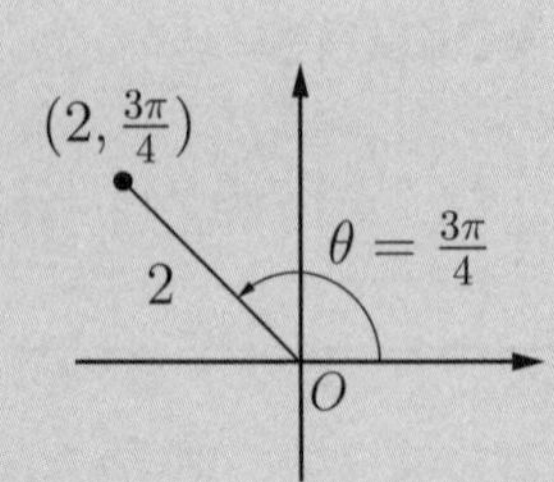

Figure 5: $(2, \frac{3\pi}{4})$

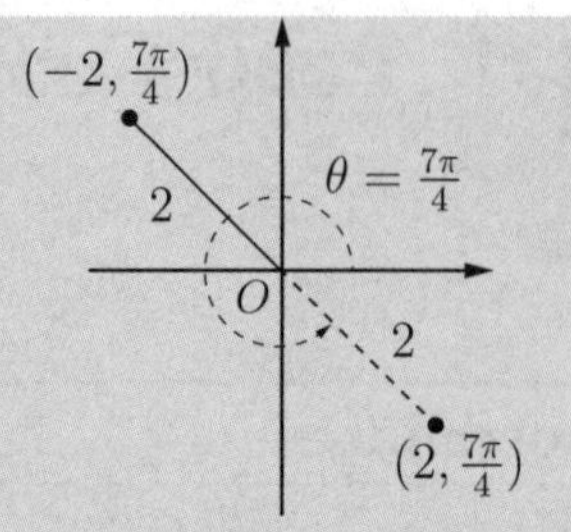

Figure 6: $(-2, \frac{7\pi}{4})$

For instance, points $(2, \frac{3\pi}{4})$ and $(2, \frac{7\pi}{4})$ are plotted in the polar coordinates shown in Figure 5 and 6. To plot $(-2, \frac{7\pi}{4})$, go to the direction $\frac{7\pi}{4}$ and then move a distance 2 in the opposite direction.

$$\left(-2, \frac{7\pi}{4}\right) = \left(2, \frac{7\pi}{4} - \pi\right) = \left(2, \frac{3\pi}{4}\right)$$

Thus, point $(-2, \frac{7\pi}{4})$ is the same point as $(2, \frac{3\pi}{4})$, and is plotted in Figure 6.

Converting from Polar Coordinates to Rectangular Coordinates

If the polar coordinates of a point is (r, θ), the rectangular coordinates of the point (x, y) are given by

$$x = r\cos\theta, \qquad y = r\sin\theta$$

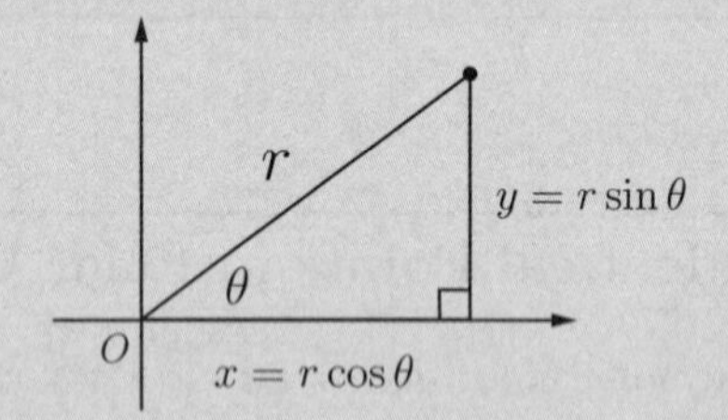

For instance, if the polar coordinates of a point is $(2, \frac{3\pi}{4})$, the rectangular coordinates of the point are as follows:

$$x = r\cos\theta \quad \Longrightarrow \quad x = 2\cos\frac{3\pi}{4} = 2\left(-\frac{\sqrt{2}}{2}\right) = -\sqrt{2}$$

$$y = r\sin\theta \quad \Longrightarrow \quad y = 2\sin\frac{3\pi}{4} = 2\left(\frac{\sqrt{2}}{2}\right) = \sqrt{2}$$

Thus, the rectangular coordinates of the point is $(-\sqrt{2}, \sqrt{2})$.

Converting from Rectangular Coordinates to Polar Coordinates

To convert from the rectangular coordinates (x, y) to the polar coordinates (r, θ), do the following three steps:

- Step 1: Plot the point (x, y) to determine the quadrant the angle θ lies in.
- Step 2: Find the distance between the point (x, y) and the origin, r, and the reference angle, β, formed by the positive x-axis and the terminal side.

$$r = \sqrt{x^2 + y^2}, \qquad \beta = \left|\tan^{-1}\frac{y}{x}\right|$$

  Note that the reference angle β is the **absolute value** of the inverse tangent function of $\frac{y}{x}$.
- Step 3: Find θ using the reference angle based on the quadrant that θ lies in.

Example 1 Converting from rectangular coordinates to polar coordinates

Find the polar coordinates of a point with the rectangular coordinates $(-1, -\sqrt{3})$.

Solution To convert from the rectangular coordinates $(-1, -\sqrt{3})$ to the polar coordinates (r, θ), do the following three steps:

- Step 1: Plot the point $(-1, -\sqrt{3})$ as shown in Figure 7. Since the point $(-1, -\sqrt{3})$ is in the third quadrant, $\pi < \theta < \frac{3\pi}{2}$.
- Step 2: Find the distance between the point $(-1, -\sqrt{3})$ and the origin, r, and the reference angle, β, formed by the positive x-axis and the terminal side as shown in Figure 8.

$$r = \sqrt{x^2 + y^2} = \sqrt{(-1)^2 + (-\sqrt{3})^2} = 2$$
$$\beta = \left|\tan^{-1}\frac{y}{x}\right| = \left|\tan^{-1}\left(\frac{-\sqrt{3}}{-1}\right)\right| = \frac{\pi}{3}$$

- Step 3: Find θ using the reference angle based on the quadrant that θ lies in as shown in Figure 9. Since $\beta = \frac{\pi}{3}$ and θ lies in the third quadrant, $\theta = \pi + \beta = \pi + \frac{\pi}{3} = \frac{4\pi}{3}$.

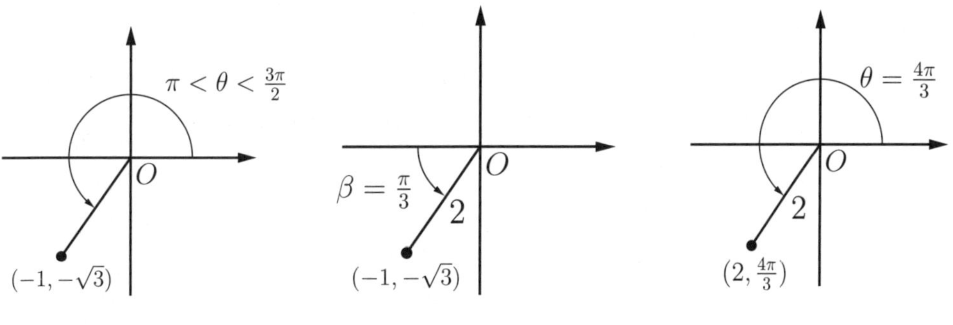

Figure 7 Figure 8 Figure 9

Therefore, the polar coordinates of a point with the rectangular coordinates $(-1, -\sqrt{3})$ are $(2, \frac{4\pi}{3})$.

Converting between Polar and Rectangular Equations

To convert from a polar equation to a rectangular equation, or vice versa, use the following conversion formulas:

$$r^2 = x^2 + y^2, \qquad x = r\cos\theta, \qquad y = r\sin\theta, \qquad \tan\theta = \frac{y}{x}$$

Example 2 Converting a polar equation to a rectangular equation

Write the polar equation $r = 2\sin\theta$ in rectangular form.

Solution To use $r^2 = x^2 + y^2$ and $y = r\sin\theta$, multiply each side of the polar equation by r.

$r = 2\sin\theta$ Multiply each side by r

$r^2 = 2r\sin\theta$ Replace r^2 with $x^2 + y^2$, and $r\sin\theta$ with y

$x^2 + y^2 = 2y$

Example 3 Converting a rectangular equation to a polar equation

Write the rectangular equation $(x-2)^2 + y^2 = 4$ in polar form.

Solution To use $r^2 = x^2 + y^2$ and $x = r\cos\theta$, expand $(x-2)^2$ and simplify the equation.

$(x-2)^2 + y^2 = 4$ Expand $(x-2)^2$

$x^2 - 4x + 4 + y^2 = 4$ Simplify

$x^2 + y^2 - 4x = 0$ Replace $x^2 + y^2$ with r^2, and x with $r\cos\theta$

$r^2 - 4r\cos\theta = 0$ Add $4r\cos\theta$ to each side

$r^2 = 4r\cos\theta$ Divide each side by r

$r = 4\cos\theta$

Vectors

A vector is a quantity that has a magnitude and a direction. If point $A(x_1, y_1)$ is the initial point and point $B(x_2, y_2)$ is the terminal point, a vector, denoted by either $\overrightarrow{AB}$ or **AB**, is defined as

$$\mathbf{AB} = \langle x_2 - x_1, y_2 - y_1 \rangle$$

and the magnitude of the vector, denoted by either $|\overrightarrow{AB}|$ or $|\mathbf{AB}|$, is defined as

$$|\mathbf{AB}| = \sqrt{(x_2 - x_1)^2 + (y_2 - y_1)^2}$$

Tip

1. The unit vector whose direction is along the positive x-axis is $\mathbf{i} = \langle 1, 0 \rangle$. The unit vector whose direction is along the positive y-axis is $\mathbf{j} = \langle 0, 1 \rangle$.
2. Let $\mathbf{v} = \langle a, b \rangle$, where a and b are the horizontal and vertical components of $\mathbf{v}$. Another way to define vector $\mathbf{v}$ using the unit vectors $\mathbf{i}$ and $\mathbf{j}$ is as follows:

$$\mathbf{v} = \langle a, b \rangle = a\langle 1, 0 \rangle + b\langle 0, 1 \rangle = a\mathbf{i} + b\mathbf{j}$$

Algebraic Operations on Vectors

If $\mathbf{u} = \langle a, b \rangle$, $\mathbf{v} = \langle c, d \rangle$, and c is a scalar (Constant), then

Addition:	$\mathbf{u} + \mathbf{v} = \langle a + c, b + d \rangle = (a + c)\mathbf{i} + (b + d)\mathbf{j}$
Subtraction:	$\mathbf{u} - \mathbf{v} = \langle a - c, b - d \rangle = (a - c)\mathbf{i} + (b - d)\mathbf{j}$
Scalar product:	$c\mathbf{u} = \langle ca, cb \rangle = ca\mathbf{i} + cb\mathbf{j}$

Example 4 Algebraic operations on vectors

If $\mathbf{v} = \langle 3, 4 \rangle$, $\mathbf{w} = \langle 2, 8 \rangle$, find $\mathbf{v} + \mathbf{w}$, $\mathbf{w} - \mathbf{v}$, $5\mathbf{v}$, and $|\mathbf{v} + \mathbf{w}|$.

Solution

$$\begin{aligned}
&\mathbf{v} + \mathbf{w} = \langle 3, 4 \rangle + \langle 2, 8 \rangle = \langle 5, 12 \rangle \\
&\mathbf{w} - \mathbf{v} = \langle 2, 8 \rangle - \langle 3, 4 \rangle = \langle -1, 4 \rangle \\
&5\mathbf{v} = 5\langle 3, 4 \rangle = \langle 15, 20 \rangle \\
&|\mathbf{v} + \mathbf{w}| = |\langle 5, 12 \rangle| = \sqrt{5^2 + 12^2} = 13
\end{aligned}$$

Graphing Vectors

Two vectors **u** and **v** are equal if they have the same magnitude and the same direction. In Figure 10 below, three vectors **u**, **v**, and **v** are drawn. Although the three vectors have different initial points and different terminal points, they are equal because they have the same magnitude and the same direction.

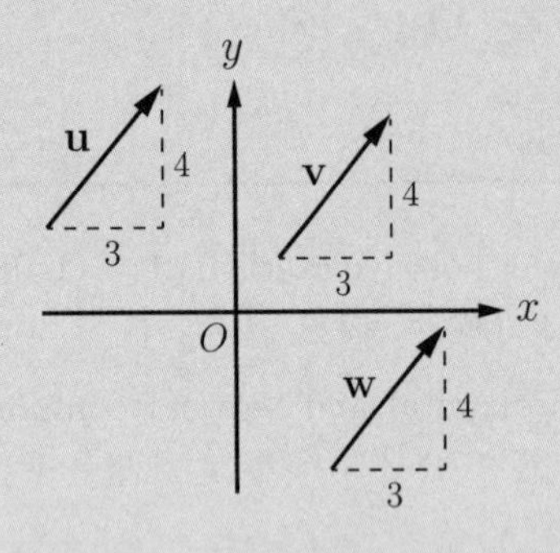

Figure 10

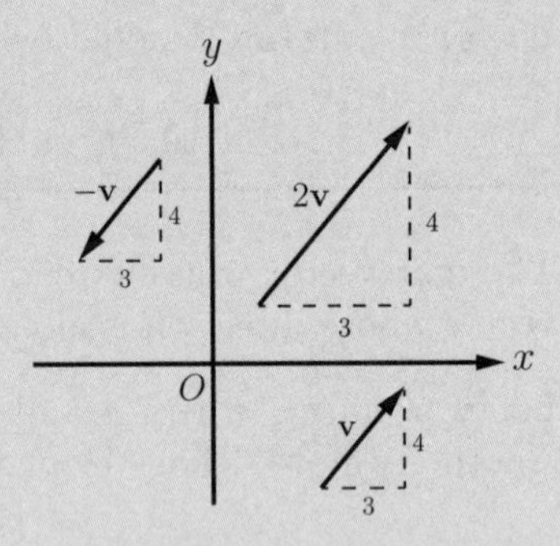

Figure 11

If **v** is a vector drawn in Figure 11, $-\mathbf{v}$ is a vector whose magnitude is the same as **v**, but whose direction is opposite to **v**. Whereas, $2\mathbf{v}$ is a vector whose magnitude is twice the magnitude of **v**, but whose direction is the same as **v**.

Drawing the resultant vectors

The sum of two vectors is called the resultant vector. The resultant vector can be drawn in the following steps:

- Step 1: Draw the first vector **u** as shown in Figure 12.
- Step 2: Draw the second vector, **v**, so that the initial point of the second vector coincides with the terminal point of the first vector **u** as shown in Figure 12.
- Step 3: Draw the resultant vector, $\mathbf{u}+\mathbf{v}$, from the initial point of the first vector **u** to the terminal point of the second vector **v** as shown in Figure 13.

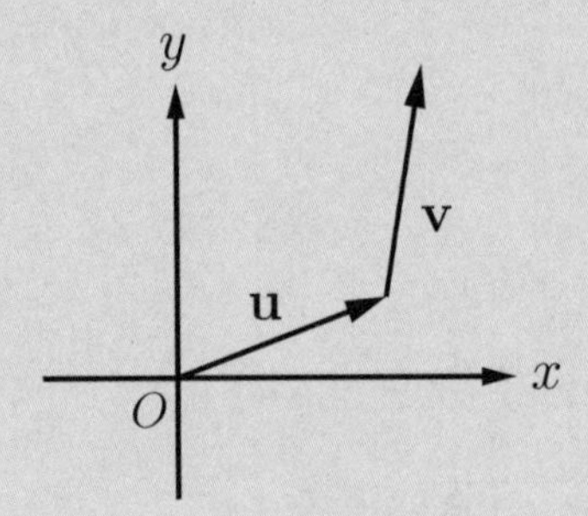

Figure 12

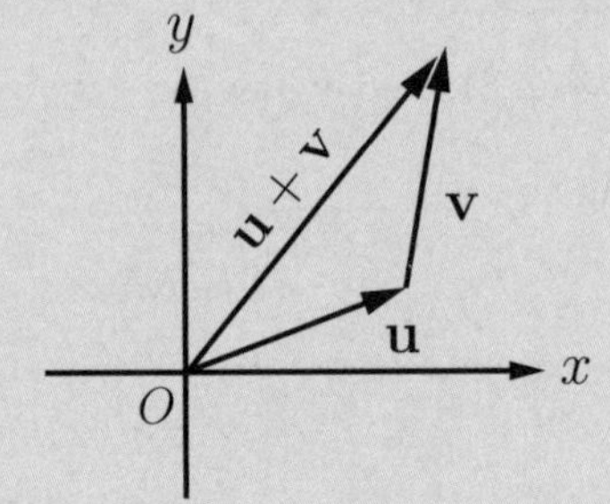

Figure 13

EXERCISES

1. What are the rectangular coordinates (x, y) that correspond with the polar coordinates $(4, \frac{5\pi}{6})$?

   (A) $(-2\sqrt{3}, 2)$

   (B) $(-2, -2\sqrt{3})$

   (C) $(2, -2\sqrt{3})$

   (D) $(2\sqrt{2}, 2\sqrt{2})$

   (E) $(2\sqrt{3}, -2)$

2. Which of the following rectangular equation is equivalent to the polar equation $r = 2\csc\theta$?

   (A) $x = 2$

   (B) $x = 2y$

   (C) $y = 2$

   (D) $y = 2x$

   (E) $x^2 + y^2 = 4$

3. If vector $\mathbf{V}$ has an initial point of $(-1, 2)$ and a terminal point of $(-3, 4)$, what is the position vector of $\mathbf{V}$?

   (A) $\langle -2, 2 \rangle$

   (B) $\langle -4, 6 \rangle$

   (C) $\langle 2, -2 \rangle$

   (D) $\langle 2, 6 \rangle$

   (E) $\langle 4, 6 \rangle$

4. Which of the following vector has the largest magnitude?

   (A) $\langle -3, -3 \rangle$

   (B) $\langle -6, 9 \rangle$

   (C) $\langle 5, 12 \rangle$

   (D) $\langle 7, -8 \rangle$

   (E) $\langle 8, 9 \rangle$

5. If $\mathbf{V} = \langle 2, 3 \rangle$ and $\mathbf{W} = \langle 6, -2 \rangle$, what is $\frac{1}{2}\mathbf{W} - 2\mathbf{V}$?

   (A) $\langle 9, 5 \rangle$

   (B) $\langle 8, 1 \rangle$

   (C) $\langle 1, 5 \rangle$

   (D) $\langle -1, -7 \rangle$

   (E) $\langle -7, 5 \rangle$

6. Which of the following point has polar coordinates equivalent to $(-3, \frac{3\pi}{4})$?

   (A) $(-3, \frac{\pi}{4})$

   (B) $(-3, \frac{5\pi}{4})$

   (C) $(3, \frac{3\pi}{4})$

   (D) $(3, \frac{5\pi}{4})$

   (E) $(3, \frac{7\pi}{4})$

7. If a point with the rectangular coordinates $(4, -4)$ is identical to a point with the polar coordinates $(4\sqrt{2}, \theta)$, what is value of θ ?

   (A) 2π

   (B) $\frac{7\pi}{4}$

   (C) $\frac{5\pi}{4}$

   (D) $\frac{3\pi}{4}$

   (E) $\frac{\pi}{4}$

8. Which of the following polar equation is equivalent to $x^2 + y^2 = 2x$?

   (A) $r = \sin 2\theta$

   (B) $r = \cos 2\theta$

   (C) $r = 2\tan\theta$

   (D) $r = 2\sin\theta$

   (E) $r = 2\cos\theta$

9. If vectors **u** and **v** are given below, what is $|\mathbf{u}+\mathbf{v}|$?

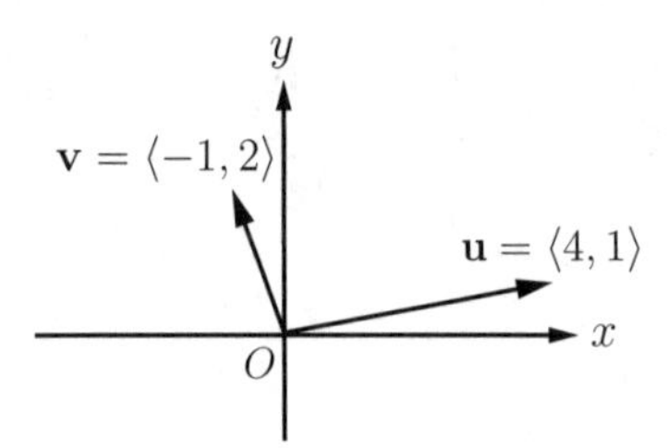

(A) 1.41

(B) 3

(C) 3.61

(D) 4.23

(E) 6

10. If vector **A** has an initial point of $(1,2)$ and a terminal point $(x,6)$, what are the possible values of x so that $|\mathbf{A}|=5$?

(A) 2 or −4

(B) 4 or −2

(C) 3 or −3

(D) 5 or 1

(E) 6 or 3

ANSWERS AND SOLUTIONS

1. (A)

Tip: If the polar coordinates of a point are (r,θ), the rectangular coordinates of the point (x,y) are given by $x=r\cos\theta$ and $y=r\sin\theta$.

The polar coordinates of the point are $(4, \frac{5\pi}{6})$. Thus, $r=4$ and $\theta=\frac{5\pi}{6}$. Since $\cos\frac{5\pi}{6}=-\cos\frac{\pi}{6}=-\frac{\sqrt{3}}{2}$, and $\sin\frac{5\pi}{6}=\sin\frac{\pi}{6}=\frac{1}{2}$,

$$x=4\cos\frac{5\pi}{6}=4\left(\frac{-\sqrt{3}}{2}\right)=-2\sqrt{3}$$

$$y=4\sin\frac{5\pi}{6}=4\left(\frac{1}{2}\right)=2$$

Therefore, the rectangular coordinates $(x,y)=(-2\sqrt{3},2)$.

2. (C)

$$r=2\csc\theta \qquad \text{Since } \csc\theta=\frac{1}{\sin\theta}$$

$$r=\frac{2}{\sin\theta} \qquad \text{Multiply each side by } \sin\theta$$

$$r\sin\theta=2 \qquad \text{Since } y=r\sin\theta$$

$$y=2$$

Therefore, the rectangular equation $y=2$ is equivalent to the polar equation $r=2\csc\theta$.

3. (A)

Tip If (x_1, y_1) is the initial point and point (x_2, y_2) is the terminal point, vector $\mathbf{V}$ is defined as $\mathbf{V} = \langle x_2 - x_1, y_2 - y_1 \rangle$.

Since the initial point is $(-1, 2)$ and the terminal point is $(-3, 4)$,

$$\mathbf{V} = \langle x_2 - x_1, y_2 - y_1 \rangle = \langle -3 - (-1), 4 - 2 \rangle = \langle -2, 2 \rangle$$

Therefore, the position vector of $\mathbf{V}$ is $\langle -2, 2 \rangle$.

4. (C)

Tip If a vector $\mathbf{V} = \langle x, y \rangle$ is given, the magnitude of the vector, denoted by $|\mathbf{V}|$, is defined as $|\mathbf{V}| = \sqrt{(x)^2 + (y)^2}$.

$$\begin{aligned}
&\text{Magnitude of } \langle -3, -3 \rangle \text{ is } \sqrt{(-3)^2 + (-3)^2} = \sqrt{18} \\
&\text{Magnitude of } \langle -6, 9 \rangle \text{ is } \sqrt{(-6)^2 + (9)^2} = \sqrt{117} \\
&\text{Magnitude of } \langle 5, -12 \rangle \text{ is } \sqrt{(5)^2 + (-12)^2} = \sqrt{169} \\
&\text{Magnitude of } \langle 7, -8 \rangle \text{ is } \sqrt{(7)^2 + (-8)^2} = \sqrt{113} \\
&\text{Magnitude of } \langle 8, 9 \rangle \text{ is } \sqrt{(8)^2 + (9)^2} = \sqrt{145}
\end{aligned}$$

Therefore, (C) is the correct answer.

5. (D)

Since $\mathbf{V} = \langle 2, 3 \rangle$, $\mathbf{W} = \langle 6, -2 \rangle$,

$$\begin{aligned}
\frac{1}{2}\mathbf{W} - 2\mathbf{V} &= \frac{1}{2}\langle 6, -2 \rangle - 2\langle 2, 3 \rangle \\
&= \langle 3, -1 \rangle - \langle 4, 6 \rangle \\
&= \langle -1, -7 \rangle
\end{aligned}$$

Therefore, $\frac{1}{2}\mathbf{W} - 2\mathbf{V}$ is $\langle -1, -7 \rangle$.

6. (E)

Tip $(-r, \theta) = (r, \theta \pm \pi)$ Note: If $\theta + \pi > 2\pi$, use $\theta - \pi$ instead.

$$\left(-3, \frac{3\pi}{4}\right) = \left(3, \frac{3\pi}{4} + \pi\right) = \left(3, \frac{7\pi}{4}\right)$$

Therefore, $(-3, \frac{3\pi}{4}) = (3, \frac{7\pi}{4})$.

7. (B)

The point $(4, -4)$ is in the fourth quadrant as shown below. Thus, θ lies in the fourth quadrant.

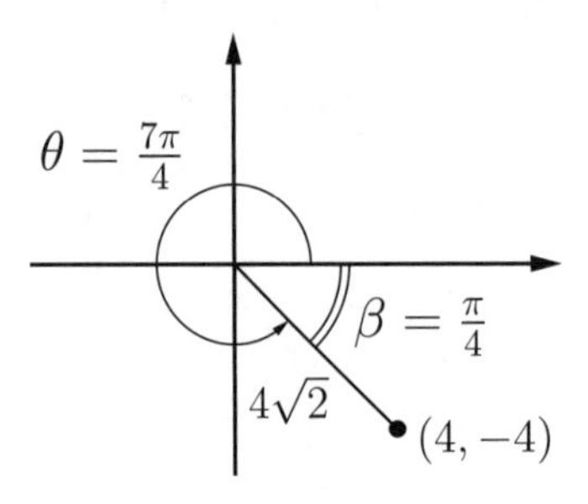

The reference angle β, formed by the positive x-axis and the terminal side, is

$$\beta = \left|\tan^{-1}\frac{y}{x}\right| = \left|\tan^{-1}\left(\frac{-4}{4}\right)\right| = \frac{\pi}{4}$$

Since $\beta = \frac{\pi}{4}$ and θ lies in the fourth quadrant, $\theta = 2\pi - \beta = 2\pi - \frac{\pi}{4} = \frac{7\pi}{4}$. Therefore, the value of θ is $\frac{7\pi}{4}$.

8. (E)

Tip $x^2 + y^2 = r^2$ and $x = r\cos\theta$

$x^2 + y^2 = 2x$	Substitute r^2 for $x^2 + y^2$ and $r\cos\theta$ for x
$r^2 = 2r\cos\theta$	Divide each side by r
$r = 2\cos\theta$	

Therefore, $r = 2\cos\theta$ is equivalent to $x^2 + y^2 = 2x$.

9. (D)

Since $\mathbf{u} = \langle 4, 1\rangle$ and $\mathbf{v} = \langle -1, 2\rangle$, $\mathbf{u} + \mathbf{v} = \langle 3, 3\rangle$. Thus, the magnitude of $\mathbf{u} + \mathbf{v}$, $|\mathbf{u} + \mathbf{v}|$, is

$$|\mathbf{u} + \mathbf{v}| = \sqrt{3^2 + 3^2} = 3\sqrt{2} = 4.23$$

Therefore, (D) is the correct answer.

10. (B)

The initial point is $(1, 2)$ and terminal point is $(x, 6)$. Thus, $\mathbf{A} = \langle x - 1, 4\rangle$. Since the magnitude of $\mathbf{A}$, $\sqrt{(x-1)^2 + (4)^2}$, is 5,

$\sqrt{(x-1)^2 + (4)^2} = 5$	Square each side
$(x-1)^2 + (4)^2 = 25$	Subtract 16 from each side
$(x-1)^2 = 9$	Take the square root of each side
$x - 1 = \pm 3$	Add 1 to each side
$x = 1 \pm 3$	
$x = 4$ or $x = -2$	

Therefore, the possible values of x are 4 or -2.

LESSON 12 Conic Sections and Parametric Equations

Identifying Conic Sections

Conic sections, or conics, consist of circles, parabolas, ellipses, and hyperbolas. The following guidelines summarize how to identify the conics from the equations given.

- When the equation contains $Ax^2 + By^2$ (Squared terms for both x and y):
  - If $A = B$, the equation defines a circle.
  - If $A \neq B$, the equation defines an ellipse.
- When the equation contains $Ax^2 - By^2$ (Squared terms have a negative sign in between them):
  - If $A = B$, the equation defines a hyperbola.
  - If $A \neq B$, the equation defines a hyperbola.
- When the equation contains $Ax + By^2$ or $Ax^2 + By$ (Squared term for only x or y):

  The equation defines a parabola when it contains the squared term for only the x or y variable.

Example 1 Identifying conic sections

Identify the conic section that each equation represents.

(a) $3x^2 + 6x + y^2 + 4y - 6 = 0$

(b) $x^2 - 2y^2 - 4x + 2y + 1 = 0$

(c) $2x^2 - 4x + 2y^2 - 8y - 2 = 0$

(d) $4y^2 + 4y + 2x + 4 = 0$

Solution

(a) The equation has squared terms $3x^2 + y^2$, where $A = 3$ and $B = 1$. Thus, the equation represents an ellipse.

(b) Since the squared terms, $x^2 - 2y^2$, have a negative sign in between them, the equation represents a hyperbola.

(c) The equation has squared terms $2x^2 + 2y^2$, where $A = 2$ and $B = 2$. Thus, the equation represents a circle.

(d) Since the equation contains the squared term for only y, the equation represents a parabola.

Circles

A circle is the set of all points that are equidistant from a fixed point called the center. The general equation of a circle is given by

$$(x-h)^2 + (y-k)^2 = r^2$$

where the point (h, k) is the center of the circle, and r is the radius of the circle.

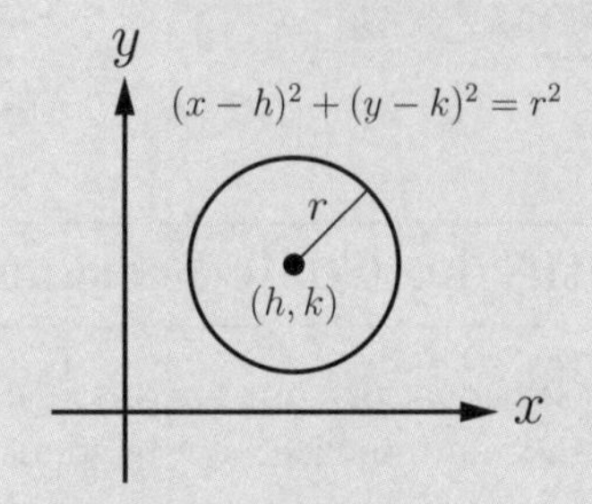

Tip In order to avoid a common mistake when finding the center of conics (a circle, an ellipse and a hyperbola), set $x - h = 0$ and $y - k = 0$ and solve for x and y. Thus, the x and y coordinates of the center of the conics are $x = h$ and $y = k$.

Ellipses

An ellipse is the set of all points in a plane such that the sum of the distances from two fixed points, called the **foci**, is a constant. The longer side of the ellipse is called the **major axis**, the shorter side of the ellipse is called the **minor axis**. The two endpoints along the major axis are the **vertices**. The other two endpoints along the minor axis are the **co-vertices**. The general equation of an ellipse is given by

$$\frac{(x-h)^2}{a^2} + \frac{(y-k)^2}{b^2} = 1$$

where the point (h, k) is the center of the ellipse. The lengths of the major axis and minor axis can be determined by the following rules.

- If $a > b$, the length of the major axis is $2a$, and the length of the minor axis is $2b$ as shown in Figure 1.
- If $a < b$, the length of the major axis is $2b$, and the length of the minor axis is $2a$ as shown in Figure 2.

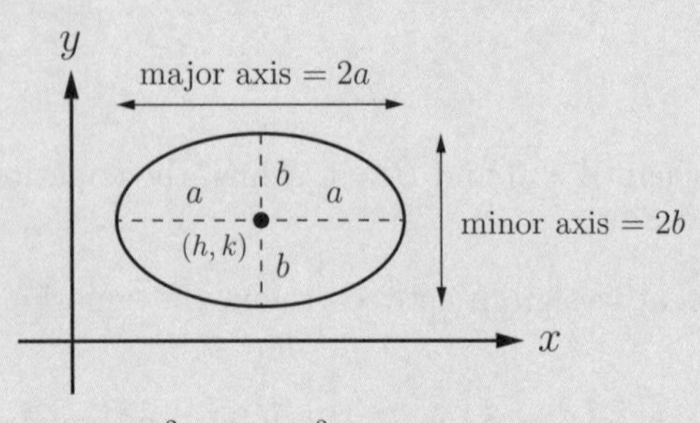

Fig. 1: $\frac{(x-h)^2}{a^2} + \frac{(y-k)^2}{b^2} = 1$, if $a > b$

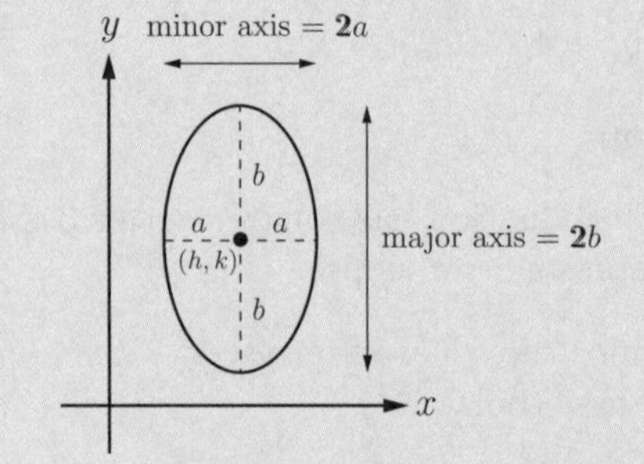

Fig. 2: $\frac{(x-h)^2}{a^2} + \frac{(y-k)^2}{b^2} = 1$, if $a < b$

Tip Let c be the distance between the center and foci of an ellipse. c can be determined by either $c^2 = a^2 - b^2$ (if $a > b$) or $c^2 = b^2 - a^2$ if (if $a < b$). However, it is less likely to find c on the SAT II Math level II test.

Example 2 Finding the center of a circle

Find the center and radius of the circle $(x-1)^2+(y+2)^2=9$.

Solution In order to find the center of the circle, set $x-1=0$ and $y+2=0$ and solve for x and y. Thus, the center of the circle is $(1,-2)$. Since $9=3^2$, the radius of the circle is 3, not 9.

Example 3 Finding the center and the length of the major axis of an ellipse

Find the center, vertices, and the lengths of the major and minor axis of the ellipse shown below.

$$4x^2+y^2+16x-6y+21=0$$

Solution In order to write a general equation of the ellipse $\frac{(x-h)^2}{a^2}+\frac{(y-k)^2}{b^2}=1$, proceed to complete the squares in x and in y.

$$4x^2+y^2+16x-6y+21=0$$

$4x^2+y^2+16x-6y=-21$	Subtract 21 from each side
$4(x^2+4x)+(y^2-6y)=-21$	Rearrange the terms
$4(x^2+4x+4)+(y^2-6y)=-5$	Add 16 to each side to complete squares in x
$4(x+2)^2+(y^2-6y+9)=4$	Add 9 to each side to complete squares in y
$4(x+2)^2+(y-3)^2=4$	Divide each side by 4
$\frac{(x+2)^2}{1^2}+\frac{(y-3)^2}{2^2}=1$	

In order to find the center of the ellipse, set $x+2=0$ and $y-3=0$ and solve for x and y. Thus, the center of the ellipse is $(-2,3)$.

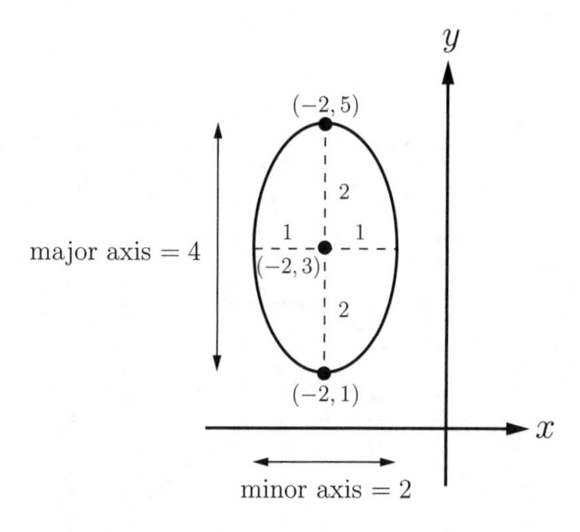

The lengths of the major axis and minor axis are 4 and 2, respectively. Since the vertices are vertically 2 units from the center, vertices are at $(-2,5)$, and $(-2,1)$.

Hyperbolas

Hyperbola is the set of all points in a plane such that the difference of the distances from two fixed points, called the foci, is a constant. The two endpoints on the hyperbola are the vertices. The midpoint of the line segment joining the vertices is called the **center** of the hyperbola. The line passes through the vertices and the center is called the **transverse axis**. The line through the center and perpendicular to the transverse axis is called the conjugate axis.

There are two types of equations of a hyperbola based on a horizontal transverse axis or a vertical transverse axis.

- An equation of a hyperbola with a horizontal transverse axis: $\dfrac{(x-h)^2}{a^2} - \dfrac{(y-k)^2}{b^2} = 1$

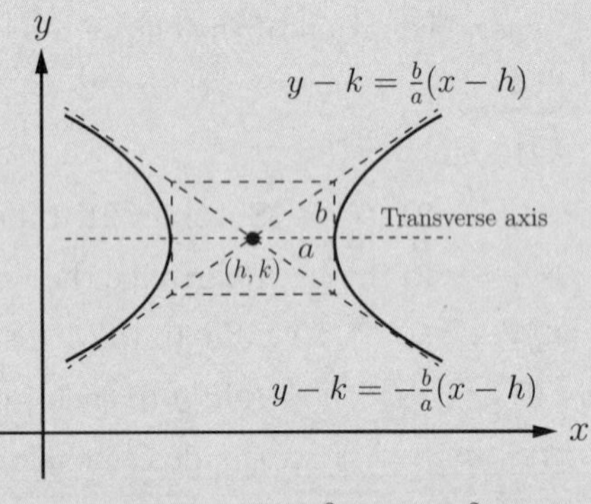

Figure 3: $\frac{(x-h)^2}{a^2} - \frac{(y-k)^2}{b^2} = 1$

  where the center of the hyperbola is (h, k). The equations of asymptotes are $y-k = \pm\frac{b}{a}(x-h)$ as shown in Figure 3.

- An equation of a hyperbola with a vertical transverse axis: $\dfrac{(y-k)^2}{a^2} - \dfrac{(x-h)^2}{b^2} = 1$

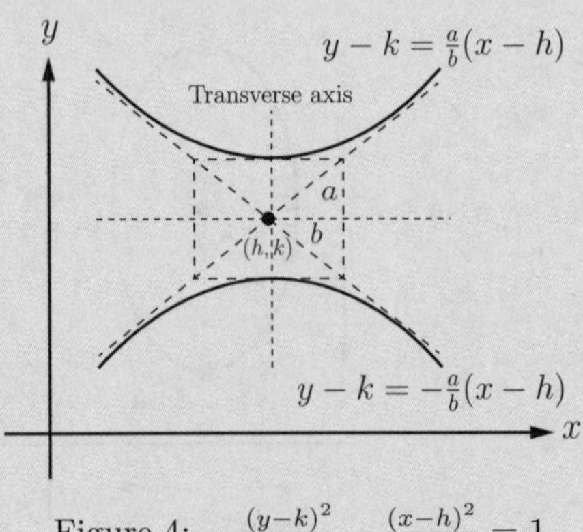

Figure 4: $\frac{(y-k)^2}{a^2} - \frac{(x-h)^2}{b^2} = 1$

  where the center of the hyperbola is (h, k). The equations of asymptotes are $y-k = \pm\frac{a}{b}(x-h)$ as shown in Figure 4.

Example 4 Finding the center and asymptotes of a hyperbola

Find the center, vertices, and asymptotes of the hyperbola shown below.

$$x^2 - 4y^2 - 6x + 5 = 0$$

Solution In order to write a general equation of the hyperbola $\frac{(x-h)^2}{a^2} - \frac{(y-k)^2}{b^2} = 1$, proceed to complete the squares in x and in y.

$x^2 - 4y^2 - 6x + 5 = 0$	
$x^2 - 4y^2 - 6x = -5$	Subtract 5 from each side
$(x^2 - 6x) - 4y^2 = -5$	Rearrange the terms
$(x^2 - 6x + 9) - 4y^2 = 4$	Add 9 to each side to complete squares in x
$(x-3)^2 - 4y^2 = 4$	Divide each side by 4
$\frac{(x-3)^2}{2^2} - \frac{(y)^2}{1^2} = 1$	

In order to find the center of the hyperbola, set $x - 3 = 0$ and $y = 0$ and solve for x and y. Thus, the center of the hyperbola is $(3, 0)$.

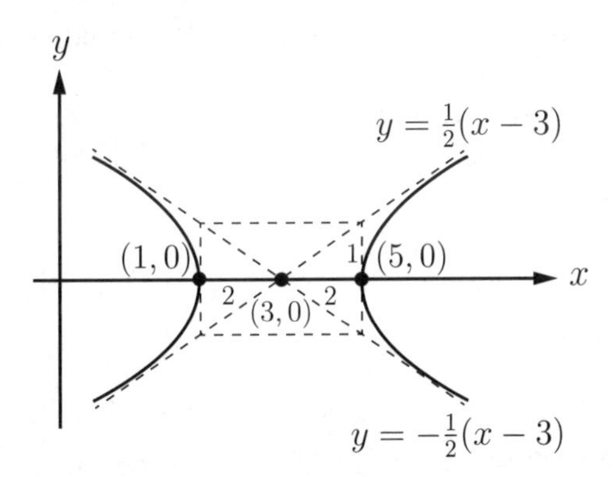

Since the vertices are horizontally 2 units from the center, the vertices are at $(1, 0)$ and $(5, 0)$. The slopes of the asymptotes are $\pm\frac{1}{2}$ and the asymptotes pass through the center $(3, 0)$. Thus, the equations of the asymptotes are $y = \frac{1}{2}(x-3)$ and $y = -\frac{1}{2}(x-3)$.

Parametric Equations

If variables x and y are related to a third variable t, called a **parameter**, the equations defined by

$$x = f(t), \qquad y = g(t)$$

are called **parametric equations**. Usually, t represents time. As a result, the parametric equations describe the movement of the graph at given time t. Below are examples of parametric equations.

$$x = t + 1, \qquad y = t^2 + 2t, \qquad -1 \le t \le 1$$

Graphing Parametric Equations

In order to graph parametric equations, do the following two steps.

- Step 1: Eliminate the parameter t so that parametric equations become a rectangular equation $y = f(x)$.
- Step 2: Graph $y = f(x)$ based on the values of t.

Let's graph the parametric equations defined by

$$x = t + 1, \qquad y = t^2 + 2t, \qquad -1 \le t \le 1$$

- Step 1: Solve for t from the first equation $x = t + 1$. We found that $t = x - 1$. Then, substitute $x - 1$ for t in the second equation $y = t^2 + 2t$ so that $y = (x-1)^2 + 2(x-1) = x^2 - 1$. Now, we eliminate the parameter t so that the parametric equations become $y = x^2 - 1$ as shown in Figure 5.
- Step 2: Since $x = t + 1$ and $-1 \le t \le 1$,

$-1 \le t \le 1$	Add 1 to each side of inequality
$0 \le t + 1 \le 2$	Since $x = t + 1$
$0 \le x \le 2$	

  the domain of the parametric equations is $0 \le x \le 2$. Thus, the rectangular equation of the parametric equations is $y = x^2 - 1$, $0 \le x \le 2$. The Figure 6 shows the graph of $y = x^2 - 1$, $0 \le x \le 2$.

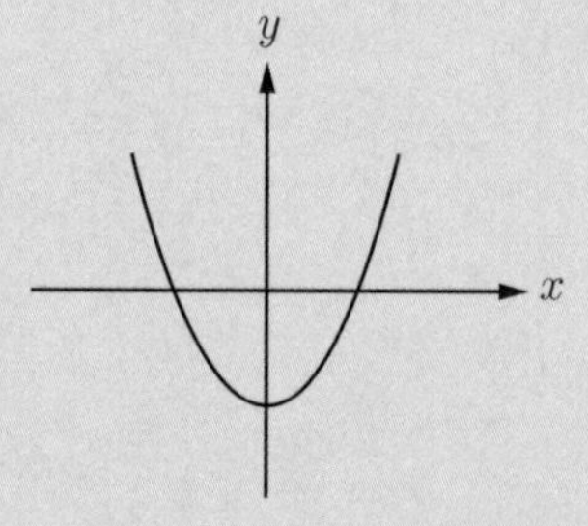

Figure 5: $y = x^2 - 1$

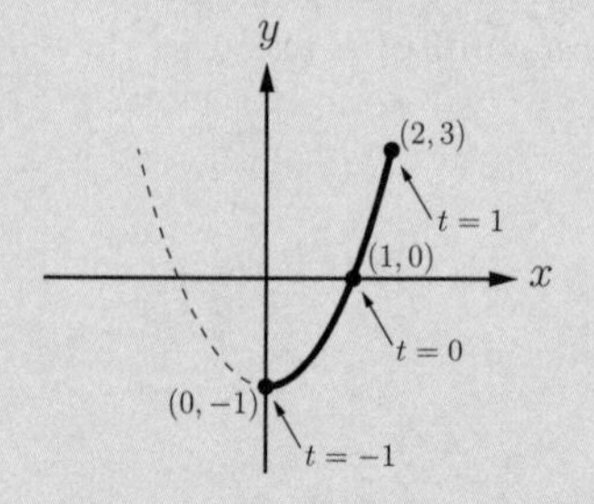

Figure 6: $y = x^2 - 1$, $0 \le x \le 2$

Solving Systems of Nonlinear Equations

A nonlinear equation represents a curve, not a straight line. Usually, the curve is one of the four conic sections: a circle, a parabola, an ellipse, and a hyperbola. A system means more than one. Thus, a **system of nonlinear equations** contains at least one curve. Below is an example of a system of nonlinear equations.

$$x^2 + y^2 = 41$$
$$y = x - 1$$

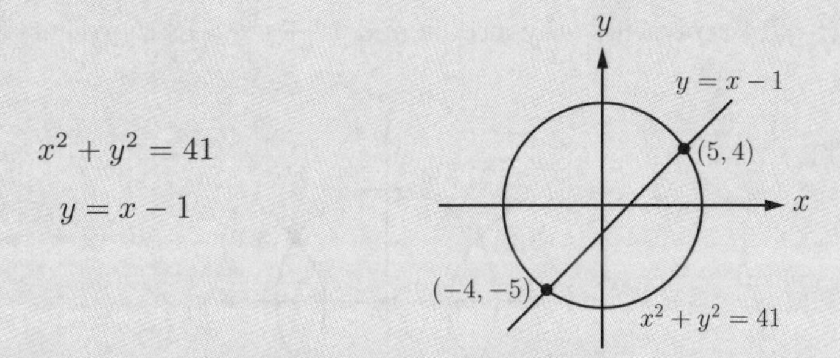

Solutions to a system of nonlinear equations are ordered pairs (x, y) that satisfy all equations in the system. In other words, solutions to a system of nonlinear equations are intersection points that lie on all graphs. In the figure above, $(5, 4)$ and $(-4, -5)$ are ordered pairs that satisfy all equations,

$$x^2 + y^2 = 41 \implies 5^2 + 4^2 = 41 \qquad x^2 + y^2 = 41 \implies (-4)^2 + (-5)^2 = 41$$
$$y = x - 1 \implies 4 = 5 - 1 \qquad y = x - 1 \implies -5 = -4 - 1$$

and are the intersection points of both graphs.

Solving a system of nonlinear equations means finding the x and y coordinates of the intersection points of both graphs. There are two methods to solve a system of nonlinear equations: **substitution** and **elimination**.

Let's solve the nonlinear equations shown below using the substitution method.

$$x^2 + y^2 = 41$$
$$y = x - 1$$

Substitute $x - 1$ for y in the first equation $x^2 + y^2 = 41$.

$x^2 + y^2 = 41$	Substitute $x - 1$ for y
$x^2 + (x-1)^2 = 41$	Simplify
$2x^2 - 2x - 40 = 0$	Divide each side by 2
$x^2 - x - 20 = 0$	Factor
$(x+4)(x-5) = 0$	Solve for x
$x = -4$ or $x = 5$	

In order to find the values of y, substitute $x = 5$ and $x = -4$ into $y = x - 1$. Thus, $y = 4$ and $y = -5$, respectively. Therefore, the solutions to the system of nonlinear equations are $(5, 4)$ and $(-4, -5)$.

Example 5 Solving a system of nonlinear equations

Solve the following system of nonlinear equations.

$$2x^2 - y^2 = 14$$
$$x^2 + y^2 = 13$$

Solution $2x^2 - y^2 = 14$ represents a hyperbola and $x^2 + y^2 = 13$ represents a circle as shown below.

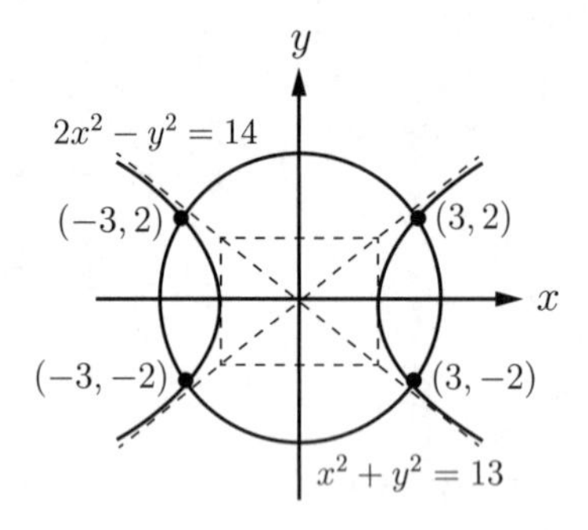

Both graphs intersect four times. In order to find the intersection points of both graphs, use the elimination method since the coefficients of y^2 are opposite.

$$\begin{aligned} 2x^2 - y^2 &= 14 \\ x^2 + y^2 &= 13 && \text{Add two equations} \\ \hline 3x^2 \quad &= 27 && \text{Divide each side by 3} \\ x^2 \quad &= 9 && \text{Solve for } x \\ x &= \pm 3 \end{aligned}$$

Substitute 3 and -3 for x in $x^2 + y^2 = 13$, respectively and solve for y. Thus, $y = \pm 2$ when $x = 3$, and $y = \pm 2$ when $x = -3$. Therefore, the solutions to the system of nonlinear equations are $(3, 2)$, $(3, -2)$, $(-3, 2)$, and $(-3, -2)$.

EXERCISES

1. Which of the following conic section does $2x^2 + 2x - 3y^2 - 12y = 7$ represent?

   (A) Line

   (B) Circle

   (C) Parabola

   (D) Hyperbola

   (E) Ellipse

2. What is the radius of the following circle $x^2 - 4x + y^2 - 6y - 3 = 0$?

   (A) 1

   (B) 2

   (C) 4

   (D) 8

   (E) 16

3. What is the rectangular equation of the following parametric equations?

$$x = \frac{1}{2}t - 1, \qquad y = 3t + 2$$

   (A) $y = \frac{3}{2}x + 1$

   (B) $y = 3x - 2$

   (C) $y = 6x + 8$

   (D) $y = 9x - 6$

   (E) $y = 12x - 8$

4. What is the length of the major axis of the ellipse $2x^2 + 4y^2 = 16$?

   (A) $8\sqrt{2}$

   (B) $4\sqrt{2}$

   (C) $2\sqrt{3}$

   (D) 4

   (E) 2

5. What are the x-intercepts of the equation $2y^2 + x^2 - 4y - x - 6 = 0$?

   (A) -2 and 3

   (B) -1 and 2

   (C) 1 and -2

   (D) 2 and -3

   (E) 2 and 3

6. If a circle with a radius of 4 does not touch nor cross both the x-axis and the y-axis, which of the following ordered pair cannot be the center of the circle?

   (A) $(6, 5)$

   (B) $(5, -6)$

   (C) $(-3, 5)$

   (D) $(-5, -6)$

   (E) $(-6, 5)$

7. Which of the following point lies outside the circle $(x-1)^2 + (y-2)^2 = 16$?

   (A) $(-2, 1)$

   (B) $(-1, 3)$

   (C) $(1, 6)$

   (D) $(3, 5)$

   (E) $(4, 5)$

8. What are the equations of the asymptotes of $9x^2 - 4y^2 = 36$?

   (A) $y = \pm\frac{9}{4}x$

   (B) $y = \pm\frac{3}{2}x$

   (C) $y = \pm\frac{2}{3}x$

   (D) $y = \pm\frac{4}{9}x$

   (E) $y = \pm\frac{2}{9}x$

9. What is the rectangular equation of the following parametric equations ?

$$x = \cos t, \qquad y = \sin t, \qquad 0 \le t \le 2\pi$$

(A) $x^2 + y^2 = 1$

(B) $x^2 - y^2 = 1$

(C) $x = 1 + y$

(D) $y = 1 - x$

(E) $y = 1 - x^2$

10. What are the solutions to the following system of nonlinear equations?

$$y = -x^2 + 3x$$
$$y = x^2 - x$$

(A) $(-2, 2)$ and $(0, 3)$

(B) $(-2, -2)$ and $(3, 0)$

(C) $(0, 0)$ and $(2, 2)$

(D) $(0, 0)$ and $(3, 0)$

(E) $(2, -2)$ and $(3, 6)$

ANSWERS AND SOLUTIONS

1. (D)

Since the squared terms, $2x^2 - 3y^2$, have a negative sign in between them, the equation $2x^2 + 2x - 3y^2 - 12y = 7$ represents a hyperbola.

2. (C)

In order to write a general equation of the circle $(x-h)^2 + (y-k)^2 = r^2$, complete the squares in x and y.

$$\begin{aligned} x^2 - 4x + y^2 - 6y - 3 &= 0 && \text{Add 3 to each side} \\ x^2 - 4x + y^2 - 6y &= 3 && \text{Add 4 to each side to complete squares in } x \\ (x^2 - 4x + 4) + y^2 - 6y &= 7 && \text{Add 9 to each side to complete squares in } y \\ (x-2)^2 + (y^2 - 6y + 9) &= 16 \\ (x-2)^2 + (y-3)^2 &= 4^2 \end{aligned}$$

Therefore, the radius of the circle $x^2 - 4x + y^2 - 6y - 3 = 0$ is 4.

3. (C)

In order to change the parametric equations to a rectangular equation $y = f(x)$, eliminate the parameter t. Solve for t from the first equation $x = \frac{1}{2}t - 1$. We found that $t = 2(x+1) = 2x + 2$. Then, substitute $2x + 2$ for t in the second equation $y = 3t + 2$ so that $y = 3(2x+2) + 2 = 6x + 8$. Therefore, (C) is the correct answer.

4. (B)

In order to find the general equation of an ellipse $\frac{(x-h)^2}{a^2} + \frac{(y-k)^2}{b^2} = 1$, divide each side of the equation by 16.

$$2x^2 + 4y^2 = 16 \qquad \text{Divide each side by 16}$$
$$\frac{x^2}{(2\sqrt{2})^2} + \frac{y^2}{2^2} = 1$$

Since $a = 2\sqrt{2}$ and $b = 2$, $a > b$. Therefore, the length of the major axis is $2a = 2(2\sqrt{2}) = 4\sqrt{2}$.

5. (A)

In order to find the x-intercepts of $2y^2 + x^2 - 4y - x - 6 = 0$, substitute 0 for y and solve for x.

$$2y^2 + x^2 - 4y - x - 6 = 0 \qquad \text{Substitute 0 for } y \text{ and solve for } x$$
$$x^2 - x - 6 = 0 \qquad \text{Factor}$$
$$(x+2)(x-3) = 0$$
$$x = -2 \quad \text{or} \quad x = 3$$

Therefore, the x-intercepts of $2y^2 + x^2 - 4y - x - 6 = 0$ are -2 and 3.

6. (C)

Let (a, b) be the center of the circle with a radius of 4. In order for the circle to neither touch nor cross the y-axis, the distance between the x-coordinate of the center, a, and the y-axis should be greater than 4. In order words, $|a| > 4$. Likewise, in order for the circle to neither touch nor cross the x-axis, the distance between the y-coordinate of the center, b, and the x-axis should be greater than 4. In order words, $|b| > 4$. Since the ordered pairs in answer choices (A), (B), (D), and (E) satisfy $|a| > 4$ and $|b| > 4$, you can eliminate them. Therefore, (C) is the correct answer.

7. (E)

Tip
1. If the distance between a point (x, y) and the center of a circle is greater than the radius, the point (x, y) lies outside the circle.
2. Distance formula: $D = \sqrt{(x_2 - x_1)^2 + (y_2 - y_1)^2}$

The center and the radius of the circle $(x-1)^2 + (y-2)^2 = 4^2$ are $(1, 2)$ and 4, respectively. In order to find a point that lies outside the circle, find the distance between the center of the circle and a point given in each answer choice.

(A) Distance between $(1, 2)$ and $(-2, 1)$: $D = \sqrt{(-2-1)^2 + (1-2)^2} = \sqrt{10}$

(B) Distance between $(1, 2)$ and $(-1, 3)$: $D = \sqrt{(-1-1)^2 + (3-2)^2} = \sqrt{5}$

(C) Distance between $(1, 2)$ and $(1, 6)$: $D = \sqrt{(1-1)^2 + (6-2)^2} = \sqrt{16} = 4$

(D) Distance between $(1, 2)$ and $(3, 5)$: $D = \sqrt{(3-1)^2 + (5-2)^2} = \sqrt{13}$

(E) Distance between $(1, 2)$ and $(4, 5)$: $D = \sqrt{(4-1)^2 + (5-2)^2} = \sqrt{18} = 3\sqrt{2}$

Since $3\sqrt{2} > 4$, (E) is the correct answer.

8. (B)

In order to have a general equation of a hyperbola $\frac{(x-h)^2}{a^2} - \frac{(y-k)^2}{b^2} = 1$, divide each side of the equation $9x^2 - 4y^2 = 36$ by 36. Thus, $\frac{x^2}{2^2} - \frac{y^2}{3^2} = 1$. The graph of the hyperbola is shown below.

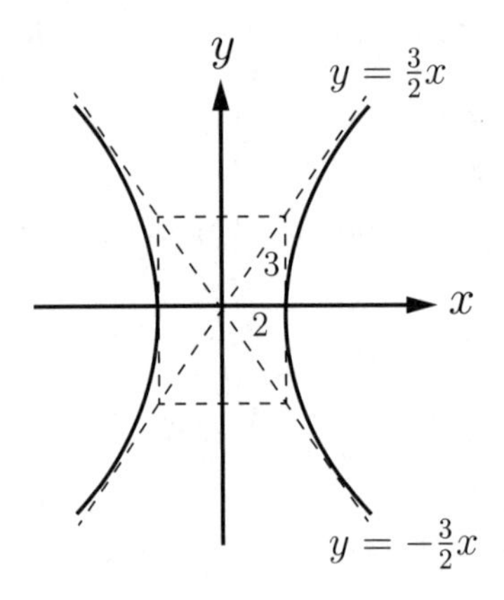

Since the slopes of the asymptotes are $\pm\frac{3}{2}$ and the asymptotes pass through the center $(0, 0)$, the equations of the asymptotes are $y = \pm\frac{3}{2}x$.

9. (A)

Tip Use the Pythagorean Identity: $\cos^2 t + \sin^2 t = 1$.

In order to change the parametric equations to a rectangular equation, square each side of parametric equations.

$$x = \cos t \implies x^2 = \cos^2 t$$
$$y = \sin t \implies y^2 = \sin^2 t$$

In order to eliminate t, add the two equations $x^2 = \cos^2 t$ and $y^2 = \sin^2 t$.

$$x^2 = \cos^2 t$$
$$y^2 = \sin^2 t \qquad \text{Add the two equations}$$
$$x^2 + y^2 = \cos^2 t + \sin^2 t \qquad \text{Since } \cos^2 t + \sin^2 t = 1$$
$$x^2 + y^2 = 1$$

Therefore, (A) is the correct answer.

10. (C)

Let's solve the nonlinear equations shown below using the substitution method.

$$y = -x^2 + 3x$$
$$y = x^2 - x$$

Substitute $x^2 - x$ for y in the first equation $y = -x^2 + 3x$.

$y = -x^2 + 3x$	Substitute $x^2 - x$ for y
$x^2 - x = -x^2 + 3x$	Add x^2 from each side
$2x^2 - x = 3x$	Subtract $3x$ from each side
$2x^2 - 4x = 0$	Factor
$2x(x - 2) = 0$	Solve for x
$x = 0 \quad \text{or} \quad x = 2$	

In order to find the values of y, substitute $x = 0$ and $x = 2$ into $y = x^2 - x$. Thus, $y = 0$ and $y = 2$, respectively. Therefore, the solutions to the system of nonlinear equations are $(0, 0)$ and $(2, 2)$.

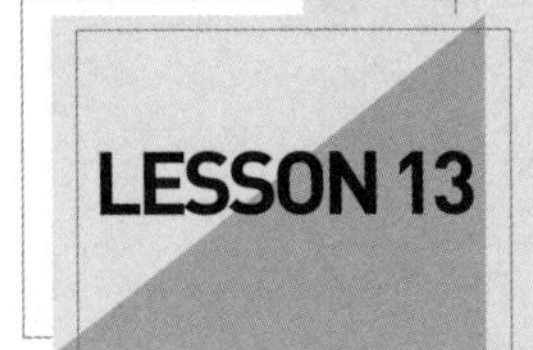

LESSON 13 Sequences and Series

Sequences

A sequence (or progression) is a list of numbers in order. The numbers in the list are called **terms** of the sequence and are denoted with subscripted letters: a_1 for the first term, a_2 for the second term, a_n for the nth term.

To evaluate the value of a_n, there are two types of formulas: the **explicit formula** and the **recursive formula**. The explicit formula evaluates a_n directly by substituting a value into the formula. Whereas, the recursive formula involves all previous terms to evaluate a_n. For instance, to evaluate the 10th term using the recursive formula, we need to evaluate the first nine terms.

Example 1 Evaluating the nth term using an explicit formula

If the sequence is defined by $a_n = 2n + 5$, evaluate the 5th term and the 11th term.

Solution In order to evaluate the 5th term and the 11th term, substitute $n = 5$ and $n = 11$ into $a_n = 2n + 5$, respectively.

$$a_n = 2n + 5 \implies a_5 = 2(5) + 5 = 15$$
$$a_n = 2n + 5 \implies a_{11} = 2(11) + 5 = 27$$

Therefore, the value of a_5 is 15 and the value of a_{11} is 27.

Example 2 Evaluating the nth term using a recursive formula

If the sequence is defined by $a_n = 2a_{n-1} + 3,\ a_1 = 4$, evaluate the 5th term.

Solution In order to evaluate the 5th term, we need to find the previous four terms as shown below.

$a_n = 2a_{n-1} + 3,\ a_1 = 4$ — Recursive formula with $a_1 = 4$

$a_2 = 2a_1 + 3 = 2(4) + 3 = 11$ — Substitute 2 for n to find a_2

$a_3 = 2a_2 + 3 = 2(11) + 3 = 25$ — Substitute 3 for n to find a_3

$a_4 = 2a_3 + 3 = 2(25) + 3 = 53$ — Substitute 4 for n to find a_4

$a_5 = 2a_4 + 3 = 2(53) + 3 = 109$ — Substitute 5 for n to find a_5

Therefore, the value of a_5 is 109.

Arithmetic Sequences and Geometric Sequences

There are two most common sequences: Arithmetic sequences and geometric sequences.

- In an **arithmetic sequence**, add or subtract the same number (common difference) to one term to get the next term.
- In a **geometric sequence**, multiply or divide one term by the same number (common ratio) to get the next term.

Type	Definition	Example	nth term
Arithmetic sequence	The common difference between any consecutive terms is constant.	$1, 3, 5, 7, \ldots$	$a_n = a_1 + (n-1)d$ where d is the common difference.
Geometric sequence	The common ratio between any consecutive terms is constant	$2, 4, 8, 16, \ldots$	$a_n = a_1 \times r^{n-1}$ where r is the common ratio.

Example 3 Writing the nth term of an arithmetic sequence

In the arithmetic sequence, if $a_9 = 51$ and $a_{17} = 99$, write an explicit formula for a_n.

Solution Write the 9th term and 17th term of the arithmetic sequence in terms of a_1 and d using the nth term formula: $a_n = a_1 + (n-1)d$.

$$a_{17} = a_1 + 16d = 99$$
$$a_9 = a_1 + 8d = 51$$

Use the linear combinations method to solve for d and a_1.

$$a_1 + 16d = 99$$
$$a_1 + 8d = 51 \qquad \text{Subtract the two equations}$$
$$8d = 48 \qquad \text{Divide both sides by 8}$$
$$d = 6$$

Substitute $d = 6$ into $a_9 = a_1 + 8d = 51$ and solve for a_1. Thus, $a_1 = 3$. Therefore, the nth term of the arithmetic sequence is $a_n = a_1 + (n-1)d = 3 + (n-1)6 = 6n - 3$.

Series

A **series** is the sum of a sequence. A series can be either a finite series, S_n or infinite series, S. The finite series S_n is the sum of a finite number of terms. Whereas, the infinite series is the sum of an infinite number of terms. Often, the finite series S_n is called the ***n*th partial sum** and the infinite series is called an **infinite sum**.

A series can be represented in a compact form, called summation notation or sigma notation $\sum$. Using the summation notation, the nth partial sum S_n and infinite sum S can be expressed as follows:

$$S_n = a_1 + a_2 + \cdots + a_n = \sum_{k=1}^{n} a_k,$$

$$S = a_1 + a_2 + a_3 \cdots = \sum_{k=1}^{\infty} a_k$$

where k is called the **index** of the sum. $k = 1$ indicates where to start the sum and $k = n$ indicates where to end the sum. For instance,

$$\sum_{k=1}^{5} k^2 = 1^2 + 2^2 + 3^2 + 4^2 + 5^2$$

Arithmetic Series and Geometric Series

An arithmetic series is the sum of an arithmetic sequence. A geometric series is the sum of a geometric sequence. For instance, $1 + 3 + 5 + 7 + \cdots$ is an arithmetic series and $\frac{1}{2} + \frac{1}{4} + \frac{1}{8} + \frac{1}{16} + \cdots$ is a geometric series.

Below summarizes the nth partial sum and infinite sum for an arithmetic series and a geometric series.

Type	Arithmetic Series	Geometric Series
nth Partial Sum	$S_n = \frac{n}{2}(a_1 + a_n)$	$S_n = \frac{a_1(1-r^n)}{1-r}$
Infinite Sum	$S = \infty$	$S = \begin{cases} \frac{a_1}{1-r}, & \lvert r\rvert < 1 \\ \infty, & \lvert r\rvert \geq 1 \end{cases}$

Tip: Note that the infinite sum S of a geometric series converges to $\frac{a_1}{1-r}$ if $|r| < 1$, where r is the common ratio of a geometric sequence.

Example 4 Finding the sum of an arithmetic series

Find the sum of the arithmetic series shown below.

$$3+7+11+\cdots+83$$

Solution $3, 7, 11, \cdots, 83$ is the arithmetic sequence with a common difference of 4. Using the nth term formula $a_n = a_1 + (n-1)d$,

$a_n = a_1 + (n-1)d$	Substitute 83 for a_n, 3 for a_1, and 4 for d
$83 = 3 + (n-1)4$	Subtract 3 from each side
$4(n-1) = 80$	Divide each side by 4
$n-1 = 20$	Add 1 to each side
$n = 21$	

we found that 83 is the 21st term of the arithmetic sequence. Thus, $3+7+11+\cdots+83$ is the sum of the first 21 terms of the arithmetic sequence.

$S_n = \dfrac{n}{2}(a_1 + a_n)$	Substitute 21 for n
$S_{21} = \dfrac{21}{2}(a_1 + a_{21})$	Substitute 3 for a_1 and 83 for a_{21}
$= \dfrac{21}{2}(3+83) = 903$	

Therefore, $3+7+11+\cdots+83 = 903$.

Example 5 Finding the sum of a geometric series

Find the sum of each geometric series.

(a) $\frac{1}{2} + \frac{3}{4} + \frac{9}{8} + \cdots$

(b) $1 - \frac{1}{2} + \frac{1}{4} - \frac{1}{8} + \cdots$

(c) $\displaystyle\sum_{k=1}^{\infty} 4\left(-\frac{2}{3}\right)^{k-1}$

Solution

(a) The common ratio is

$$r = \frac{\frac{3}{4}}{\frac{1}{2}} = \frac{3}{2}$$

Since $|r| > 1$, the infinite sum S of the geometric series diverges. In other words, $S = \infty$.

(b) The common ratio is $-\frac{1}{2}$. Since $|r| < 1$,

$$S = \frac{a_1}{1-r} = \frac{1}{1-(-\frac{1}{2})} = \frac{1}{\frac{3}{2}} = \frac{2}{3}$$

Therefore, the infinite sum S of the geometric series is $\frac{2}{3}$.

(c) Substitute $k = 1$, $k = 2$, and $k = 3$ to write out the sum.

$$\sum_{k=1}^{\infty} 4\left(-\frac{2}{3}\right)^{k-1} = 4 - \frac{8}{3} + \frac{16}{9} + \cdots$$

The common ratio is $-\frac{2}{3}$. Since $|r| < 1$,

$$S = \frac{a_1}{1-r} = \frac{4}{1-(-\frac{2}{3})} = \frac{4}{\frac{5}{3}} = \frac{12}{5}$$

Therefore, the infinite sum S of the geometric series is $\frac{12}{5}$.

EXERCISES

1. If the sequence is defined by $a_n = \frac{1}{a_{n-1}}$, $a_1 = 2$, where $n \geq 2$, what is the 12th term?

   (A) $\frac{1}{4}$

   (B) $\frac{1}{2}$

   (C) 2

   (D) 12

   (E) 26

2. If the sequence is defined by $a_n = 3n + 1$, what is 25th term?

   (A) 72

   (B) 73

   (C) 74

   (D) 75

   (E) 76

3. What is the sum of $\sum_{n=1}^{10}(-1)^n n$?

(A) -15

(B) -9

(C) 5

(D) 36

(E) 55

4. If the three terms, x, y, and z, form a geometric sequence, which of the following equation best describes the relationship between x, y, and z ?

(A) $y = \frac{x-z}{2}$

(B) $y = \frac{x+z}{2}$

(C) $y = xz$

(D) $y^2 = x + z$

(E) $y^2 = xz$

5. In the arithmetic sequence $5, 12, 19, 26, \cdots$, what is the value of the 19th term?

(A) 110

(B) 117

(C) 124

(D) 131

(E) 138

6. $\frac{2}{3} + 1 + \frac{3}{2} + \frac{9}{4} + \cdots =$

(A) $-\frac{1}{3}$

(B) $\frac{1}{3}$

(C) 5

(D) 7

(E) ∞

7. In the arithmetic sequence, the 5th term is 13 and the 24th term is 89. What is the 16th term?

(A) 57

(B) 61

(C) 65

(D) 69

(E) 73

8. $\sum_{k=1}^{\infty}\left(-\frac{1}{2}\right)^k$?

(A) $-\frac{1}{4}$

(B) $-\frac{1}{3}$

(C) $\frac{1}{4}$

(D) $\frac{1}{3}$

(E) $\frac{1}{2}$

9. If the sequence is defined as shown below, what is the 4th term?

$$a_n = \sqrt{a_{n-1} + n}, \quad a_1 = 2, \quad \text{where } n \geq 2$$

(A) 3.18

(B) 2.96

(C) 2.73

(D) 2.50

(E) 2.24

10. $11 + 14 + 17 + \cdots + 98 =$

(A) 1481

(B) 1558

(C) 1635

(D) 1712

(E) 1789

ANSWERS AND SOLUTIONS

1. (B)

Substitute $n = 2, 3, 4$, and 5 into the recursive formula to find a pattern.

$a_n = \dfrac{1}{a_{n-1}},\ a_1 = 2$	Recursive formula with $a_1 = 2$
$a_2 = \dfrac{1}{a_1} = \dfrac{1}{2}$	Substitute 2 for n to find a_2
$a_3 = \dfrac{1}{a_2} = \dfrac{1}{\frac{1}{2}} = 2$	Substitute 3 for n to find a_3
$a_4 = \dfrac{1}{a_3} = \dfrac{1}{2}$	Substitute 4 for n to find a_4
$a_5 = \dfrac{1}{a_4} = \dfrac{1}{\frac{1}{2}} = 2$	Substitute 5 for n to find a_5
$a_6 = \dfrac{1}{a_5} = \dfrac{1}{2}$	Substitute 6 for n to find a_6

The result above suggests a pattern such that all the even numbered terms, $a_2, a_4, a_6, \cdots a_{12}$, are $\frac{1}{2}$. Therefore, the 12th term is $\frac{1}{2}$.

2. (E)

In order to find the 25th term, substitute $n = 25$ into $a_n = 3n + 1$.

$$a_n = 3n + 1 \qquad \text{Substitute 25 for } n$$
$$a_{25} = 3(25) + 1 = 76$$

Therefore, the 25th term of the sequence is 76.

3. (C)

Substitute $n = 1$, $n = 2$, $\cdots$, $n = 10$ to write out the sum.

$$\sum_{n=1}^{10} (-1)^n n = -1 + 2 - 3 + 4 - 5 + 6 - 7 + 8 - 9 + 10 = 5$$

Therefore, $\displaystyle\sum_{n=1}^{10} (-1)^n n = 5$.

4. (E)

Since the three terms, x, y, and z form a geometric sequence, the common ratios, $\frac{y}{x}$ and $\frac{z}{y}$, are the same.

$$\frac{y}{x} = \frac{z}{y} \qquad \text{Cross multiply}$$
$$y^2 = xz$$

Therefore, the equation that best describes the relationship between x, y, and z is $y^2 = xz$.

5. (D)

In the arithmetic sequence $5, 12, 19, 26, \cdots$, the first term is 5 and the common difference, d, is 7. Use the nth term formula to find the 19th term.

$$a_n = a_1 + (n-1)d \qquad \text{Substitute 19 for } n\text{, 5 for } a_1\text{, and 7 for } d$$
$$a_{19} = 5 + (19-1)7 = 131$$

Therefore, the 19th term of the arithmetic sequence is 131.

6. (E)

$\frac{2}{3} + 1 + \frac{3}{2} + \frac{9}{4} + \cdots$ is the geometric series. The common ratio, r, is $\frac{1}{\frac{2}{3}} = \frac{3}{2}$. Since $|r| > 1$, the infinite sum of the geometric series diverges. Therefore, $\frac{2}{3} + 1 + \frac{3}{2} + \frac{9}{4} + \cdots = \infty$.

7. (A)

Write the 5th term and 24th term of the arithmetic sequence in terms of a_1 and d using the nth term formula: $a_n = a_1 + (n-1)d$.

$$a_{24} = a_1 + 23d = 89$$
$$a_5 = a_1 + 4d = 13$$

Use the linear combinations method to solve for d and a_1.

$$a_1 + 23d = 89$$
$$a_1 + 4d = 13 \qquad \text{Subtract the two equations}$$
$$19d = 76 \qquad \text{Divide both sides by 19}$$
$$d = 4$$

Substitute $d = 4$ in $a_5 = a_1 + 4d = 13$ and solve for a_1. Thus, $a_1 = -3$. Therefore, the 16th term of the arithmetic sequence is $a_{16} = a_1 + 15d = -3 + 15(4) = 57$.

8. (B)

Substitute $k = 1$, $k = 2$, and $k = 3$ to write out the sum.

$$\sum_{k=1}^{\infty}\left(-\frac{1}{2}\right)^k = -\frac{1}{2} + \frac{1}{4} - \frac{1}{8} + \cdots$$

The common ratio is $-\frac{1}{2}$. Since $|r| < 1$,

$$S = \frac{a_1}{1-r} = \frac{-\frac{1}{2}}{1-(-\frac{1}{2})} = \frac{-\frac{1}{2}}{\frac{3}{2}} = -\frac{1}{3}$$

Therefore, the infinite sum of $\sum_{k=1}^{\infty}\left(-\frac{1}{2}\right)^k$ is $-\frac{1}{3}$.

9. (D)

In order to evaluate the 4th term, we need to find the previous three terms as shown below.

$a_n = \sqrt{a_{n-1} + n}, \quad a_1 = 2$	Recursive formula with $a_1 = 2$
$a_2 = \sqrt{a_1 + 2} = \sqrt{2+2} = 2$	Substitute 2 for n to find a_2
$a_3 = \sqrt{a_2 + 3} = \sqrt{2+3} = 2.24$	Substitute 3 for n to find a_3
$a_4 = \sqrt{a_3 + 4} = \sqrt{2.24+4} = 2.50$	Substitute 4 for n to find a_4

Therefore, the 4th term is 2.50.

10. (C)

$11, 14, 17, \cdots, 98$ is the arithmetic sequence with a common difference of 3. Using the nth term formula $a_n = a_1 + (n-1)d$,

$a_n = a_1 + (n-1)d$	Substitute 98 for a_n, 11 for a_1, and 3 for d
$98 = 11 + (n-1)3$	Subtract 11 from each side
$3(n-1) = 87$	Divide each side by 3
$n - 1 = 29$	Add 1 to each side
$n = 30$	

we found that 98 is the 30th term of the arithmetic sequence. Thus, $11 + 14 + 17 + \cdots + 98$ is the sum of the first 30 terms of the arithmetic sequence.

$S_n = \frac{n}{2}(a_1 + a_n)$	Substitute 30 for n
$S_{30} = \frac{30}{2}(a_1 + a_{30})$	Substitute 11 for a_1 and 98 for a_{30}
$= \frac{30}{2}(11 + 98)$	
$= 1635$	

Therefore, $11 + 14 + 17 + \cdots + 98 = 1635$.

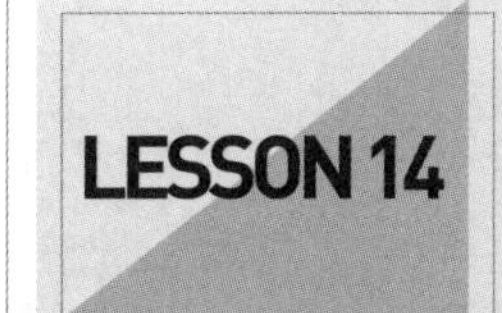

LESSON 14 Probability and Statistics

Counting

Counting integers

How many positive integers are there between 42 and 97 inclusive? Are there 54, 55, or 56 integers? Even in this simple counting problem, many students are not sure what the right answer is. A rule for counting integers is as follows:

$$\text{The number of integers} = \text{Greatest integer} - \text{Least integer} + 1$$

According to this rule, the number of integers between 42 and 97 inclusive is $97 - 42 + 1 = 56$ integers.

Venn Diagram

A venn diagram is very useful in counting. It helps you count numbers correctly.

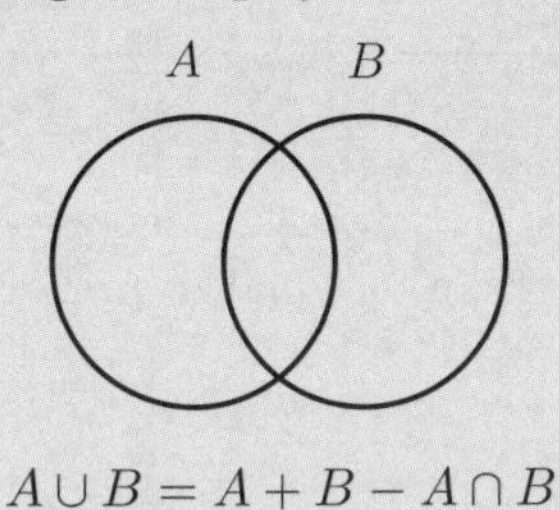

$$A \cup B = A + B - A \cap B$$

In the figure above, $A \cup B$ represents the combined area of two circles A and B. $A \cap B$ represents the common area where the two circles overlap. The venn diagram suggests that the combined area $(A \cup B)$ equals the sum of areas of circles $(A + B)$ minus the common area $(A \cap B)$.
In counting, each circle A and B represents a set of numbers. $n(A)$ and $n(B)$ represent the number of elements in set A and B, respectively. For instance, $A = \{2, 4, 6, 8, 10\}$ and $n(A) = 5$. Thus, the total number of elements that belong to either set A or set B, $n(A \cup B)$, can be counted as follows:

$$n(A \cup B) = n(A) + n(B) - n(A \cap B)$$

Let's find out how many positive integers less than or equal to 20 are divisible by 2 or 3. Define A as the set of numbers divisible by 2 and B as the set of numbers divisible by 3.

$$\begin{aligned} A &= \{2, 4, 6, \cdots, 18, 20\}, & n(A) &= 10 \\ B &= \{3, 6, 9, 12, 15, 18\}, & n(B) &= 6 \\ A \cap B &= \{6, 12, 18\}, & n(A \cap B) &= 3 \end{aligned}$$

Notice that $A \cap B = \{6, 12, 18\}$ are multiples of 2 and multiples of 3. They are counted twice so they must be excluded in counting. Thus,

$$\begin{aligned} n(A \cup B) &= n(A) + n(B) - n(A \cap B) \\ &= 10 + 6 - 3 \\ &= 13 \end{aligned}$$

Therefore, the total number of positive integers less than or equal to 20 that are divisible by 2 or 3 is 13.

The Fundamental Counting Principle

If one event can occur in m ways and another event can occur in n ways, then the number of ways both events can occur is $m \times n$. For instance, Jason has three shirts and four pairs of jeans. He can dress up in $3 \times 4 = 12$ different ways.

Permutation and Combination

Factorial notation

n factorial, denoted by $n!$, is defined as $n! = n(n-1)(n-2)\cdots 3 \cdot 2 \cdot 1$. In other words, n factorial is the product of all positive integers less than or equal to n. For instance, $3! = 3 \cdot 2 \cdot 1 = 6$. Below are the properties of factorials.

1. $0! = 1$ and $1! = 1$.
2. $n! = n \times (n-1)!$ or $n! = n \cdot (n-1) \cdot (n-2)!$.
   For instance, $5! = 5 \cdot 4!$, or $5! = 5 \cdot 4 \cdot 3!$.

Permutations without repetition

A permutation, denoted by ${}_nP_r$, represents a number of ways to select r objects from the total number of objects n where the order is important. The permutation ${}_nP_r$ is given by

$${}_nP_r = \frac{n!}{(n-r)!}, \qquad \text{where } r \leq n$$

For instance, how many words can be formed using all the letters in the word ABCDE?

Since all the letters A, B, C, D, and E are distinguishable, the order is important. Thus, this is a permutation problem.

$${}_5P_5 = \frac{5!}{(5-5)!} = \frac{5!}{0!} = 120$$

Therefore, the number of different words can be formed using the letters in word ABCDE is 120.

Permutations with repetition

The number of permutations of n objects, where there are n_1 indistinguishable objects of one kind, and n_2 indistinguishable objects of a second kind, is given by

$$\text{Permutations with repetition} = \frac{n!}{n_1! \cdot n_2!}$$

For instance, how many words can be formed using all the letters in the word AABBB?

Since letters A and B are distinguishable, the order is important. However, there are 2 A's and 3 B's out of 5 letters.

$$\text{Permutations with repetition} = \frac{5!}{2! \cdot 3!} = \frac{5 \cdot 4 \cdot \cancel{3!}}{2! \cdot \cancel{3!}} = 10$$

Therefore, the number of different words can be formed using the letters in word AABBB is 10.

Combinations

A combination, denoted by ${}_nC_r$ or $\binom{n}{r}$, represents a number of ways to select r objects from the total number of objects n where the order is NOT important. The combination $\binom{n}{r}$ is given by

$$\binom{n}{r} = \frac{n!}{(n-r)! \cdot r!}, \qquad \text{where } r \leq n$$

For instance, how many 2 different books can be selected from a list of 10 books?

Since 2 books are indistinguishable, the order is not important. Thus, this is a combination problem.

$$\binom{10}{2} = \frac{10!}{8! \cdot 2!} = \frac{10 \cdot 9 \cdot \cancel{8!}}{\cancel{8!} \cdot 2!} = 45$$

Therefore, the number of selecting 2 different books from a list of 10 books is 45.

Example 1 Permutations and Combinations

(a) How many ways can a group of 10 people elect a president and a vice president?

(b) How many different committees of 2 officers can be formed from a group of 10 people?

Solution

(a) Since a president and a vice president are distinguishable, the order is important. Thus, this is a permutation problem.

$$_{10}P_2 = \frac{10!}{8!} = \frac{10 \cdot 9 \cdot \cancel{8!}}{\cancel{8!}} = 90$$

Therefore, the number of ways to elect a president and a vice present from a group of 10 people is 90.

(b) Since 2 officers are indistinguishable, the order is not important. Thus, this is a combination problem.

$$\binom{10}{2} = \frac{10!}{8! \cdot 2!} = \frac{10 \cdot 9 \cdot \cancel{8!}}{\cancel{8!} \cdot 2!} = 45$$

Therefore, the number of different committees of 2 officers can be formed from a group of 10 people is 45.

Probability

The definition of probability of an event, E, is as follows:

$$\text{Probability(E)} = \frac{\text{The number of outcomes event } E \text{ that can happen}}{\text{The total number of possible outcomes}}$$

Probability is a measure of how likely an event will happen. Probability can be expressed as a fraction, a decimal, and a percent, and is measured on scale from 0 to 1. Probability can not be less than 0 nor greater than 1.

- Probability equals 0 means an event will never happen.
- Probability equals 1 means an event will always happen.
- Higher the probability, higher chance an event will happen.

For instance, what is the probability of selecting a prime number at random from 1 to 5? In this problem, the event E is selecting a prime number from three possible prime numbers: 2, 3, and 5. The total possible outcomes are numbers from 1 to 5. Thus, the probability of selecting a prime number is $P(E) = \frac{\{2,3,5\}}{\{1,2,3,4,5\}} = \frac{3}{5}$.

Geometric probability

Geometric probability involves the length or area of the geometric figures. The definition of the geometric probability is as follows:

$$\text{Geometric probability} = \frac{\text{Area of desired region}}{\text{Total area}}$$

In the figure below, a circle is inscribed in the square with side length of 10. Assuming that a dart always lands inside the square, what is the probability that a dart lands on a region that lies outside the circle and inside the square?

The area of the square is $10^2 = 100$, and the area of the circle is $\pi(5)^2 = 25\pi$. Thus, the area of desired region is $100 - 25\pi$.

$$\text{Geometric probability} = \frac{\text{Area of desired region}}{\text{Total area}} = \frac{100 - 25\pi}{100}$$
$$= \frac{25(4-\pi)}{100} = \frac{4-\pi}{4}$$

Probability of independent events

Two events, A and B, are said to be independent if the outcome of A does not affect the outcome of B. If two events, A and B, are independent, the probability of both occurring is as follows:

$$\text{P}(A \text{ and } B) = P(A) \times P(B)$$

Example 2 Calculating probability

Two standard dice are rolled. What is the probability that the sum of the two numbers on top of each die is 6?

Solution The first and the second die have 6 possible outcomes each: 1, 2, 3, 4, 5, and 6, which are shown in the second row and the second column of the table below. There are a total number of $6 \times 6 = 36$ possible outcomes. Each of the 36 outcomes represents the sum of the two numbers on the top of the first and the second die. For instance, when 4 is on the first die and 2 is on the second die, expressed as $(4, 2)$, the sum of the two numbers is 6.

		1st die					
		1	2	3	4	5	6
2nd die	1					6	
	2				6		
	3			6			
	4		6				
	5	6					
	6						

There are 5 outcomes for which the sum of the two numbers is 6: $(1,5)$, $(2,4)$, $(3,3)$, $(4,2)$, and $(5,1)$. Therefore, the probability that the sum of the two numbers on the top of each die is 6 is $\frac{5}{36}$.

Statistics

Mean, or **Average**, is the sum of all elements in a set divided by the number of elements in the set. For instance, if there are the numbers 3, 7, and 11 in a set, the mean $= \frac{3+7+11}{3} = 7$.

Median is the middle number when a set of numbers is arranged from least to greatest.

- If there is an odd number of numbers (represented by n) in a set, the median is the middle number which is the $(\frac{n+1}{2})^{\text{th}}$ number in the set. For instance, if there are 3, 2, 5, 7, and 10 in a set, arrange the numbers in the set from least to greatest: 2, 3, 5, 7, and 10. Since there are 5 numbers in the set, the median is the $\frac{5+1}{2} = 3^{\text{rd}}$ number in the set. Thus, the median is 5.
- If there is an even number of numbers (represented by n) in a set, the median is the average of the two middle numbers which are the $(\frac{n}{2})^{\text{th}}$ and $(\frac{n}{2}+1)^{\text{th}}$ numbers. For instance, if there are 1, 4, 6, 8, 9, and 11 in a set, the median is the average of the 3rd number and the 4th number in the set. Thus, the median is $\frac{6+8}{2} = 7$.

Mode is a number that appears most frequently in a set. It is possible to have more than one mode or no mode in a set.

Range is the difference between the greatest number and the smallest number in a set.

Standard Deviation

The **standard deviation**, denoted by σ, measures the amount of variation or dispersion from the mean. In other words, it is a measure of how spread out the numbers are. A small standard deviation indicates that the numbers tend to be very close to the mean. Whereas, a large standard deviation indicates that the numbers are spread out over a large range of values.

How to calculate standard deviation from a TI-84 calculator:

Do the following procedures to find the standard deviation from a data set $\{2, 5, 8, 9, 15\}$.

1. Press [STAT] [ENTER] to enter the statistics list editor.
2. Enter the numbers 2, 5, 8, 9, and 15 of the data set into L1, pressing [ENTER] after each entry.
3. Press [STAT], arrow over to CALC, and press 1:1-Var Stats.
4. After displaying 1-Var Stats screen, arrow down to Calculate and press [ENTER].
5. The standard deviation of the data set is $\sigma_x = 4.35$.

Properties of the Standard Deviation

1. The standard deviation is always positive.

2. The standard deviation is zero if all the numbers in the data set are the same. For instance, If a data set is $\{3, 3, 3, 3, 3\}$, the standard deviation of the data set is zero.

3. If all numbers in the data set are added by the same number, k, the standard deviation does not change. For instance, let $A = \{2, 5, 8, 9, 15\}$ and $\sigma_A = 4.35$. If each element in set $B = \{4, 7, 10, 11, 17\}$ is 2 more than each element in set A, the standard deviation of set B, σ_B, is $\sigma_B = \sigma_A = 4.35$.

4. If all numbers in the data set are subtracted by the same number, k, the standard deviation does not change. For instance, let $A = \{2, 5, 8, 9, 15\}$ and $\sigma_A = 4.35$. If each element in set $B = \{-1, 2, 5, 6, 12\}$ is 3 less than each element in set A, the standard deviation of set B is $\sigma_B = \sigma_A = 4.35$.

5. If all numbers in the data set are multiplied by the same number, k, the standard deviation is multiplied by k. For instance, let $A = \{2, 5, 8, 9, 15\}$ and $\sigma_A = 4.35$. If each element in set $B = \{6, 15, 24, 27, 45\}$ is three times each element in set A, the standard deviation of set B is $\sigma_B = 3\sigma_A = 3(4.35) = 13.05$.

6. If all numbers in the data set are divided by the same number, k, the standard deviation is divided by k. For instance, let $A = \{2, 5, 8, 9, 15\}$ and $\sigma_A = 4.35$. If each element in set $B = \{1, 2.5, 4, 4.5, 7.5\}$ is one-half each element in set A, the standard deviation of set B is $\sigma_B = \frac{\sigma_A}{2} = \frac{4.35}{2} = 2.175$.

Example 3 Finding mean, median and standard deviation

Let the mean, median, and standard deviation of set A is 10, 8, and 6.72, respectively. If 5 is added to each element in set A, find the new mean, new median, and new standard deviation.

Solution Let $A = \{1, 6, 8, 15, 20\}$ so that the mean is 10, the median is 8, and the standard deviation is 6.72. Let $B = \{6, 11, 13, 20, 25\}$ so that each element in set B is 5 more than each element in set A. Therefore, the mean of set B is 15, the median of set B is 13, and the standard deviation of set B is the same as standard deviation of set A, which is 6.72. That is, both the new mean and median are 5 more than the old mean and old median. However, the new standard deviation is the same as the old standard deviation.

EXERCISES

1. $\dfrac{n!}{(n-2)!} =$

   (A) 2

   (B) $\frac{n}{2}$

   (C) $n - 1$

   (D) $n^2 - n$

   (E) n^2

2. There are five types of bread, two types of meat, and four types of cheese. Assuming you have to select one of each category, how many different sandwiches can you make?

   (A) 56

   (B) 40

   (C) 32

   (D) 24

   (E) 11

3. There are seven points on a circle. How many different chords can be drawn? (A chord is a line segment whose endpoints both lie on a circle.)

   (A) 16

   (B) 21

   (C) 30

   (D) 35

   (E) 42

4. There are two red chairs and two blue chairs in a room. How many ways can you arrange these four chairs in a row?

   (A) 2

   (B) 4

   (C) 6

   (D) 18

   (E) 24

5. If an integer is selected at random from 9 to 38 inclusive, what is the probability that the selected integer is a prime number?

   (A) $\frac{1}{3}$

   (B) $\frac{7}{29}$

   (C) $\frac{8}{29}$

   (D) $\frac{3}{15}$

   (E) $\frac{4}{15}$

6. Set $A = \{2, 5, 7, 12, 13\}$ has a standard deviation of 4.17. If 5 is added to each number in the set, what is the new standard deviation of the set?

   (A) 20.85

   (B) 9.17

   (C) 4.17

   (D) 1.86

   (E) 0.83

7. Two numbers are selected without replacement from the set $\{3, 5, 7, 11\}$ to form a fraction. What is the probability that the fraction formed is an improper fraction?

(A) $\frac{4}{5}$

(B) $\frac{3}{4}$

(C) $\frac{2}{3}$

(D) $\frac{1}{2}$

(E) $\frac{1}{3}$

8. If you toss a coin three times, what is the probability that one head will be shown?

(A) $\frac{3}{4}$

(B) $\frac{2}{3}$

(C) $\frac{5}{8}$

(D) $\frac{1}{2}$

(E) $\frac{3}{8}$

9. In a set of five positive integers, the mode is 4, the median is 5, and the mean is 6. What is the greatest of these integers?

(A) 11

(B) 10

(C) 9

(D) 8

(E) 7

10. There are 4 boys and 5 girls in Mr. Rhee's class. Mr. Rhee wants to select a team of 2 boys and 2 girls from his class. How many different teams can Mr. Rhee can select?

(A) 80

(B) 60

(C) 50

(D) 40

(E) 30

ANSWERS AND SOLUTIONS

1. (D)

$$\frac{n!}{(n-2)!} = \frac{n \cdot (n-1) \cdot (n-2)!}{(n-2)!} = n(n-1) = n^2 - n$$

2. (B)

Tip: The fundamental counting principle: If one event can occur in m ways and another event can occur in n ways, then the number of ways both events can occur is $m \times n$.

Event 1, event 2, and event 3 are selecting 1 out of 5 types of bread, 1 out of 2 types of meat, and 1 out of 4 types of cheese, respectively. According to the fundamental counting principle, you can make $5 \times 2 \times 4 = 40$ different sandwiches.

3. (B)

Since the 7 points on the circle are indistinguishable, this is a combination problem. In order to draw a chord, out of the 7 points, two points must be selected. Thus,

$$\binom{7}{2} = \frac{7!}{5! \cdot 2!} = \frac{7 \cdot 6 \cdot \cancel{5!}}{\cancel{5!} \cdot 2!} = 21$$

Therefore, the number of different chords can be drawn from the 7 points is 21.

4. (C)

Let R and B represent a red chair and a blue chair, respectively. There are 2 red chairs and 2 blue chairs. Thus, this problem is exactly the same as counting the number of different words that can be formed using all the letters in the word RRBB. Since letters R and B are distinguishable and there are 2 R's and 2 B's out of 4 letters,

$$\text{Permutations with repetition} = \frac{4!}{2!\cdot 2!} = \frac{4\cdot 3\cdot \cancel{2!}}{2!\cdot \cancel{2!}} = 6$$

Therefore, the number of different arrangements of the two red chairs and two blue chairs in a row is 6.

5. (E)

The number of integers from 9 to 38 inclusive is $38-9+1=30$. Since the prime numbers between 9 and 38 are 11, 13, 17, 19, 23, 29, 31, and 37, the probability that the selected integer is a prime number is $\frac{8}{30}=\frac{4}{15}$.

6. (C)

> Tip: If all numbers in the data set are added by the same number, the standard deviation does not change.

Since 5 is added to each number in the set, the new standard deviation does not change. Therefore, the new standard deviation is 4.17.

7. (D)

Two numbers are selected at random without replacement from the set $\{3, 5, 7, 11\}$ to form a fraction: one number for the numerator and another number for the denominator. If the numerator of the fraction is 3, there are three possible numbers for the denominator: 5, 7, and 11. Thus, there are three fractions that can be formed using these numbers: $\frac{3}{5}$, $\frac{3}{7}$, $\frac{3}{11}$. All of the fractions are proper fractions. The table below shows a list of all possible fractions that can be formed when the numerator of fractions are 3, 5, 7, or 11.

Numerator	Denominator	Fraction	Improper fraction
3	5	$\frac{3}{5}$	
3	7	$\frac{3}{7}$	
3	11	$\frac{3}{11}$	
5	3	$\frac{5}{3}$	✓
5	7	$\frac{5}{7}$	
5	11	$\frac{5}{11}$	
7	3	$\frac{7}{3}$	✓
7	5	$\frac{7}{5}$	✓
7	11	$\frac{7}{11}$	
11	3	$\frac{11}{3}$	✓
11	5	$\frac{11}{5}$	✓
11	7	$\frac{11}{7}$	✓

Out of 12 fractions, there are 6 improper fractions. Therefore, the probability that the fraction formed is an improper fraction is $\frac{6}{12}$ or $\frac{1}{2}$.

8. (E)

When a coin is tossed, there are two possible outcomes: head or tail. If you toss a coin three times, according to the fundamental counting principle, the total number of the possible outcomes is $2 \times 2 \times 2 = 8$. The table below shows the 8 possible outcomes.

H	H	H	
H	H	T	
H	T	H	
H	T	T	✓
T	H	H	
T	H	T	✓
T	T	H	✓
T	T	T	

Out of 8 possible outcomes, there are 3 outcomes that have one head: $H\ T\ T$, $T\ H\ T$, and $T\ T\ H$. Therefore, the probability that one head will be shown is $\frac{3}{8}$.

9. (A)

Since the mean of the five positive integers is 6, the sum of the five positive integers is $5 \times 6 = 30$. Define x as the second greatest integer and y as the greatest integer in the set. Let's consider the three cases shown below. In case 1, the median is 4 because there are three 4's. This doesn't satisfy the given information such that the median is 5. Thus, case 1 is false.

Case 1:	$4+4+4+x+y=30$	mode=4, median=4: It doesn't work
Case 2:	$4+4+5+5+y=30$	mode=4 and 5, median=5: It doesn't work
Case 3:	$4+4+5+x+y=30$	mode=4, median=5: It works

In case 2, there are two 4's and two 5's in which the modes are both 4 and 5. This doesn't satisfy the given information such that the mode is 4. Thus, case 2 is false. Finally, in case 3, there are two 4's and one 5. If x is greater than 5, the mode is 4 and the median is 5 which satisfies the given information. x must be the smallest positive integer greater than 5 so that y will have the greatest possible value. Thus, $x = 6$ and $y = 11$. Therefore, the greatest of these integers is 11.

10. (B)

Event 1 is selecting 2 boys out of 4 boys. Since 4 boys are indistinguishable, the number of ways to select 2 boys out of 4 boys is $\binom{4}{2} = 6$. Event 2 is selecting 2 girls out of 5 girls. Since 5 girls are indistinguishable, the number of ways to select 2 girls out of 5 girls is $\binom{5}{2} = 10$. Each team that Mr. Rhee selects consists of 2 boys and 2 girls. Therefore, according to the fundamental counting principle, the number of different teams that Mr. Rhee can select is $\binom{4}{2} \times \binom{5}{2} = 6 \times 10 = 60$.

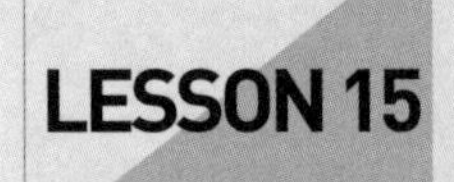

LESSON 15 Miscellaneous

Simplifying Expressions

Simplifying numerical expressions

To simplify numerical expressions, use the **order of operations** (PEMDAS): P (Parenthesis), E (Exponent), M (Multiplication), D (Division), A (Addition), and S (Subtraction). The order of operations suggests to first perform any calculations inside parentheses. Afterwards, evaluate any exponents. Next, perform all multiplications and divisions working from left to right. Finally, do additions and subtractions from left to right.

Simplifying and evaluating algebraic expressions

Like terms are terms that have same variables and same exponents; only the coefficients could be different or the same. Knowing like terms is essential when you simplify algebraic expressions. For instance,

- $2x$ and $3x$: (Like terms)
- $2x$ and $3x^2$: (Not like terms since the two expressions have different exponents)
- 2 and 3: (Like terms)

Use the **distributive property** to expand an algebraic expression that has a parenthesis.

$$x(y+z) = x \times y + x \times z$$

To simplify an algebraic expression, expand the expression using the distributive property. Then group the like terms and simplify them. For instance,

$$\begin{aligned} 2(-x+2)+3x+5 &= -2x+4+3x+5 && \text{Use distributive property to expand} \\ &= (-2x+3x)+(4+5) && \text{Group the like terms and simplify} \\ &= x+9 \end{aligned}$$

To evaluate an algebraic expression, substitute the numerical value into the variable. When substituting a negative numerical value, make sure to use a **parenthesis** to avoid a mistake. For instance, let $f(x) = x^2 - 2x$. In order to evaluate $f(-2)$, substitute -2 for x.

$$\begin{aligned} f(x) &= x^2 - 2x && \text{Substitute } -2 \text{ for } x \\ f(-2) &= (-2)^2 - 2(-2) = 4 \end{aligned}$$

Simplifying rational expressions

To multiply two rational expressions, factor the numerator and denominator of each rational expression. Then, cancels out any common factors to both the numerator and denominator. For instance,

$$\frac{x-1}{x^2-4}\cdot\frac{x+2}{x-1}=\frac{\cancel{(x-1)}}{\cancel{(x+2)}(x-2)}\cdot\frac{\cancel{(x+2)}}{\cancel{(x-1)}}=\frac{1}{x-2}$$

To divide two rational expressions, multiply the first rational expression by the reciprocal of the second rational expression. For instance,

$$\frac{x-1}{x+2}\div\frac{x-1}{x^2+5x+6}=\frac{\cancel{(x-1)}}{\cancel{(x+2)}}\cdot\frac{\cancel{(x+2)}(x+3)}{\cancel{(x-1)}}=x+3$$

To add (or subtract) two rational expressions with like denominators, simply add (or subtract) their numerators and put the result over the common denominator. For instance,

$$\frac{2}{x-3}+\frac{x}{x-3}=\frac{x+2}{x-3}$$

To add (or subtract) two rational expressions with unlike denominators, find the least common denominator of the rational expressions. Then, rewrite each expression as an equivalent expression using the least common denominator. Finally, add (or subtract) their numerators and put the result over the least common denominator. For instance,

$$\frac{3}{x+1}+\frac{2}{x-2}=\frac{3(x-2)}{(x+1)(x-2)}+\frac{2(x+1)}{(x+1)(x-2)}=\frac{5x-4}{(x+1)(x-2)}$$

Solving Equations

Solving an equation is finding the value of the variable that makes the equation true. In order to solve an equation, use the rule called SADMEP with inverse operations (SADMEP is the reverse order of the order of operations, PEMDAS). Inverse operations are the operations that cancel each other. Addition and subtraction, and multiplication and division are good examples.

SADMEP suggests to first cancel subtraction or addition. Then, cancel division or multiplication next. Finally, cancel exponent and parenthesis by applying corresponding inverse operation. Below is an example that shows you how to solve $\sqrt[3]{x+1}-2=1$, which involves subtraction, exponent, and parenthesis.

	✓ ✓ ✓
$\sqrt[3]{x+1}-2=1$	$S\ A\ D\ M\ E\ P$
$+2=+2$	Add 2 to cancel substraction
$(x+1)^{\frac{1}{3}}=3$	Raise each side to the power of 3 to cancel the exponent
$(x+1)=27$	Subtract 1 from each side
$x=26$	

Solving radical equations

An equation that contains a radical ($\sqrt{x}$) is called a radical equation. Often, solving a radical equation involves squaring a binomial. The binomial expansion formulas are shown below.

$$(x+y)^2 = x^2 + 2xy + y^2 \qquad \text{Common mistake: } (x+y)^2 \neq x^2 + y^2$$
$$(x-y)^2 = x^2 - 2xy + y^2 \qquad \text{Common mistake: } (x-y)^2 \neq x^2 - y^2$$

To solve a radical equation, square both sides of the equation to eliminate the square root. Then, solve for the variable. Once you get the solution of the radical equation, you need to substitute the solution in the original equation to check the solution. If the solution doesn't make the equation true, it is called an extraneous solution and is disregarded. Below shows how to solve a radical equation.

$$\begin{aligned}
x - 1 &= \sqrt{x+5} && \text{Square both sides} \\
(x-1)^2 &= x + 5 && \text{Use the binomial expansion formula} \\
x^2 - 2x + 1 &= x + 5 && \text{Subtract } x+5 \text{ from each side} \\
x^2 - 3x - 4 &= 0 && \text{Factor the quadratic expression} \\
(x+1)(x-4) &= 0 && \text{Use the zero product property: If } ab = 0, \text{ then } a = 0 \text{ or } b = 0. \\
x = -1 \quad &\text{or} \quad x = 4
\end{aligned}$$

Substitute -1 and 4 for x in the original equation to check the solutions.

$$(-1) - 1 = \sqrt{-1+5} \qquad\qquad (4) - 1 = \sqrt{4+5}$$
$$-2 \neq 2 \quad \text{(Not a solution)} \qquad\qquad 3 = 3 \quad \checkmark \text{ (Solution)}$$

Solving rational equations

To solve a rational equation, multiply each side of the equation by the least common denominator (LCD) of the rational expressions. Then, simplify and solve the resulting polynomial equation. Once you get the solution of the rational equation, you need to substitute the solution in the original equation to check the solution. Below shows how to solve a rational equation.

$$\frac{x-1}{x-2} + \frac{2}{x-4} = \frac{4}{x^2 - 6x + 8}$$
$$(x-2)(x-4)\left(\frac{x-1}{x-2} + \frac{2}{x-4}\right) = \left(\frac{4}{x^2 - 6x + 8}\right)(x-2)(x-4) \qquad \text{LCD: } (x-2)(x-4)$$
$$(x-1)(x-4) + 2(x-2) = 4$$
$$x^2 - 3x - 4 = 0$$
$$(x+1)(x-4) = 0$$
$$x = -1 \quad \text{or} \quad x = 4$$

Substitute -1 and 4 for x in the original equation to check the solutions. Thus, the only solution to the rational equation is $x = -1$.

Distance Formula in Three-Dimensional Space

If the points (x_1, y_1, z_1) and (x_2, y_2, z_2) are given in three-dimensional space, the distance D between the two points is defined as follows:

$$D = \sqrt{(x_2 - x_1)^2 + (y_2 - y_1)^2 + (z_2 - z_1)^2}$$

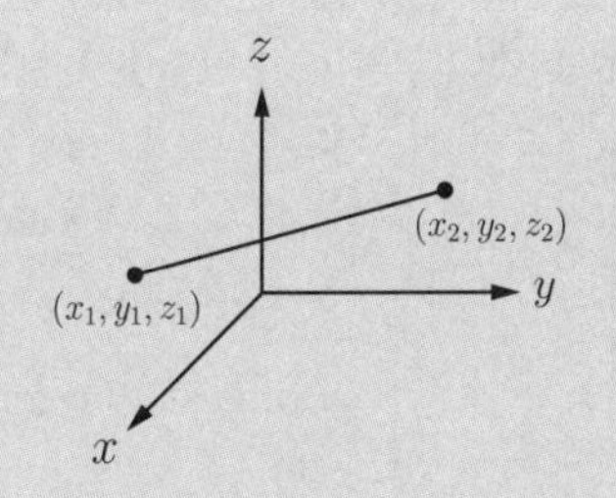

Example 1 Finding the distance between two points in 3-D

If two points, $A(1, -2, 3)$ and $B(-2, 2, 15)$, are in three-dimensional space, find the distance between the two points.

Solution

$$\begin{aligned} D &= \sqrt{(x_2 - x_1)^2 + (y_2 - y_1)^2 + (z_2 - z_1)^2} \\ &= \sqrt{(-2 - 1)^2 + (2 - (-2))^2 + (15 - 3)^2} \\ &= \sqrt{(-3)^2 + 4^2 + 12^2} \\ &= 13 \end{aligned}$$

Therefore, the distance between two points is 13.

Three-Dimensional Figures

Rectangular box

The volume V, surface area A, and longest diagonal AB of a rectangular box with side lengths a, b, and c are defined as follows:

Volume: $V = abc$

Surface area: $A = 2(ab + bc + ca)$

Longest diagonal: $AB = \sqrt{a^2 + b^2 + c^2}$

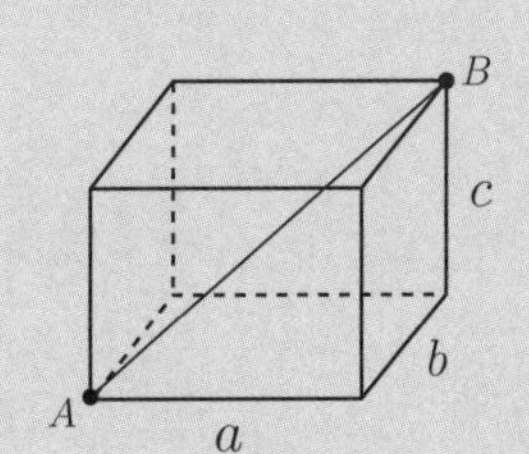

Cube

The volume V, surface area A, and longest diagonal AB of a cube with side lengths x are defined as follows:

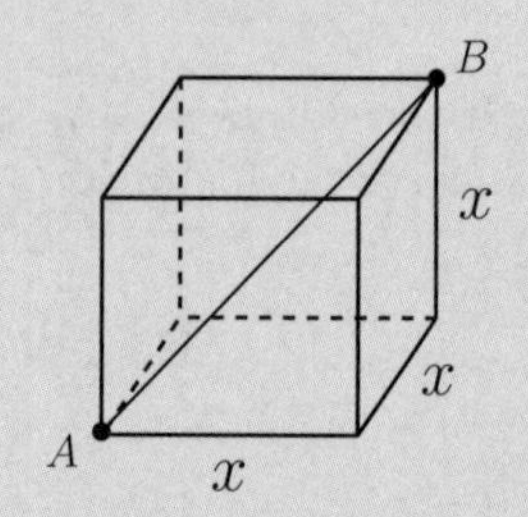

Volume: $V = x^3$

Surface area: $A = 6x^2$

Longest diagonal: $AB = \sqrt{x^2 + x^2 + x^2} = x\sqrt{3}$

Rectangular pyramid

The volume V of a rectangular pyramid is one-third of the area of the base B times the height h.

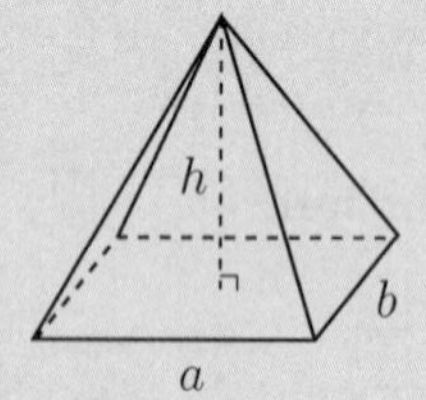

Volume: $V = \frac{1}{3}Bh = \frac{1}{3}abh$

Cylinder

The volume V of a cylinder with radius r is the area of the base B times the height h.

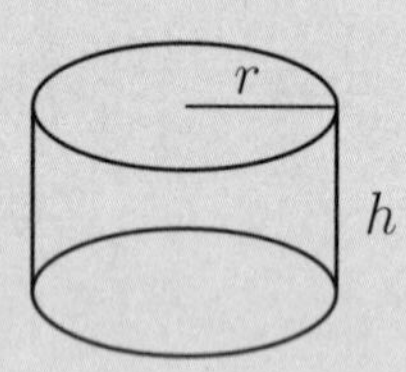

Volume: $V = Bh = \pi r^2 h$

Right circular cone

The volume V of a right circular cone with radius r is one-third of the area of the base B times the height h. The lateral area A of a right circular cone with circumference of the base c and slant height l is half times the product of the circumference of the base and the slant height.

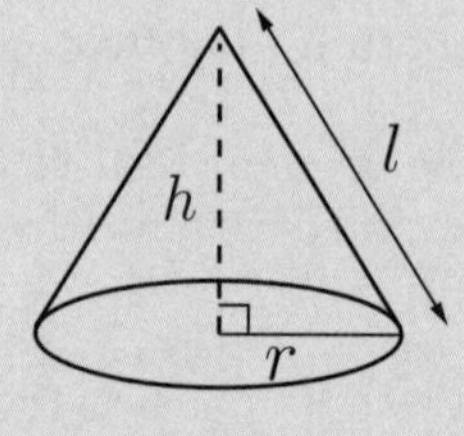

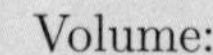

Volume: $V = \frac{1}{3}Bh = \frac{1}{3}\pi r^2 h$

Lateral area: $A = \frac{1}{2}cl = \pi rl$

Sphere

The general equation of a sphere with center (x_0, y_0, z_0) and radius r is $(x-x_0)^2+(y-y_0)^2+(z-z_0)^2=r^2$. The volume V and surface area A of a sphere with radius r are as follows:

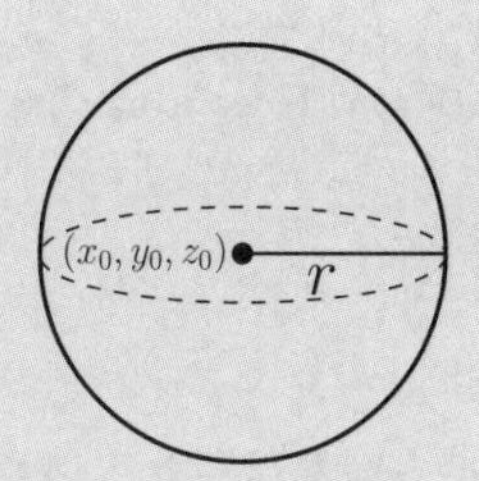

Volume: $V=\frac{4}{3}\pi r^3$

Surface area: $A=4\pi r^2$

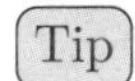

1. When a sphere is inscribed in a cube with side length x, the diameter of the sphere is the same as the length of the cube. Thus, the radius of the sphere is half the side length of the cube, or $\frac{x}{2}$.
2. When a cube with side length x is inscribed in a sphere, the diameter of the sphere is the same as longest diagonal of the cube. Thus, the diameter of the sphere is $x\sqrt{3}$ and the radius of the sphere is $\frac{x}{2}\sqrt{3}$.

Example 2 Finding the center, volume, and surface area of a sphere

If the equation of a sphere is $(x-2)^2+(y+1)^2+(z+3)^2=4$, find the center, volume, and surface area of the sphere.

Solution In order to find the center of the sphere, set expressions inside the parenthesis equal to zero and solve for each variable: $x-2=0$, $y+1=0$, and $z+3=0$. We found that $x=2$, $y=-1$, and $z=-3$. Thus, the center of the sphere is $(2,-1,-3)$. Since the radius of the sphere is 2, the volume and the surface area of the sphere are $\frac{32\pi}{4}$ and 16π, respectively.

$$V=\frac{4}{3}\pi r^3=\frac{4}{3}\pi(2)^3=\frac{32\pi}{4}$$
$$A=4\pi r^2=4\pi(2)^2=16\pi$$

Indirect Proof: Contrapositive

A conditional statement consists of two parts, a hypothesis in "if" clause and a conclusion in "then" clause. For instance, a conditional statement "if $x=2$, then $x^2=4$" is given, "if $x=2$" is the hypothesis, and "then $x^2=4$" is the conclusion.

When you interchange the hypothesis and the conclusion and negate both, you form the **contrapositive**. For instance, the contrapositive of "if $x=2$, then $x^2=4$" is "if $x^2\neq 4$, then $x\neq 2$.

Sometimes, it is difficult to prove directly whether a conditional statement is true. Instead, mathematicians use an indirect proof, contrapositive, to prove it easily. Logically, a conditional statement and its contrapositive are equivalent statements. That is, if the conditional statement is true, the contrapositive is also true. If the conditional statement is false, the contrapositive is also false.

The Product of Two Matrices

A matrix is a rectangular arrangement of numbers in rows and columns. For instance, matrix A below has two rows and three columns. The dimension of matrix A is 2×3 (is read as 2 by 3).

$$A = \begin{bmatrix} 1 & 3 & 5 \\ 2 & 4 & 6 \end{bmatrix}$$

Let A be a $m \times n$ matrix and B be a $n \times p$ matrix. The product of the two matrices, AB, is defined if the number of columns in matrix A is equal to the number of rows in matrix B, and is a $m \times p$ matrix.

For instance, A is a 3×2 matrix and B is a 2×1 matrix. Since the number of columns in matrix A is equal to the number of rows in matrix B, the product AB is defined and is a 3×1 matrix. However, the product BA is undefined since the number of columns in matrix B, 1, is not the same as the number of rows in matrix A, 3.

Variations

Direct variation

If y varies directly with x or y is proportional to x, then x and y have a relationship called **direct variation**. This means that as the x value gets larger, the y value gets larger. Additionally, as the x value gets smaller, the y value gets smaller. A direct variation is expressed as

$$y = kx \qquad \text{or} \qquad \frac{y}{x} = k$$

where k is the constant of variation. In most cases, you need to solve the constant of variation, k, from the given information.

Inverse variation

If y varies inversely with x or y is inversely proportional to x, then x and y have a relationship called **inverse variation**. This means that as the x value gets larger, the y value gets smaller. Furthermore, as the x value gets smaller, the y value gets larger. An inverse variation is expressed as

$$y = \frac{k}{x} \qquad \text{or} \qquad xy = k$$

where k is the constant of variation.

Joint variation

Joint variation is a direct variation that involves two or more variables. For instance, if z varies jointly with x and square of y, z can be expressed as

$$z = kxy^2$$

where k is the constant of variation.

Combined variation

Combined variation involves both direct variation (or joint variation) and inverse variation. For instance, if z varies directly with square root of x and inversely with y, z can be expressed as

$$z = \frac{k\sqrt{x}}{y}$$

where k is the constant of variation.

Example 3 Writing a combined variation equation

z varies directly with x and inversely with square root of y. When $x = 8$ and $y = 4$, $z = 16$. find z when $x = -2$ and $y = 16$.

Solution Since z varies directly with x and inversely with square root of y, start with $z = \frac{kx}{\sqrt{y}}$. Substitute 8 for x, 4 for y, and 16 for z to find the value of k.

$$z = \frac{kx}{\sqrt{y}} \qquad \text{Substitute 8 for } x\text{, 4 for } y\text{, and 16 for } z$$

$$16 = \frac{8k}{\sqrt{4}} \qquad \text{Solve for } k$$

$$k = 4$$

Thus, the equation that relates x, y, and z is $z = \frac{4x}{\sqrt{y}}$. Substitute -2 for x and 16 for y to find the value of z.

$$z = \frac{4x}{\sqrt{y}} \qquad \text{Substitute } -2 \text{ for } x \text{ and 16 for } y$$

$$z = \frac{4(-2)}{\sqrt{16}} \qquad \text{Solve for } z$$

$$z = -2$$

Therefore, the value of z when $x = -2$ and $y = 16$ is -2.

EXERCISES

1. Solve: $(x-2)^3+1=3$

(A) 2.76

(B) 3.26

(C) 3.96

(D) 4.46

(E) 4.96

2. $\dfrac{2x-3}{x-2}+\dfrac{3x-7}{x-2}=$

(A) 1

(B) 2

(C) 3

(D) 4

(E) 5

3. If the length of the longest diagonal of a cube is $6\sqrt{3}$, what is the volume of the cube?

(A) 512

(B) 216

(C) 125

(D) 64

(E) 27

4. A is a 3×4 matrix and B is a 4×3 matrix. What are the dimensions of the product of two matrices, BA ?

(A) 4×4

(B) 4×3

(C) 3×4

(D) 3×3

(E) Undefined

5. If the distance between two points, $(3,1,2)$ and $(x,1,2)$, is 4, what is the sum of the possible values of x ?

(A) -6

(B) -3

(C) 3

(D) 6

(E) 8

6. If both the volume and the surface area of a sphere are the same, what is the radius of the sphere?

(A) 1

(B) 2

(C) 3

(D) 4

(E) 6

7. Solve: $\sqrt{2x-1}-3\sqrt{x+2}=0$

(A) $\frac{17}{7}$

(B) $\frac{9}{5}$

(C) $\frac{3}{7}$

(D) $-\frac{13}{5}$

(E) $-\frac{19}{7}$

8. $\dfrac{1}{\frac{1}{x}+\frac{1}{y}}=$

(A) $\frac{xy}{x+y}$

(B) $\frac{x+y}{xy}$

(C) $xy(x+y)$

(D) xy

(E) $x+y$

9. z varies jointly with y and the square of x. When $x = -2$ and $y = 3$, $z = 4$. What is the value of z when $x = 3$ and $y = -2$?

(A) 12

(B) 9

(C) -6

(D) -9

(E) -12

10. Solve: $\dfrac{x}{x-3} + \dfrac{4x}{x^2-9} = \dfrac{x+4}{x+3}$

(A) 2

(B) 1

(C) -1

(D) -2

(E) -3

ANSWERS AND SOLUTIONS

1. (B)

$(x-2)^3 + 1 = 3$	Subtract 1 from each side
$(x-2)^3 = 2$	Take the cube root of each side
$x - 2 = \sqrt[3]{2}$	Add 2 to each side
$x = 3.26$	

Therefore, the value of x for which $(x-2)^3 + 1 = 3$ is 3.26.

2. (E)

In order to add two rational expressions with like denominators, simply add their numerators and put the result over the common denominator.

$$\frac{2x-3}{x-2} + \frac{3x-7}{x-2} = \frac{5x-10}{x-2} = \frac{5(x-2)}{x-2} = 5$$

Therefore, $\frac{2x-3}{x-2} + \frac{3x-7}{x-2} = 5$.

3. (B)

Tip The length of the longest diagonal of a cube with side length x is $x\sqrt{3}$.

Since the length of the longest diagonal of the cube is $6\sqrt{3}$, the length of the cube is 6. Therefore, the volume of the cube is 6^3 or 216.

4. (A)

Since the number of columns in matrix B is the same as the number of rows in matrix A, the product of the two matrices, BA, is defined as a 4×4 matrix.

5. (D)

Tip

1. The distance D between two points (x_1, y_1, z_1) and (x_2, y_2, z_2) is $D = \sqrt{(x_2 - x_1)^2 + (y_2 - y_1)^2 + (z_2 - z_1)^2}$.
2. Since $\sqrt{x^2} = |x|$, $\sqrt{(x-3)^2} = |x-3|$

In order to find the possible values of x, use the distance formula between the two points $(3, 1, 2)$ and $(x, 1, 2)$.

$$\begin{aligned} D &= \sqrt{(x_2 - x_1)^2 + (y_2 - y_1)^2 + (z_2 - z_1)^2} \\ &= \sqrt{(x-3)^2 + (1-1)^2 + (2-2)^2} \\ &= \sqrt{(x-3)^2} \\ &= |x-3| \end{aligned}$$

Since the distance between the two points is 4, $\sqrt{(x-3)^2} = |x-3| = 4$. Thus,

$$\begin{aligned} |x-3| &= 4 \\ x - 3 &= \pm 4 \\ x &= 3 \pm 4 \\ x = 7 \quad &\text{or} \quad x = -1 \end{aligned}$$

Therefore, the sum of the possible values of x is $7 + (-1) = 6$.

6. (C)

Tip

1. The volume of a sphere: $V = \frac{4}{3}\pi r^3$
2. The surface area of a sphere: $A = 4\pi r^2$

Since the volume and the surface area of the sphere are the same, set $V = \frac{4}{3}\pi r^3$ and $A = 4\pi r^2$ equal to each other and solve for r.

$$\frac{4}{3}\pi r^3 = 4\pi r^2 \qquad \text{Divide each side by } 4\pi r^2$$

$$\frac{1}{3}r = 1 \qquad \text{Multiply each side by 3}$$

$$r = 3$$

Therefore, the radius of the sphere is 3.

7. (E)

$$\sqrt{2x-1} - 3\sqrt{x+2} = 0 \qquad \text{Add } 3\sqrt{x+2} \text{ to each side}$$

$$\sqrt{2x-1} = 3\sqrt{x+2} \qquad \text{Square both sides}$$

$$2x - 1 = 9(x+2) \qquad \text{Simplify}$$

$$7x = -19 \qquad \text{Divide each side by 7}$$

$$x = -\frac{19}{7}$$

Therefore, the value of x for which $\sqrt{2x-1} - 3\sqrt{x+2} = 0$ is $-\frac{19}{7}$.

8. (A)

Simplify the denominator of the two rational expressions. The least common denominator of the two rational expressions is xy.

$$\frac{1}{x}+\frac{1}{y}=\frac{y}{xy}+\frac{x}{xy}=\frac{x+y}{xy}$$

Thus,

$$\frac{1}{\frac{1}{x}+\frac{1}{y}}=\frac{1}{\frac{x+y}{xy}}=\frac{xy}{x+y}$$

Therefore, the correct answer is (A).

9. (C)

Since z varies jointly with y and the square of x, start with $z=kx^2y$. Substitute -2 for x, 3 for y, and 4 for z to find the value of k.

$$\begin{aligned} z&=kx^2y && \text{Substitute } -2 \text{ for } x, 3 \text{ for } y, \text{ and } 4 \text{ for } z\\ 4&=k(-2)^2(3) && \text{Solve for } k\\ k&=\frac{1}{3} \end{aligned}$$

Thus, the equation that relates x, y, and z is $z=\frac{1}{3}x^2y$. Substitute 3 for x and -2 for y to find the value of z.

$$\begin{aligned} z&=\frac{1}{3}x^2y && \text{Substitute } 3 \text{ for } x \text{ and } -2 \text{ for } y\\ z&=\frac{1}{3}(3)^2(-2) && \text{Solve for } z\\ z&=-6 \end{aligned}$$

Therefore, the value of z when $x=3$ and $y=-2$ is -6.

10. (D)

The least common denominator of the rational expressions is $(x+3)(x-3)$.

$$\begin{aligned} \frac{x}{x-3}+\frac{4x}{x^2-9}&=\frac{x+4}{x+3}\\ (x+3)(x-3)\left(\frac{x}{x-3}+\frac{4x}{x^2-9}\right)&=\left(\frac{x+4}{x+3}\right)(x+3)(x-3) && \text{LCD: } (x+3)(x-3)\\ x(x+3)+4x&=(x+4)(x-3)\\ x^2+3x+4x&=x^2-3x+4x-12\\ x^2+7x&=x^2+x-12\\ 6x&=-12\\ x&=-2 \end{aligned}$$

Substitute -2 in the original equation to check the solution. Therefore, the solution to the rational equation is $x=-2$.

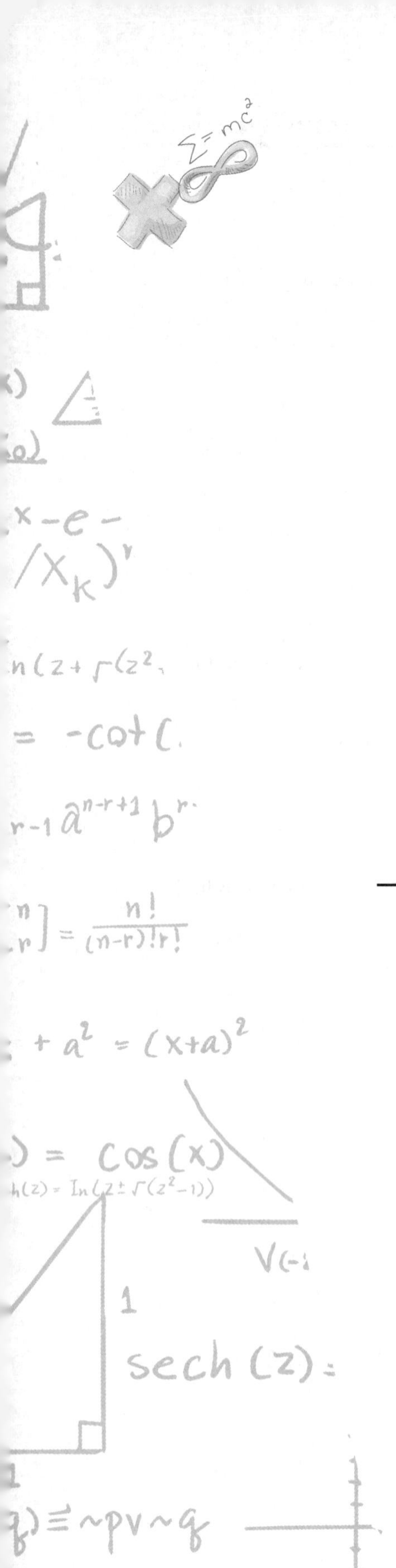

SAT II MATH LEVEL 2 TEST 1

ANSWERS AND SOLUTIONS

SAT II MATH LEVEL 2 TEST 2

ANSWERS AND SOLUTIONS

SAT II MATH LEVEL 2 TEST 3

ANSWERS AND SOLUTIONS

SAT II MATH LEVEL 2 TEST 4

ANSWERS AND SOLUTIONS

SAT II MATH LEVEL 2 TEST 5

ANSWERS AND SOLUTIONS

SAT II MATH LEVEL 2 TEST 6

ANSWERS AND SOLUTIONS

SAT II MATH LEVEL 2 TEST 1

Directions: Among the given answer choices, choose the BEST answer for each problem. If the exact numerical value is not within the given answer choices, select the answer that best approximates this value. Afterwards, fill in the corresponding oval on the answer sheet.

Notes:

1. A calculator may be required to answer some of the questions. Scientific and graphing calculators are allowed during this test.

2. For some questions, it is up to you to decide whether the calculator should be in degree mode or radian mode.

3. Provided figures for questions are drawn as accurately as possible UNLESS otherwise specified by the problem. Unless otherwise indicated, all figures are assumed to lie in a plane.

4. Unless otherwise specified, it can be assumed that the domain of any function f is to be the set of all real numbers x for which $f(x)$ is a real number.

Reference Information: Use the following information and formulas as a reference in answering questions on this test.

1. If the radius and height of a right circular cone are r and h respectively, then the Volume V of the cone is $V = \frac{1}{3}\pi r^2 h$.

2. If the circumference and slant height of a right circular cone are c and ℓ respectively, then the Lateral Area A of the cone is $A = \frac{1}{2}c\ell$.

3. At any given radius r, the Volume V of a sphere is $V = \frac{4}{3}\pi r^3$.

4. At any given radius r, the Surface Area A of a sphere is $A = 4\pi r^2$.

5. The Volume V of a pyramid is $V = \frac{1}{3}Bh$; given that B and h represent the base area and height of the pyramid respectively.

USE THIS SPACE FOR SCRATCH WORK

1. If $2(1-x)=3(1-x)$, what is the value of x ?

   (A) -2

   (B) -1

   (C) 0

   (D) 1

   (E) 2

2. If the parametric equations, $x = 2t+1$ and $y=\frac{t}{2}-1$, represent a line, what is the slope of the line?

   (A) $\frac{1}{6}$

   (B) $\frac{1}{5}$

   (C) $\frac{1}{4}$

   (D) $\frac{1}{3}$

   (E) $\frac{1}{2}$

3. The graph of the hyperbola $\frac{x^2}{9}-\frac{y^2}{4}=1$ intersects the x-axis. What are the x-intercepts?

   (A) 9 and -9

   (B) 6 and -6

   (C) 4 and -4

   (D) 3 and -3

   (E) 2 and -2

USE THIS SPACE FOR SCRATCH WORK

4. What is the probability of selecting a prime number at random from the first ten positive integers?

(A) $\frac{4}{5}$

(B) $\frac{7}{10}$

(C) $\frac{3}{5}$

(D) $\frac{1}{2}$

(E) $\frac{2}{5}$

5. What are the rectangular coordinates (x, y) that correspond with the polar coordinates $(2, \frac{\pi}{3})$?

(A) $(\sqrt{3}, 1)$

(B) $(1, \sqrt{3})$

(C) $(2, 2\sqrt{3})$

(D) $(-\sqrt{3}, 1)$

(E) $(-1, -\sqrt{3})$

6. A is a 2×4 matrix and B is a 4×3 matrix. What are the dimensions of the product of the two matrices, AB ?

(A) 2×2

(B) 2×3

(C) 3×2

(D) 4×3

(E) Undefined

USE THIS SPACE FOR SCRATCH WORK

7. In $\triangle ABC$, $m\angle C = 90°$. If $AB = 10$ and $BC = 8$, what is the measure of $\angle BAC$?

(A) 53.13°

(B) 51.34°

(C) 48.85°

(D) 42.92°

(E) 36.87°

8. Let $f(x) = |x|$. If $g(x)$ is obtained by translating $f(x)$ 3 units to the left and 2 units down, what is the value of $g(-4)$?

(A) 9

(B) 5

(C) 3

(D) −1

(E) −3

9. $\lim_{x \to 2} x^3 - 4x + 4 =$

(A) 0

(B) 2

(C) 4

(D) 5

(E) 8

USE THIS SPACE FOR SCRATCH WORK

10. If $f(x) = \frac{x+2}{x+1}$, for what value of x is f undefined?

(A) -2

(B) -1

(C) 0

(D) 1

(E) 2

11. If a car travels at 40 miles per hour for the first half an hour and travels at 60 miles per hour for the next 20 minutes, what is the total distance, in miles, that the car would travel?

(A) 100 miles

(B) 80 miles

(C) 60 miles

(D) 50 miles

(E) 40 miles

12. $\sin^2(-\theta) + \cos^2(-\theta) =$

(A) 2

(B) 1

(C) 0

(D) -1

(E) -2

USE THIS SPACE FOR SCRATCH WORK

13. What is the inverse function of $y = \log_3 x$?

(A) $\log_{\frac{1}{3}} x$

(B) $\log_3 \frac{1}{x}$

(C) 3^x

(D) $\left(\frac{1}{3}\right)^x$

(E) x^3

14. In order to use an indirect proof of the statement "If $x = 9$, then $x^2 = 81$", which of the following assumption could you begin with?

(A) If x is a positive integer

(B) If $x^2 = 81$

(C) If $x^2 \neq 81$

(D) If $x = 9$

(E) If $x \neq 9$

15. You purchased a car for $\$20,000$. If the value of the car depreciates at a rate of 15% per year, how much is the car worth after 5 years?

(A) $\$13,862.92$

(B) $\$12,474.58$

(C) $\$10,440.10$

(D) $\$8,874.11$

(E) $\$7,542.99$

USE THIS SPACE FOR SCRATCH WORK

16. If $(-7, -24)$ is on the terminal side of an angle θ in standard position, what is the value of $\csc\theta$?

(A) $\frac{7}{24}$

(B) $\frac{25}{7}$

(C) $\frac{25}{24}$

(D) $-\frac{25}{24}$

(E) $-\frac{25}{7}$

17. If $z_1 = 2 + i$ and $z_2 = 1 + 3i$, which of the following letter in the complex plane shown in Figure 1 represents the product of the two complex numbers, $z_1 z_2$?

(A) A

(B) B

(C) C

(D) D

(E) E

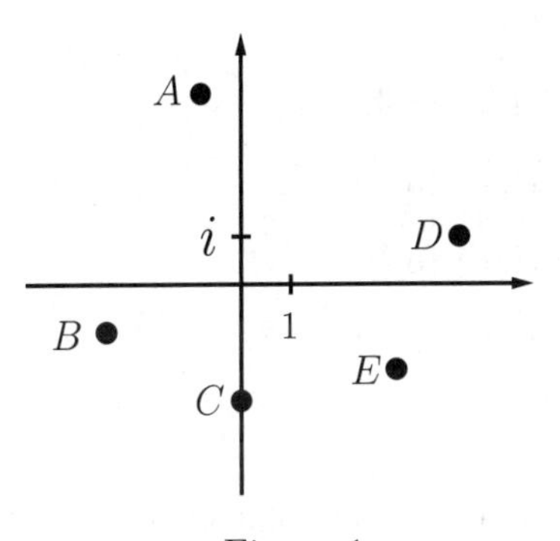

Figure 1

18. Solve the inequality: $\cos\theta < 0$, where $0 \leq \theta < 2\pi$.

(A) $0 \leq \theta < \pi$

(B) $\frac{\pi}{2} < \theta < \frac{3\pi}{2}$

(C) $\frac{\pi}{4} < \theta < \frac{7\pi}{4}$

(D) $\pi < \theta < 2\pi$

(E) $\frac{4\pi}{3} < \theta < \frac{5\pi}{3}$

USE THIS SPACE FOR SCRATCH WORK

19. If $f(x) = \log x$ and $g(x) = 10^x$, what is the value of $f(g(3))$?

(A) 30

(B) 10

(C) 3

(D) 1

(E) $\frac{1}{3}$

20. What is the vertex of the quadratic function $y = -2(x-3)^2 + 1$?

(A) $(-3, 2)$

(B) $(-3, 1)$

(C) $(3, 1)$

(D) $(3, 2)$

(E) $(6, 3)$

21. What is the range of the following function f ?

$$f(x) = \begin{cases} 3x + 2, & x < 0 \\ e^x, & x \geq 0 \end{cases}$$

(A) All real numbers

(B) $y \geq 0$

(C) $y < 0$

(D) Rational numbers

(E) Integers

USE THIS SPACE FOR SCRATCH WORK

22. If $f(x) = |x - 5| + 2$, $1 \leq x \leq 7$, what is the maximum value of f ?

(A) 7

(B) 6

(C) 5

(D) 4

(E) 3

23. If $p < 0$ and $pq - pr > 0$, which of the following inequality must be true?

(A) $\frac{q}{r} > 1$

(B) $qr < 0$

(C) $qr > 0$

(D) $r < q$

(E) $q < r$

24. The maximum height H of a projectile is given by

$$H = \frac{v_0^2 \sin^2 \theta}{2g}$$

where v_0 is an initial velocity in feet per second, θ is an angle to the horizontal in degrees, and g is the acceleration due to gravity and is 32. If the projectile is launched at an angle of 25° to the horizontal with an initial velocity of 100 feet per second, what is the maximum height of the projectile?

(A) 27.91 feet

(B) 34.25 feet

(C) 47.28 feet

(D) 54.33 feet

(E) 65.86 feet

USE THIS SPACE FOR SCRATCH WORK

25. If $\sin\theta\cos\theta = 0.3$, what is the value of $\dfrac{3}{\sin 2\theta}$?

(A) 12

(B) 10

(C) 8

(D) 6

(E) 5

26. At what value of x does the trigonometric function $y = -2\cos 2x$ have the maximum value?

(A) π

(B) $\dfrac{5\pi}{6}$

(C) $\dfrac{3\pi}{4}$

(D) $\dfrac{2\pi}{3}$

(E) $\dfrac{\pi}{2}$

27. There are nine points on a plane. Assuming that any three points are not collinear, how many different triangles can be formed by connecting three points on the plane?

(A) 36

(B) 42

(C) 60

(D) 72

(E) 84

USE THIS SPACE FOR SCRATCH WORK

28. The sequence is defined by $a_n = -2a_{n-1} - 1$ for $n \geq 2$ and $a_1 = -3$. What is the fifth term of the sequence?

    (A) 85

    (B) 31

    (C) -14

    (D) -43

    (E) -61

29. For the following system of linear equations, what must be the value of a so that the system has no solution?

$$2x + 6y = 5$$
$$x + ay = 1$$

    (A) -3

    (B) -1

    (C) 3

    (D) 6

    (E) 12

30. If $\tan(30° + x) = \cot(70° - 3x)$, what is the value of x ?

    (A) 20°

    (B) 15°

    (C) 10°

    (D) 5°

    (E) 1°

USE THIS SPACE FOR SCRATCH WORK

31. A particle can move either inside or on the surface of a cube with side length 10 feet as shown in Figure 2. The particle moves at a rate of $\frac{1}{2}$ feet per second. How much time does the particle save if it moves from point A to point B directly, rather than to move from point A to point C to point B ?

(A) 6.82 seconds

(B) 13.64 seconds

(C) 21.52 seconds

(D) 25.31 seconds

(E) 34.75 seconds

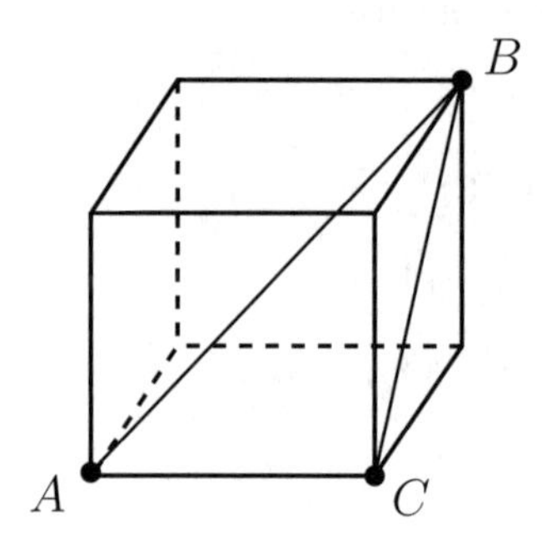

Figure 2

32. What is the domain of the logarithmic function $y = \ln(x^2 - 9)$?

(A) $0 < x < 3$

(B) $-3 < x < 3$

(C) $-3 \leq x \leq 3$

(D) $x < -3$ or $x > 3$

(E) $x \leq -3$ or $x \geq 3$

33. If the two graphs, $y = \sqrt{x-1}$ and $x = 4$, intersect on the coordinate plane, what is the y-coordinate of the intersection point?

(A) 1.04

(B) 1.27

(C) 1.41

(D) 1.65

(E) 1.73

USE THIS SPACE FOR SCRATCH WORK

34. In $\triangle ABC$, shown in Figure 3, $m\angle C = 40°$, $m\angle A = 60°$, and $AB = 6$. What is the length of $\overline{AC}$?

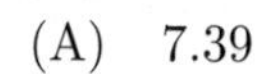

(A) 7.39

(B) 8.07

(C) 9.19

(D) 10.95

(E) 12.28

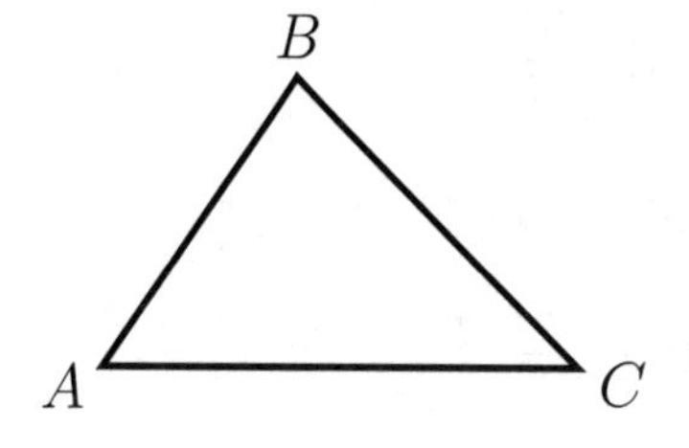

Figure 3

35. If $5^x = 14$, which of the following CANNOT be the solution of the equation?

(A) $\ln 14 - \ln 5$

(B) $\log_5 14$

(C) $\dfrac{\ln 14}{\ln 5}$

(D) $\dfrac{\log 14}{\log 5}$

(E) $\dfrac{\log_2 14}{\log_2 5}$

36. Which of the following expression is equal to $2^{\frac{1}{2}\log_2 x}$?

(A) $\dfrac{x}{2}$

(B) $\dfrac{1}{2x}$

(C) $\sqrt{x}$

(D) $x^{\frac{3}{2}}$

(E) x^2

USE THIS SPACE FOR SCRATCH WORK

37. If $\sin x = -\frac{2}{5}$, what is the exact value of $\cos 2x$?

(A) $\frac{21}{25}$

(B) $\frac{17}{25}$

(C) $\frac{12}{25}$

(D) $\frac{8}{25}$

(E) $\frac{4}{25}$

38. Which of the following statement CANNOT be true about the polynomial function $f(x) = x^3 - 3x - 18$?

(A) The real zero of f is only 3.

(B) For $x \geq 0$, $f(x) \geq -18$.

(C) f is decreasing in the interval $(-1, 1)$.

(D) The y-intercept of f is -18.

(E) As x increases without bound, $f(x)$ increases without bound.

39. If $f(x) = 2x^3 + 7x^2 + kx - 3$ has a factor of $x + 3$, what is the value of k ?

(A) 2

(B) 3

(C) 5

(D) 7

(E) 8

USE THIS SPACE FOR SCRATCH WORK

40. What is the inverse function of $f(x) = x^2,\ x \leq 0$?

(A) $\dfrac{1}{x^2}$

(B) $\sqrt{x}$

(C) $-x^2$

(D) $-\sqrt{x}$

(E) $-\dfrac{1}{x^2}$

41. What is the set of all ordered pairs (x, y) in the first quadrant such that the ratio of the shortest distance to the y-axis to the shortest distance to the x-axis is $1 : 3$?

(A) $y = \sqrt[3]{x}$

(B) $y = \dfrac{3}{x}$

(C) $y = x^3$

(D) $y = \dfrac{1}{3}x$

(E) $y = 3x$

42. In Figure 4, the three vectors **A**, **B**, and **C** are shown. Which of the following equation best represents the relationship between the three vectors?

(A) $\mathbf{C} = \mathbf{A} - \mathbf{B}$

(B) $\mathbf{C} = \mathbf{A} + \mathbf{B}$

(C) $\mathbf{A} = \mathbf{B} - \mathbf{C}$

(D) $\mathbf{B} = \mathbf{A} + \mathbf{C}$

(E) $\mathbf{B} = \mathbf{C} - \mathbf{A}$

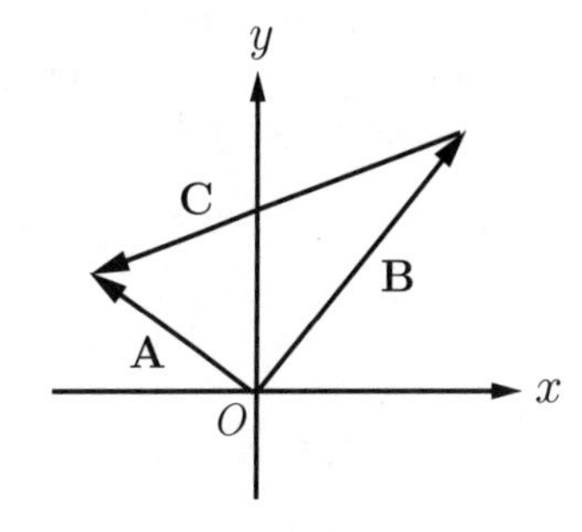

Figure 4

USE THIS SPACE FOR SCRATCH WORK

43. The quadratic function $f(x) = -2x^2 + 3x + 7$ contains the two points $(-1, f(-1))$ and $(1, f(1))$. What is the slope of a line that passes through the two points?

(A) 3

(B) 1

(C) $\frac{1}{3}$

(D) -1

(E) -3

44. Figure 5 shows a portion of the graph of $y = 2^{-x}$. Let the area of the shaded region A equal the sum of the areas of an infinite number of rectangles, of which only five rectangles are shown in Figure 5. What is the value of A ?

(A) 16

(B) 8

(C) 6

(D) 4

(E) 2

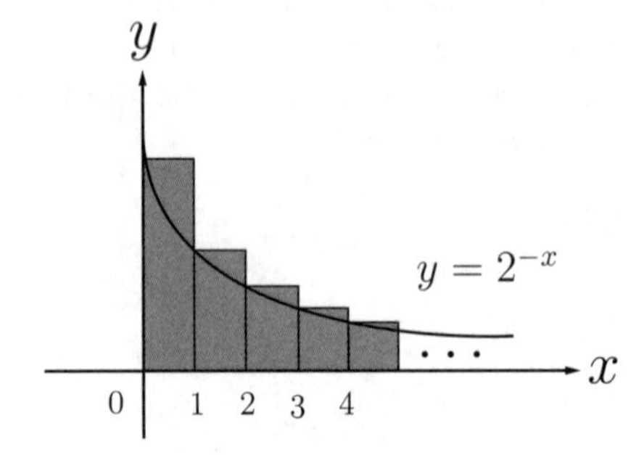

Figure 5

Note: Figure not drawn to scale.

45. Suppose $f(x)$ is an even function and $g(x)$ is an odd function. If $f(-2) = 5$ and $g(-5) = 2$, what is the value of $g(f(2))$?

(A) -5

(B) -2

(C) 2

(D) 5

(E) 7

USE THIS SPACE FOR SCRATCH WORK

46. Which of the following statement must be true?

(A) A function f is said to be increasing if $f(x_1) < f(x_2)$ when $x_1 > x_2$.

(B) A function f is said be constant if $f(x_1) \neq f(x_2)$ when $x_1 \neq x_2$.

(C) All polynomial functions are continuous.

(D) A function f is said to be decreasing if $f(x_1) < f(x_2)$ when $x_1 < x_2$.

(E) All vertical lines are functions.

47. What is the value of $\sin(\cos^{-1}(0.45))$?

(A) 0.27

(B) 0.41

(C) 0.53

(D) 0.76

(E) 0.89

48. A theater has twenty rows of seats. The first row has 12 seats. Each successive row has two more seats than the prior row. What is the total number of seats in the theater?

(A) 600

(B) 610

(C) 620

(D) 630

(E) 640

USE THIS SPACE FOR SCRATCH WORK

49. $i + i^2 + i^3 + \cdots + i^{11} + i^{12} =$

(A) -1

(B) $-i$

(C) 0

(D) 1

(E) i

50. There are 4 boys and 6 girls in Mr. Rhee's class. If Mr. Rhee selects a team of five students in his class, what is the probability that he selects a team of 2 boys and 3 girls?

(A) $\frac{\binom{4}{2}}{\binom{10}{2}} \times \frac{\binom{6}{3}}{\binom{10}{3}}$

(B) $\frac{\binom{4}{2}}{\binom{10}{2}} + \frac{\binom{6}{3}}{\binom{10}{3}}$

(C) $\frac{\binom{4}{3}}{\binom{10}{4}} \times \frac{\binom{6}{2}}{\binom{10}{6}}$

(D) $\frac{\binom{4}{2} \times \binom{6}{3}}{\binom{10}{5}}$

(E) $\frac{\binom{4}{2} + \binom{6}{3}}{\binom{10}{5}}$

STOP

Mathematics Scoring Worksheet

Directions: In order to calculate your score correctly, fill out the table below. After calculating your raw score, round the raw score to the nearest whole number. The scaled score can be determined using the "Math Test Score Conversion Table".

Mathematics Score			
A. Number Correct		**B.** Number Incorrect ÷ 4	
Total Unrounded Raw Score $A - B$		**Total Rounded Raw Score** Round to nearest whole number	

Math Test Score Conversion Table					
Raw Score	Scaled Score	Raw Score	Scaled Score	Raw Score	Scaled Score
50	800	29	660	8	490
49	800	28	650	7	480
48	800	27	640	6	480
47	800	26	630	5	470
46	800	25	630	4	460
45	800	24	620	3	450
44	800	23	610	2	440
43	800	22	600	1	430
42	790	21	590	0	410
41	780	20	580	−1	390
40	770	19	570	−2	370
39	760	18	560	−3	360
38	750	17	560	−4	340
37	740	16	550	−5	340
36	730	15	540	−6	330
35	720	14	530	−7	320
34	710	13	530	−8	320
33	700	12	520	−9	320
32	690	11	510	−10	320
31	680	10	500	−11	310
30	670	9	500	−12	310

ANSWERS AND SOLUTIONS

Answers

1. D	11. E	21. A	31. B	41. E
2. C	12. B	22. B	32. D	42. A
3. D	13. C	23. E	33. E	43. A
4. E	14. C	24. A	34. C	44. E
5. B	15. D	25. E	35. A	45. B
6. B	16. D	26. E	36. C	46. C
7. A	17. A	27. E	37. B	47. E
8. D	18. B	28. D	38. B	48. C
9. C	19. C	29. C	39. A	49. C
10. B	20. C	30. D	40. D	50. D

Solutions

1. (D)

Expand the expressions inside the parenthesis on the left and right sides of the equation.

$$\begin{aligned} 2(1-x) &= 3(1-x) && \text{Expand each side the equation} \\ 2-2x &= 3-3x && \text{Add } 3x \text{ to each side} \\ x+2 &= 3 && \text{Subtract 2 from each side} \\ x &= 1 \end{aligned}$$

Therefore, the value of x for which $2(1-x)=3(1-x)$ is 1.

2. (C)

In order to find the slope of the line, eliminate the parameter t. Solve for t from the first equation $x = 2t+1$. We found that $t = \dfrac{x-1}{2}$. Then, substitute $\dfrac{x-1}{2}$ for t in the second equation $y = \dfrac{1}{2}t - 1$ so that $y = \dfrac{1}{2}\left(\dfrac{x-1}{2}\right) - 1 = \dfrac{1}{4}x - \dfrac{5}{4}$. Therefore, the slope of the line is $\dfrac{1}{4}$.

3. (D)

In order to find the x-intercepts, substitute 0 for y and solve for x.

$$\frac{x^2}{9} - \frac{y^2}{4} = 1 \qquad \text{Substitute 0 for } y$$
$$\frac{x^2}{9} = 1 \qquad \text{Multiply each side by 9}$$
$$x^2 = 9 \qquad \text{Take the square root of both sides}$$
$$x = 3 \quad \text{or} \quad x = -3$$

Therefore, the x-intercepts are 3 and -3.

4. (E)

There are 4 prime numbers out of the first 10 positive integers: 2, 3, 5, and 7. Therefore, the probability of selecting a prime number at random from the first 10 positive integers is $\frac{4}{10}$ or $\frac{2}{5}$.

5. (B)

Tip	If the polar coordinates of a point are (r, θ), the rectangular coordinates of the point (x, y) are given by $x = r\cos\theta$ and $y = r\sin\theta$.

The polar coordinates of the point are $\left(2, \frac{\pi}{3}\right)$. Thus, $r = 2$ and $\theta = \frac{\pi}{3}$. Since $\cos\frac{\pi}{3} = \frac{1}{2}$ and $\sin\frac{\pi}{3} = \frac{\sqrt{3}}{2}$,

$$x = 2\cos\frac{\pi}{3} = 2 \cdot \frac{1}{2} = 1$$
$$y = 2\sin\frac{\pi}{3} = 2 \cdot \frac{\sqrt{3}}{2} = \sqrt{3}$$

Therefore, the rectangular coordinates that correspond with the polar coordinates $\left(2, \frac{\pi}{3}\right)$ are $(1, \sqrt{3})$.

6. (B)

Tip	Let A be a $m \times n$ matrix and B be a $n \times p$ matrix. The product of the two matrices, AB, is defined if the number of columns in matrix A is equal to the number of rows in matrix B, and is a $m \times p$ matrix.

Since the number of columns in matrix A is the same as the number of rows in matrix B, the product of two matrices, AB, is defined and is a 2×3 matrix.

7. (A)

Use the Pythagorean theorem to find AC: $10^2 = 8^2 + AC^2$. Thus, $AC = 6$.

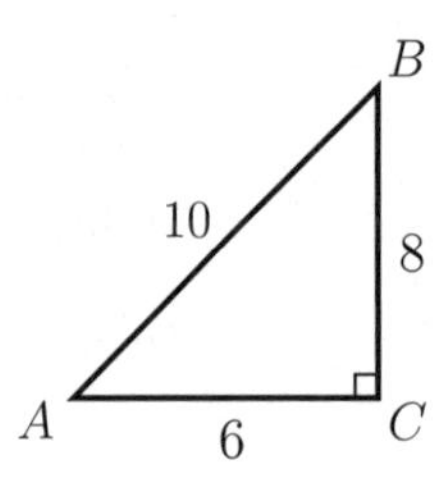

Since $\triangle ABC$ is a right triangle, use the definition of the tangent function to find the measure of $\angle BAC$.

$$\tan\angle BAC = \frac{8}{6} \quad \Longrightarrow \quad \angle BAC = \tan^{-1}\left(\frac{8}{6}\right) = 53.13°$$

Therefore, the measure of $\angle BAC$ is $53.13°$.

8. (D)

Since $g(x)$ is obtained by translating $f(x)$ 3 units to the left and 2 units down, $g(x) = f(x+3) - 2$ or $g(x) = |x+3| - 2$. In order to evaluate $g(-4)$, substitute -4 for x in $g(x)$.

$$g(x) = |x+3| - 2 \qquad \text{Substitute } -4 \text{ for } x$$
$$g(-4) = |-4+3| - 2 = -1$$

Therefore, the value of $g(-4)$ is -1.

9. (C)

> Tip: If $f(x)$ is a polynomial function, $\lim_{x\to c} f(x) = f(c)$.

$f(x) = x^3 - 4x + 4$ is a polynomial function. Thus,

$$\begin{aligned}\lim_{x\to 2} x^3 - 4x + 4 &= f(2)\\ &= (2)^3 - 4(2) + 4\\ &= 4\end{aligned}$$

Therefore, $\lim_{x\to 2} x^3 - 4x + 4 = 4$.

10. (B)

If the denominator of the rational function $f(x) = \dfrac{x+2}{x+1}$ is zero, f is undefined. Therefore, f is undefined when $x = -1$.

11. (E)

> Tip: Distance = Rate × Time

20 minutes is equal to $\frac{20\,\text{min}}{60\,\text{min}} = \frac{1}{3}$ hour. Since the car travels at 40 miles per hour for the first $\frac{1}{2}$ hour and travels at 60 miles per hour for the next $\frac{1}{3}$ hour, the total distance that the car travels is $40 \times \frac{1}{2} + 60 \times \frac{1}{3} = 40$ miles.

12. (B)

> Tip
> 1. Odd property: $\sin(-\theta) = -\sin\theta$, Even property: $\cos(-\theta) = \cos\theta$
> 2. $\sin^2\theta + \cos^2\theta = 1$

$$\begin{aligned}\sin^2(-\theta) + \cos^2(-\theta) &= (\sin(-\theta))^2 + (\cos(-\theta))^2\\ &= (-\sin\theta)^2 + (\cos\theta)^2\\ &= \sin^2\theta + \cos^2\theta\\ &= 1\end{aligned}$$

13. (C)

In order to find the inverse function, switch the x and y variables and solve for y.

$y = \log_3 x$	Switch the x and y variables
$x = \log_3 y$	Convert the equation to an exponential equation
$y = 3^x$	

Therefore, the inverse function of $y = \log_3 x$ is $f^{-1}(x) = 3^x$.

14. (C)

> Tip
> 1. Conditional statement: If P, then Q
> 2. Contrapositive: If not Q, then not P.

Note that a conditional statement and its contrapositive are logically equivalent. Since the conditional statement is "If $x = 9$, then $x^2 = 81$", the contrapositive is "If $x^2 \neq 81$, then $x \neq 9$". Thus, in order to use an indirect proof, you could begin with "If $x^2 \neq 81$". Therefore, (C) is the correct answer.

15. (D)

The value of the car depreciates 15% per year. The value of the car for the first five years is shown below.

The value of the car after 1 year: $\$20,000(1 - 0.15) = \$17,000$

The value of the car after 2 year: $\$20,000(1 - 0.15)^2 = \$14,450$

The value of the car after 3 year: $\$20,000(1 - 0.15)^3 = \$12,282.50$

The value of the car after 4 year: $\$20,000(1 - 0.15)^4 = \$10,440.10$

The value of the car after 5 year: $\$20,000(1 - 0.15)^5 = \$8,874.11$

Therefore, the value of the car after 5 years is $\$8,874.11$.

16. (D)

The point $(-7, -24)$ lies in the 3rd quadrant. Thus, the angle θ lies in the 3rd quadrant. Use the Pythagorean theorem to find the length of the hypotenuse: $c^2 = (-7)^2 + (-24)^2$. Thus, the length of the hypotenuse is 25.

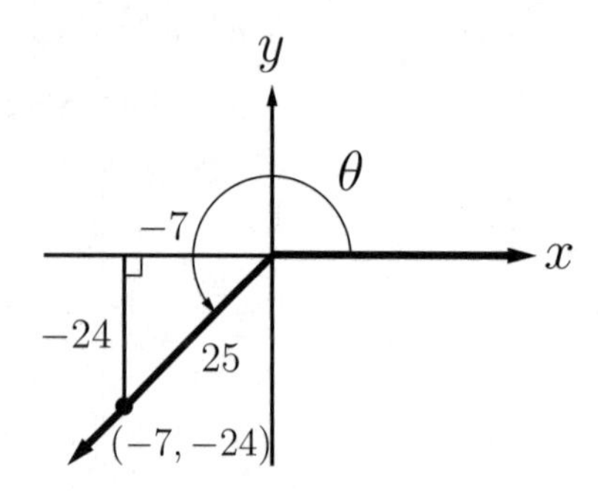

Since $\sin\theta = -\dfrac{24}{25}$, $\csc\theta = \dfrac{1}{\sin\theta} = -\dfrac{25}{24}$.

17. (A)

Tip $\quad i^2 = -1$.

$$\begin{aligned} z_1 z_2 &= (2+i)(1+3i) \\ &= 2 + 6i + i + 3i^2 \\ &= 2 - 3 + 6i + i \\ &= -1 + 7i \end{aligned}$$

Therefore, choice A best represents the product of two complex numbers, $z_1 z_2$.

18. (B)

Solving $\cos\theta < 0$ means finding the angle θ for which the cosine function lies below the x-axis.

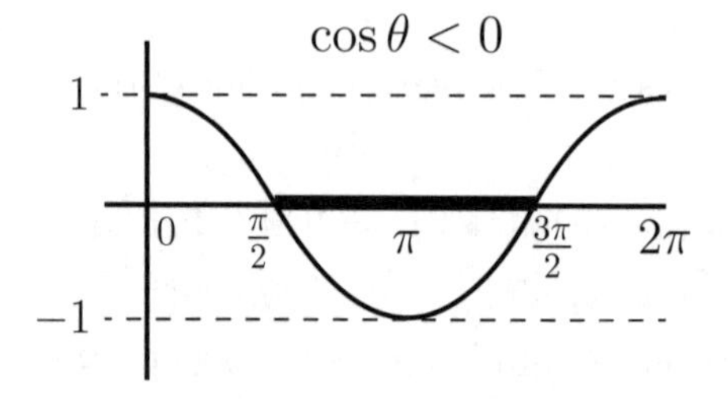

As shown in the figure above, the cosine function lies below the x-axis when $\dfrac{\pi}{2} < \theta < \dfrac{3\pi}{2}$. Therefore, the solution to $\cos\theta < 0$ is $\dfrac{\pi}{2} < \theta < \dfrac{3\pi}{2}$.

19. (C)

Tip
1. $\log_a x^n = n\log_a x$
2. $\log_a a = 1$

Note that $\log x = \log_{10} x$. Since $g(3) = 10^3$,

$$f(g(3)) = f(10^3) = \log_{10} 10^3 = 3\log_{10} 10 = 3$$

Therefore, $f(g(3)) = 3$.

20. (C)

Since the quadratic function $y = -2(x-3)^2 + 1$ is written in vertex form, the vertex of the quadratic function is $(3, 1)$.

21. (A)

The graph of the piecewise function f is shown below.

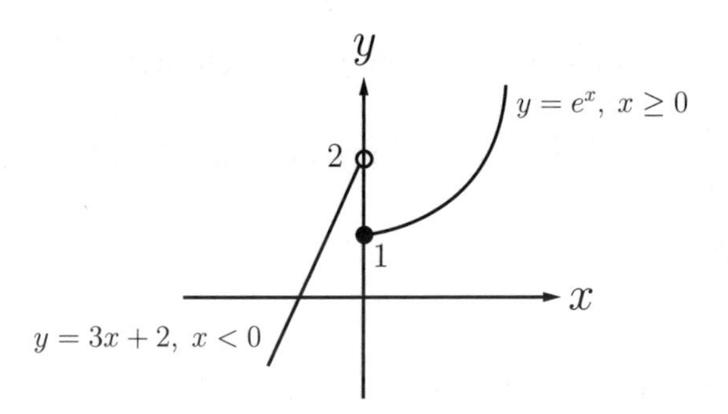

Therefore, the range of the piecewise function is all real numbers.

22. (B)

The graph of the absolute value function $f(x) = |x-5| + 2$, $1 \le x \le 7$ is shown below.

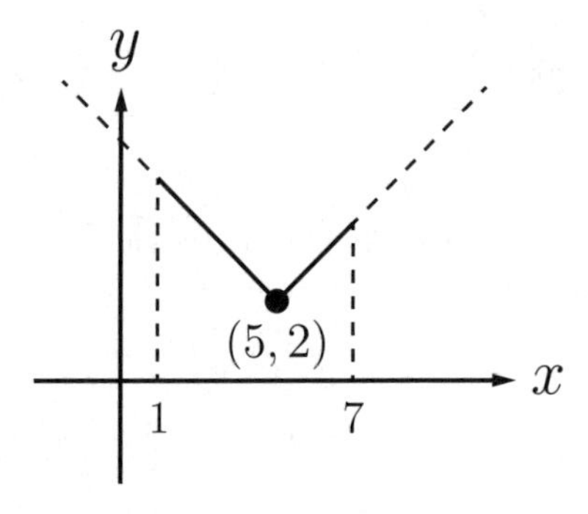

$f(1) = |1-5| + 2 = 6$ and $f(7) = |7-5| + 2 = 4$. Therefore, the maximum value of f is 6.

23. (E)

> Tip | If you multiply or divide each side of the inequality by a negative number, the inequality symbol must be reversed.

Since $p < 0$, p is a negative number.

$$pq - pr > 0 \qquad \text{Add } pr \text{ to each side}$$
$$pq > pr \qquad \text{Divide each side by } p$$
$$q < r \qquad \text{Reverse the inequality symbol}$$

Therefore, (E) is the correct answer.

24. (A)

> Tip | $\sin^2\theta = (\sin\theta)^2$

In order to find the maximum height of the projectile, substitute $v_0 = 100$, $\theta = 25°$, and $g = 32$.

$$H = \frac{v_0^2 \sin^2\theta}{2g} \qquad \text{Substitute } v_0 = 100,\ \theta = 25°, \text{ and } g = 32$$
$$H = \frac{100^2 \cdot (\sin 25°)^2}{2(32)}$$
$$= 27.91$$

Therefore, the maximum height of the projectile is 27.91 feet.

25. (E)

> Tip | $\sin 2\theta = 2\sin\theta\cos\theta$

$$\frac{3}{\sin 2\theta} = \frac{3}{2\sin\theta\cos\theta}$$
$$= \frac{3}{2(0.3)}$$
$$= 5$$

Therefore, the value of $\dfrac{3}{\sin 2\theta}$ is 5.

26. (E)

> Tip The period P of $y = -2\cos Bx$ is $P = \dfrac{2\pi}{B}$.

The amplitude and period of the trigonometric function $y = -2\cos 2x$ are 2 and $\frac{2\pi}{2} = \pi$, respectively, as shown below.

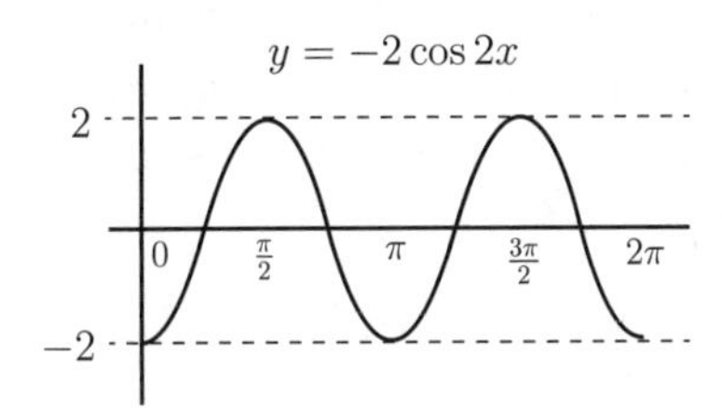

Therefore, $y = -2\cos 2x$ has a maximum value at $x = \frac{\pi}{2}$.

27. (E)

In order to form a triangle, select any three non-collinear points on the plane. Since the three points are indistinguishable, the order is not important. Thus, this is a combination problem. Therefore, the number of different triangles that can be formed by connecting the three points is $\binom{9}{3} = 84$.

28. (D)

In order to evaluate the fifth term, we need to find all previous four terms as shown below.

$a_n = -2a_{n-1} - 1,\ a_1 = -3$	Recursive formula with $a_1 = -3$
$a_2 = -2a_1 - 1 = -2(-3) - 1 = 5$	Substitute 2 for n to find a_2
$a_3 = -2a_2 - 1 = -2(5) - 1 = -11$	Substitute 3 for n to find a_3
$a_4 = -2a_3 - 1 = -2(-11) - 1 = 21$	Substitute 4 for n to find a_4
$a_5 = -2a_4 - 1 = -2(21) - 1 = -43$	Substitute 5 for n to find a_5

Therefore, the value of a_5 is -43.

29. (C)

> Tip In the system of linear equations given below,
>
> $$ax + by = c$$
> $$dx + ey = f$$
>
> the number of solutions to the system is zero if $\dfrac{a}{d} = \dfrac{b}{e} \neq \dfrac{c}{f}$.

The number of solutions to the system of linear equations below is zero if $\frac{2}{1} = \frac{6}{a}$. Thus, $a = 3$.

$$2x + 6y = 5$$
$$x + ay = -1$$

Therefore, (C) is the correct answer.

30. (D)

Tip Cofunction Identity: $\tan(90° - \theta) = \cot\theta$

The value of the tangent function is equal to the value of the cotangent function when two angles are complementary (their sum is 90°).

$$\begin{aligned} 30 + x + 70 - 3x &= 90 && \text{Sum of the two angles is } 90^\circ \\ -2x + 100 &= 90 && \text{Subtract 100 from each side} \\ -2x &= -10 && \text{Divide each side by } -2 \\ x &= 5 \end{aligned}$$

Therefore, the value of x is 5°.

31. (B)

Tip The longest diagonal of a cube with side length x is $x\sqrt{3}$.

The particle moves at a rate of $\frac{1}{2}$ feet per second, which means that it moves 1 foot in 2 seconds. Since the length of the cube is 10, the length of the path from point A to point B directly or the length of the longest diagonal is $10\sqrt{3}$ feet as shown below. It takes $2 \times 10\sqrt{3} = 34.64$ seconds for the particle to move from point A to point B directly.

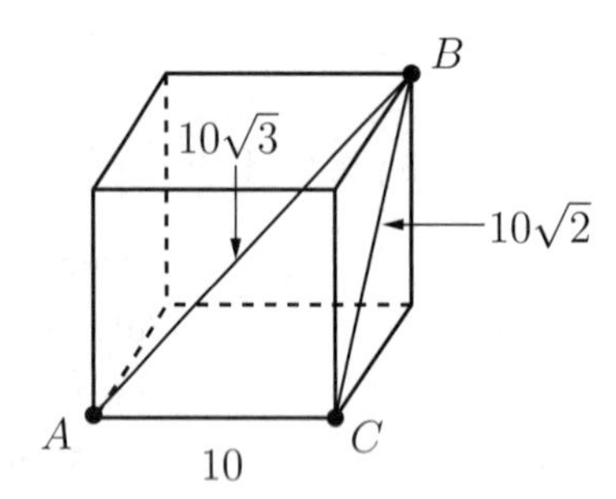

Whereas, the length of a path from point A to point C to point B is $10 + 10\sqrt{2}$ feet. It takes $2 \times (10 + 10\sqrt{2}) = 48.28$ seconds for the particle to move from point A to point C to point B. Therefore, the particle saves $48.28 - 34.64 = 13.64$ seconds if it moves from point A to point B directly.

32. (D)

Tip The general solutions of the inequality $x^2 > a^2$ are $x < -a$ or $x > a$.

The domain of the logarithmic function $y = \ln(x^2 - 9)$ is a set of all x-values for which $x^2 - 9 > 0$. Thus,

$$\begin{aligned} x^2 - 9 &> 0 && \text{Add 9 each side} \\ x^2 &> 9 && \text{Solve the inequality} \\ x < -3 \quad &\text{or} \quad x > 3 \end{aligned}$$

Therefore, the domain of $y = \ln(x^2 - 9)$ is $x < -3$ or $x > 3$.

33. (E)

The two graphs, $y = \sqrt{x-1}$ and $x = 4$, intersect on the coordinate plane as shown below.

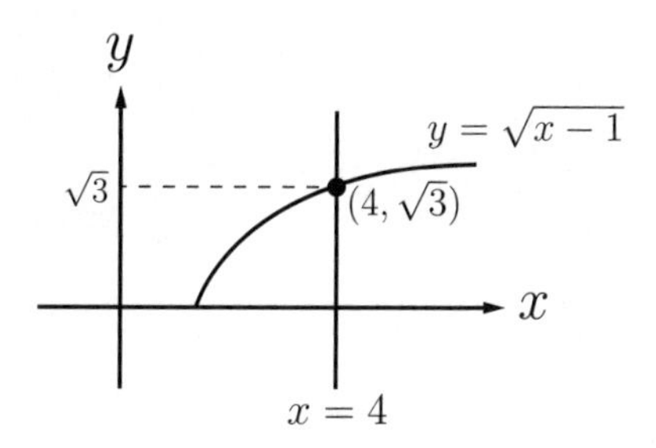

In order to find the y-coordinate of the intersection point, substitute 4 for x in $y = \sqrt{x-1}$. Thus, $y = \sqrt{4-1} = \sqrt{3}$. Therefore, the y-coordinate of the intersection point is $\sqrt{3}$ or 1.73.

34. (C)

The sum of the measures of interior angles is 180°. Thus, $m\angle B = 80°$ as shown below.

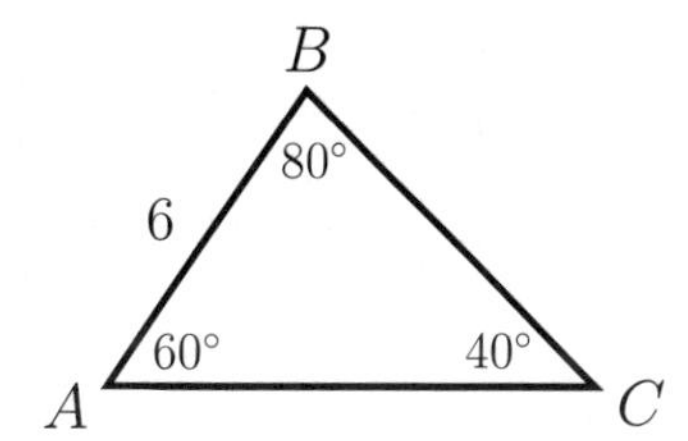

Use the Law of Sines to find AC.

$$\frac{6}{\sin 40°} = \frac{AC}{\sin 80°} \quad \Longrightarrow \quad AC = \frac{6\sin 80°}{\sin 40°} = 9.19$$

Therefore, the length of $\overline{AC}$ is 9.19.

35. (A)

Tip — Change-of-base formula: $\log_a b = \dfrac{\log_c b}{\log_c a} = \dfrac{\log_{10} b}{\log_{10} a} = \dfrac{\ln b}{\ln a}$

Since each side of the equation has a different base (the left side has a base of 5 and the right side has a base of 14), convert the exponential equation to a logarithmic equation.

$5^x = 14$ — Convert the equation to a logarithmic equation

$x = \log_5 14$

Since the solution $\log_5 14 = \dfrac{\log_2 14}{\log_2 5} = \dfrac{\log 14}{\log 5} = \dfrac{\ln 14}{\ln 5}$, (A) is the correct answer.

36. (C)

Tip Use the following properties of logarithms.

1. $\log_a a = 1$
2. $n \log_a x = \log_a x^n$
3. $a^{\log_a x} = x^{\log_a a} = x$

Since $\frac{1}{2}\log_2 x = \log_2 x^{\frac{1}{2}} = \log_2 \sqrt{x}$,

$$2^{\frac{1}{2}\log_2 x} = 2^{\log_2 \sqrt{x}} = (\sqrt{x})^{\log_2 2} = \sqrt{x}$$

Therefore, $2^{\frac{1}{2}\log_2 x} = \sqrt{x}$.

37. (B)

Tip $\cos 2x = \cos^2 x - \sin^2 x = 2\cos^2 x - 1 = 1 - 2\sin^2 x$

Since $\sin x = -\frac{2}{5}$ is given, let's use the $\cos 2x = 1 - 2\sin^2 x$ formula.

$$\cos 2x = 1 - 2\sin^2 x$$
$$\cos 2x = 1 - 2\left(-\frac{2}{5}\right)^2 = 1 - 2\left(\frac{4}{25}\right) = \frac{17}{25}$$

Therefore, the exact value of $\cos 2x$ is $\frac{17}{25}$.

38. (B)

The graph of $f(x) = x^3 - 3x - 18$ is shown below.

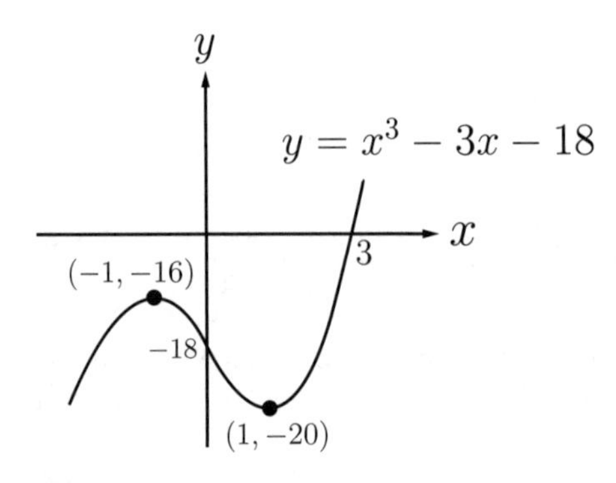

The polynomial function f has a real zero at $x = 3$ and has a y-intercept of -18. Furthermore, f is decreasing in the interval $(-1, 1)$ and $f(x)$ increases without bound as x increases without bound. However, the minimum value of f for $x \geq 0$ is -20, which can be expressed as $f(x) \geq -20$ for $x \geq 0$. Therefore, (B) is the correct answer.

39. (A)

> **Tip** Factor Theorem: If $x-k$ is a factor of $f(x)$, then the remainder $r=f(k)=0$.

Since $x+3$ is the factor of f, the remainder $r=f(-3)=0$.

$$\begin{aligned} f(x) &= 2x^3+7x^2+kx-3 && \text{Substitute } -3 \text{ for } x \\ f(-3) &= 2(-3)^3+7(-3)^2+k(-3)-3 && \text{Simplify} \\ &= 6-3k \end{aligned}$$

Since $f(-3)=0$, set $6-3k=0$ and solve for k. Therefore, the value of k is 2.

40. (D)

The domain of $f(x)=x^2,\ x\le 0$ is $x\le 0$. Since the domain of f is the range of the inverse function, the range of the inverse function is $y\le 0$. In order to find the inverse function algebraically, switch the x and y variables and solve for y.

$$\begin{aligned} y &= x^2 && \text{Switch the } x \text{ and } y \text{ variables} \\ x &= y^2 && \text{Take the square root of both sides} \\ y &= \pm\sqrt{x} && \text{Select the inverse function} \\ y &= -\sqrt{x} && \text{Since the range of the inverse function is } y\le 0 \end{aligned}$$

Therefore, the inverse function of $f(x)=x^2,\ x\le 0$ is $f^{-1}(x)=-\sqrt{x}$.

41. (E)

In the figure below, two points in the first quadrant, $(1,3)$ and $(2,6)$, are selected so that the ratio of the shortest distance to the y-axis to the shortest distance to the x-axis is $1:3$.

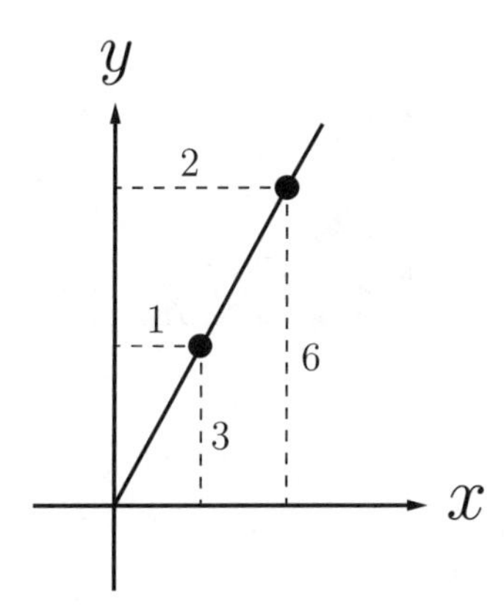

The slope and the y-intercept of the line that passes through the two points $(1,3)$ and $(2,6)$ is 3. Since the line passes through the origin, the equation of the line is $y=3x$. Therefore, a set of all ordered pairs (x,y) in the first quadrant such that the ratio of the shortest distance to the y-axis to the shortest distance to the x-axis is $1:3$ is $y=3x$.

42. (A)

In the figure below, the initial point of vector **A** is the same as the initial point of the vector **B**, and terminal point of vector **A** is the same as the terminal point of the vector **C**.

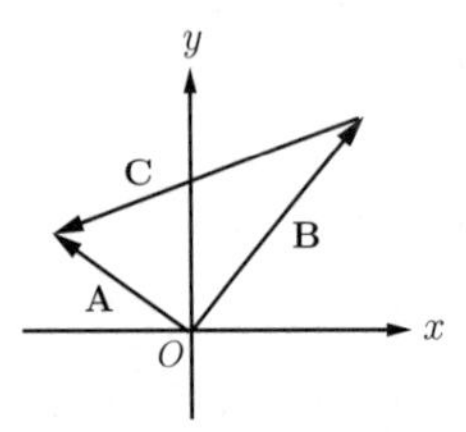

Thus, the vector A is the resultant vector. The equation that relates vectors **A**, **B**, and **C** are either $\mathbf{A} = \mathbf{B} + \mathbf{C}$ or $\mathbf{C} = \mathbf{A} - \mathbf{B}$. Therefore, (A) is the correct answer.

43. (A)

Substitute 1 and -1 for x to evaluate $f(1)$ and $f(-1)$, respectively.

$$f(1) = -2(1)^2 + 3(1) + 7 = 8, \qquad f(-1) = -2(-1)^2 + 3(-1) + 7 = 2$$

Therefore, the slope of the line that passes through the two points $(-1, f(-1))$ and $(1, f(1))$ is $\dfrac{f(1) - f(-1)}{1 - (-1)} = \dfrac{8-2}{2} = 3.$

44. (E)

Tip | Infinite sum S of geometric series: $S = \dfrac{a_1}{1-r}$, if $|r| < 1$

In order to find the heights of the first five rectangles, substitute $x = 0$, $x = 1$, $x = 2$, $x = 3$, and $x = 4$ into 2^{-x}. Thus, the heights of the first five rectangles are 1, $\dfrac{1}{2}$, $\dfrac{1}{4}$, $\dfrac{1}{8}$, and $\dfrac{1}{16}$ as shown below.

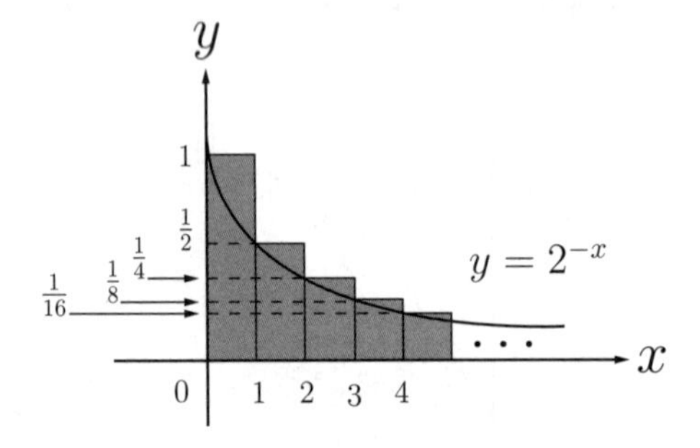

Since the length of each rectangle is 1, the sum S of the areas of the infinite number of rectangles is the geometric series with the common ration $r = \dfrac{1}{2}$.

$$S = 1 + \frac{1}{2} + \frac{1}{4} + \frac{1}{8} + \frac{1}{16} + \cdots = \frac{1}{1 - \frac{1}{2}} = 2$$

Therefore, the area of the shaded region A which is equal to the sum of the areas of the infinite number of rectangles is 2.

45. (B)

Tip	Even property: $f(-x) = f(x)$,	Odd property: $g(-x) = -g(x)$

f is an even function. So, $f(-2) = f(2) = 5$. Whereas, g is an odd function. So, $g(-5) = -g(5)$. We found that $g(5) = -g(-5) = -2$. Thus, $g(f(2)) = g(5) = -2$. Therefore, the value of $g(f(2))$ is -2.

46. (C)

A function f is said to be increasing if $f(x_1) > f(x_2)$ when $x_1 > x_2$. So eliminate answer choice (A). A function f is said be constant if $f(x_1) = f(x_2)$ when $x_1 \neq x_2$. So eliminate answer choice (B). A function f is said to be decreasing if $f(x_1) > f(x_2)$ when $x_1 < x_2$. So eliminate answer choice (D). All vertical lines are not functions. So eliminate answer choice (E). Therefore, (C) is the correct answer.

47. (E)

Use a calculator to evaluate $\cos^{-1}(0.45)$. Thus, $\cos^{-1}(0.45) = 63.26°$. Thus,

$$\sin(\cos^{-1}(0.45)) = \sin 63.26° = 0.89$$

Therefore, the value of $\sin(\cos^{-1}(0.45))$ is 0.89.

48. (C)

Tip	
	1. nth term of arithmetic sequence: $a_n = a_1 + (n-1)d$
	2. nth partial sum of arithmetic series: $S_n = \frac{n}{2}(a_1 + a_n)$

The first row has 12 seats. The second row has 14 seats, the third row has 16 seats. Thus, $12, 14, 16, \cdots$ is the arithmetic sequence with the common difference of 2. In order to find out how many seats the 20th row has, find a_{20} using the nth term formula.

$$\begin{aligned} a_n &= a_1 + (n-1)d && \text{Substitute 20 for } n\text{, 12 for } a_1\text{, and 2 for } d \\ a_{20} &= 12 + (20-1)2 \\ a_{20} &= 50 \end{aligned}$$

Thus, the 20th row has 50 seats. Use the nth partial sum of arithmetic series formula to find the total number of seats in the theater, $12 + 14 + \cdots + 50$.

$$\begin{aligned} S_n &= \frac{n}{2}(a_1 + a_n) && \text{Substitute 20 for } n \\ S_{20} &= \frac{20}{2}(a_1 + a_{20}) && \text{Substitute 12 for } a_1 \text{ and 50 for } a_{20} \\ &= \frac{20}{2}(12 + 50) \\ &= 620 \end{aligned}$$

Therefore, the total number of seats in the theater is 620.

49. (C)

Tip

The table below shows the powers of i.

Powers of i	i	i^2	i^3	i^4	i^5	i^6	i^7	i^8	i^9	i^{10}	i^{11}	i^{12}
Value	i	-1	$-i$	1	i	-1	$-i$	1	i	-1	$-i$	1

The powers of i repeat in a pattern: i, -1, $-i$, and 1. Thus, $i+i^2+i^3+i^4=0$, $i^5+i^6+i^7+i^8=0$, and $i^9+i^{10}+i^{11}+i^{12}=0$. Therefore, $i+i^2+i^3+\cdots+i^{11}+i^{12}=0$.

50. (D)

Out of 10 students, Mr. Rhee selects 5 students. Since 10 students are indistinguishable, the number of ways to select 5 students out of 10 students is $\binom{10}{5}$. Define event 1 as selecting 2 boys out of 4 boys. Since 4 boys are indistinguishable, the number of ways to select 2 boys out of 4 boys is $\binom{4}{2}$. Define event 2 as selecting 3 girls out of 6 girls. Since 6 girls are indistinguishable, the number of ways to select 3 girls out of 6 girls is $\binom{6}{3}$. Each team that Mr. Rhee selects consists of 2 boys and 3 girls. Thus, according to the fundamental counting principle, the number of different teams of 2 boys and 3 girls that Mr. Rhee can select is $\binom{4}{2}\times\binom{6}{3}$. Therefore, the probability that Mr. Rhee selects a team of 2 boys and 3 girls in his class is $\dfrac{\binom{4}{2}\times\binom{6}{3}}{\binom{10}{5}}$.

SAT II MATH LEVEL 2 TEST 2

Directions: Among the given answer choices, choose the BEST answer for each problem. If the exact numerical value is not within the given answer choices, select the answer that best approximates this value. Afterwards, fill in the corresponding oval on the answer sheet.

Notes:

1. A calculator may be required to answer some of the questions. Scientific and graphing calculators are allowed during this test.
2. For some questions, it is up to you to decide whether the calculator should be in degree mode or radian mode.
3. Provided figures for questions are drawn as accurately as possible UNLESS otherwise specified by the problem. Unless otherwise indicated, all figures are assumed to lie in a plane.
4. Unless otherwise specified, it can be assumed that the domain of any function f is to be the set of all real numbers x for which $f(x)$ is a real number.

Reference Information: Use the following information and formulas as a reference in answering questions on this test.

1. If the radius and height of a right circular cone are r and h respectively, then the Volume V of the cone is $V = \frac{1}{3}\pi r^2 h$.
2. If the circumference and slant height of a right circular cone are c and ℓ respectively, then the Lateral Area A of the cone is $A = \frac{1}{2}c\ell$.
3. At any given radius r, the Volume V of a sphere is $V = \frac{4}{3}\pi r^3$.
4. At any given radius r, the Surface Area A of a sphere is $A = 4\pi r^2$.
5. The Volume V of a pyramid is $V = \frac{1}{3}Bh$; given that B and h represent the base area and height of the pyramid respectively.

USE THIS SPACE FOR SCRATCH WORK

1. If x and y are positive real numbers and $\frac{x}{y} < 1$, which of the following inequality must be true?

(A) $x - y < 0$

(B) $x + y < 1$

(C) $y - x < 0$

(D) $xy < 1$

(E) $\frac{y}{x} < 1$

2. If $f(x) = -x^2 + 1$, which of the following expression is equal to $f(x+h) - f(x)$?

(A) $2xh + h^2$

(B) $2xh - h^2$

(C) $-h^2$

(D) $-2xh - h^2$

(E) $-2xh + h^2$

3. Which of the following function is symmetric with respect to the y-axis?

(A) $\sin x$

(B) $\cos x$

(C) $\tan x$

(D) x^3

(E) $x^2 + 2x + 3$

USE THIS SPACE FOR SCRATCH WORK

4. Which of the following measures in radians is equal to $105°$?

(A) $\frac{2\pi}{3}$

(B) $\frac{5\pi}{12}$

(C) $\frac{3\pi}{4}$

(D) $\frac{7\pi}{12}$

(E) $\frac{5\pi}{6}$

5. $\frac{(n+1)!}{(n-1)!} =$

(A) $\frac{1}{n(n+1)}$

(B) $\frac{1}{n^2}$

(C) $n(n-1)$

(D) n^2

(E) $n(n+1)$

6. The equation $x^2 + y^2 + 4x - 2y - 3 = 0$ represents a circle. What is the center of the circle?

(A) $(4, -2)$

(B) $(2, -1)$

(C) $(-4, -2)$

(D) $(-2, -1)$

(E) $(-2, 1)$

USE THIS SPACE FOR SCRATCH WORK

7. A number N of bacteria is defined by the function $N(t) = N_0 e^{0.35t}$, where N_0 is the initial number of bacteria, and t is a time in hours. If the initial number of bacteria is 100, what is the number of bacteria after 90 minutes?

   (A) 135

   (B) 142

   (C) 169

   (D) 197

   (E) 213

8. If $f(x) = 6x + 4$ and $g(f(x)) = 3x + 1$, which of the following function must be $g(x)$?

   (A) $\frac{x}{2} - 1$

   (B) $\frac{x}{3} + 2$

   (C) $\frac{x-1}{x+1}$

   (D) $2x + 1$

   (E) $3x - 2$

9. Which of the following quadratic function has a zero of $2 - \sqrt{3}$?

   (A) $x^2 - 2x + 3$

   (B) $x^2 + 2x - 3$

   (C) $x^2 + 4x + 3$

   (D) $x^2 - 4x + 1$

   (E) $x^2 - 4x - 1$

USE THIS SPACE FOR SCRATCH WORK

10. Which of the following exponential function best represents the graph in Figure 1?

(A) $y = \left(\frac{1}{2}\right)^{x-1} - 3$

(B) $y = \left(\frac{1}{3}\right)^{x+1} + 2$

(C) $y = 2^{x-1} - 3$

(D) $y = 2^{x+1} - 3$

(E) $y = 3^{x-1} - 2$

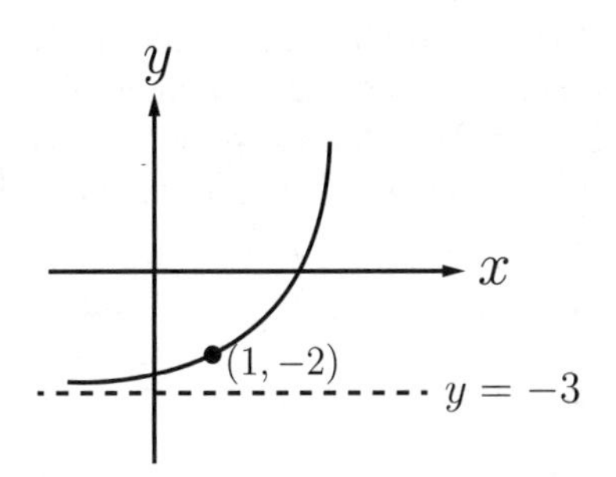

Figure 1

11. $\lim_{x \to \infty} \dfrac{2x^2 - x - 1}{x^2 - 3} =$

(A) $y = 3$

(B) $y = 2$

(C) $y = \frac{1}{2}$

(D) $y = \frac{1}{3}$

(E) $y = 0$

12. A statistician asked a random sample of 187 U.S. adults whether they own a tablet computer. Of the respondents, 79 said "Yes." The statistician used the sample to estimate the total number of U.S. adults who own a tablet computer. If the population of U.S. adults is 192 million, which of the following best approximates the total number of U.S. adults who own a tablet computer?

(A) 75 million

(B) 77 million

(C) 79 million

(D) 81 million

(E) 83 million

USE THIS SPACE FOR SCRATCH WORK

13. The greatest integer function $f(x) = [\,x\,]$ is the function that assigns each number to the greatest integer less than or equal to that number. If a part of the graph of f is shown in Figure 2, what is the domain of f ?

    (A) Positive integers

    (B) Whole numbers

    (C) Integers

    (D) Rational numbers

    (E) All real numbers

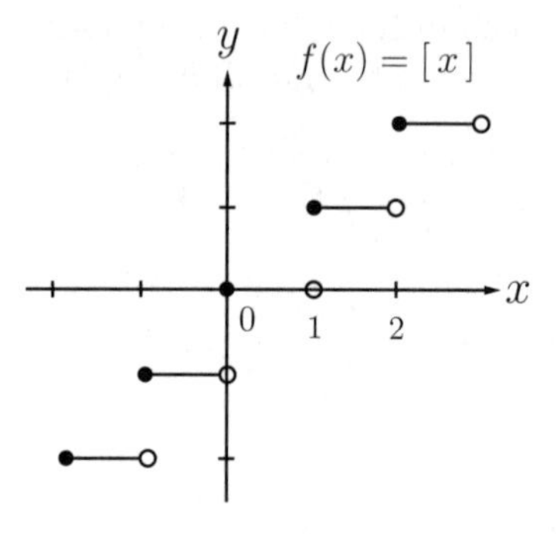

Figure 2

14. Which of the following lines are the asymptotes of the rational function $y = \dfrac{x-2}{x^2-4}$?

    I. $x = 2$

    II. $x = -2$

    III. $y = 0$

    (A) I only

    (B) I and II only

    (C) II and III only

    (D) I and III only

    (E) I, II, and III

15. Which of the following equation is perpendicular to the line that passes through the points $(-1, 1)$ and $(2, -5)$?

    (A) $x + 2y = 6$

    (B) $x - 2y = 6$

    (C) $2x + y = -6$

    (D) $2x - y = -6$

    (E) $3x + 2y = -6$

USE THIS SPACE FOR SCRATCH WORK

16. A particle is moving along a curve defined by the parametric equations $x = \sin t$, $y = \cos t$, where $0 \le t \le 2\pi$. Where is the particle located at $t = \dfrac{5\pi}{6}$?

(A) $(0.87, 0.5)$

(B) $(0.87, -0.5)$

(C) $(0.5, -0.87)$

(D) $(-0.5, 0.87)$

(E) $(-0.5, -0.87)$

17. In $\triangle ABC$, shown in Figure 3, $m\angle A = 64°$ and $m\angle B = 32°$. If $AC = 1$, what is BC ?

(A) 1.64

(B) 1.67

(C) 1.70

(D) 1.73

(E) 1.76

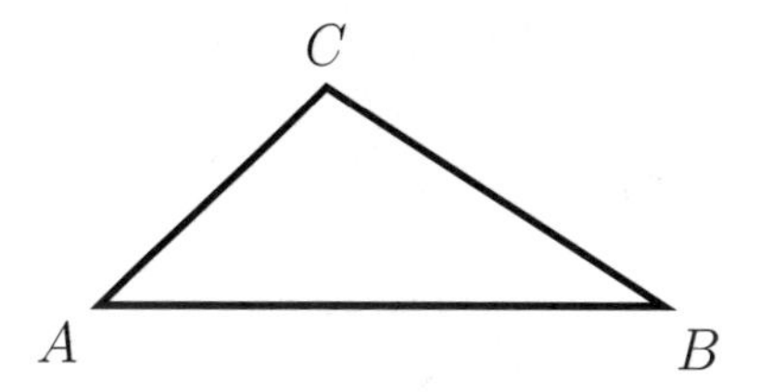

Figure 3

18. $\tan x + \cot x =$

(A) $\csc x \sec x$

(B) $\csc x \cos x$

(C) $\sin x \sec x$

(D) $\sin x \cos x$

(E) 1

USE THIS SPACE FOR SCRATCH WORK

19. The graph of f is shown in Figure 4. Which of the following graph represents $|f(x)|$?

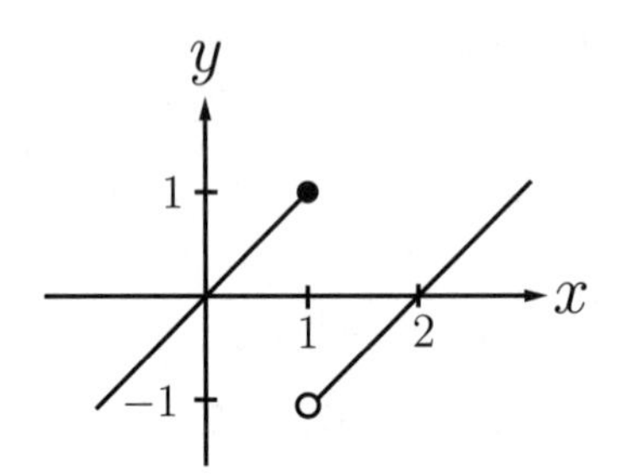

Figure 4

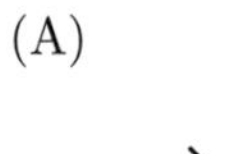
(A)

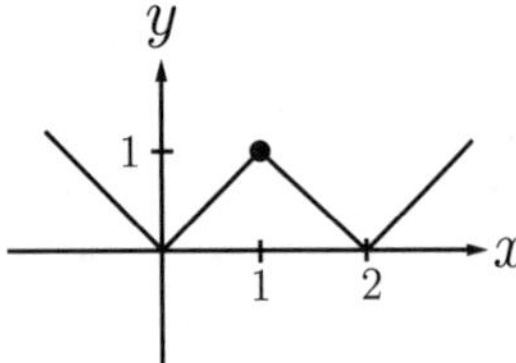

(B)

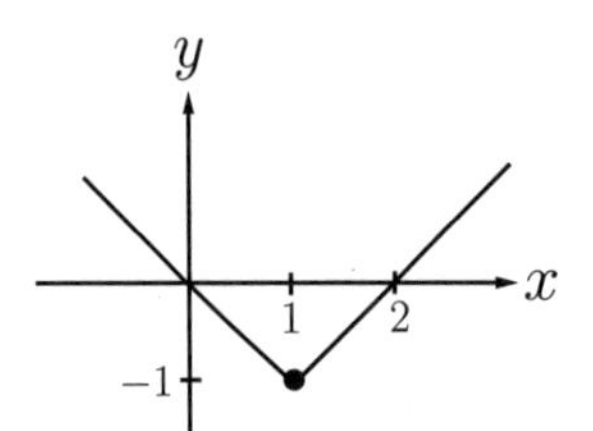

(C)

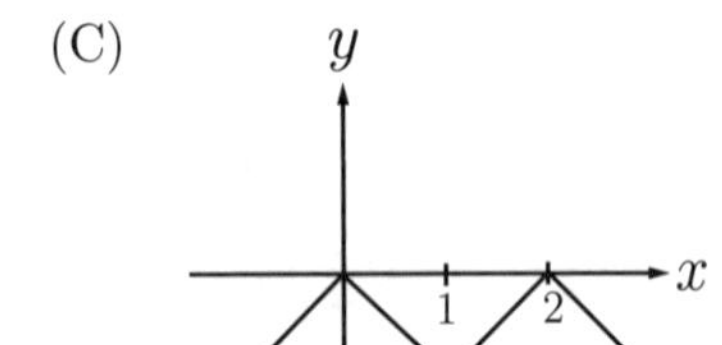

(D)

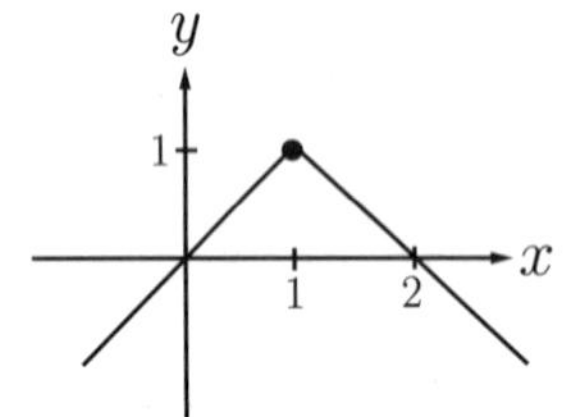

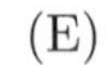
(E)

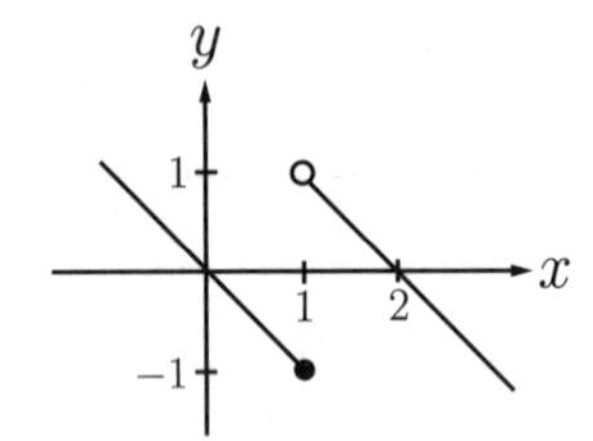

20. What are the solutions to $3x^2 + 2x = 4$?

(A) $x = 1.54$ or $x = 0.87$

(B) $x = 1.27$ or $x = 0.63$

(C) $x = 0.96$ or $x = -1.29$

(D) $x = 0.87$ or $x = -1.54$

(E) $x = 0.62$ or $x = -2.21$

USE THIS SPACE FOR SCRATCH WORK

21. $\dfrac{1}{x-3} - \dfrac{6}{x^2-9} =$

(A) $\dfrac{5}{x^2-9}$

(B) $\dfrac{3-x}{x^2-9}$

(C) $\dfrac{x+9}{x^2-9}$

(D) $\dfrac{1}{x-3}$

(E) $\dfrac{1}{x+3}$

22. $\displaystyle\sum_{n=1}^{10}(-1)^n 2n =$

(A) 6

(B) 8

(C) 10

(D) 24

(E) 42

23. From the equation of the ellipse $\dfrac{x^2}{a^2} + \dfrac{y^2}{b^2} = 1$, the area A of the ellipse can be determined by $A = \pi ab$. If an ellipse is defined by $16x^2 + 9y^2 = 144$, what is the area of the ellipse?

(A) 6π

(B) 12π

(C) 48π

(D) 72π

(E) 144π

USE THIS SPACE FOR SCRATCH WORK

24. What is the inverse function of $y = 7^{x-3}$?

(A) $\log_7(x+3)$

(B) $\log_7 x + 3$

(C) $\log_3 x - 7$

(D) $\log_{\frac{1}{3}}(x+7)$

(E) $\log_{\frac{1}{7}} x + 3$

25. The two complex numbers $z_1 = 2y + (x+6)i$ and $z_2 = (3-x) + yi$ are equal. If $z_3 = x + yi$, what is the value of z_3 ?

(A) $6 + 3i$

(B) $3 - 3i$

(C) $-3 + 3i$

(D) $-3 - 3i$

(E) $-6 + 3i$

26. Figure 5 shows the graph of a cubic function whose zeros are -1 and 3. Which of the following cubic function best represents the graph in Figure 5?

(A) $(x+1)(x-1)(x-3)$

(B) $(x+1)^2(x-3)$

(C) $-(x-1)(x+3)^2$

(D) $-(x+1)^2(x-3)$

(E) $-(x+1)(x-3)^2$

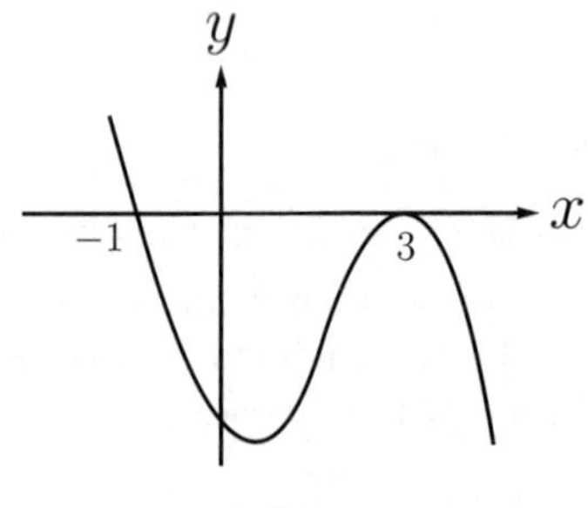

Figure 5

USE THIS SPACE FOR SCRATCH WORK

27. $\log_2 \sqrt{2} - \log_4 2 + \log_8 2 =$

(A) -2

(B) $-\dfrac{1}{2}$

(C) $\dfrac{1}{3}$

(D) $\sqrt{2}$

(E) 3

28. In Figure 6, $\triangle ABC$ is a right triangle. If $BC = a$, $AC = b$, and $m\angle ACB = \theta$, what is $\sin\theta$ in terms of a and b ?

(A) $\dfrac{\sqrt{b^2 - a^2}}{a}$

(B) $\dfrac{\sqrt{b^2 - a^2}}{b}$

(C) $\dfrac{\sqrt{b^2 + a^2}}{b}$

(D) $\dfrac{a}{\sqrt{b^2 - a^2}}$

(E) $\dfrac{b}{\sqrt{b^2 + a^2}}$

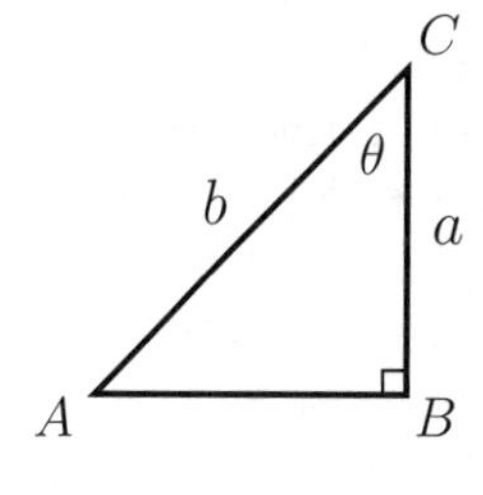

Figure 6

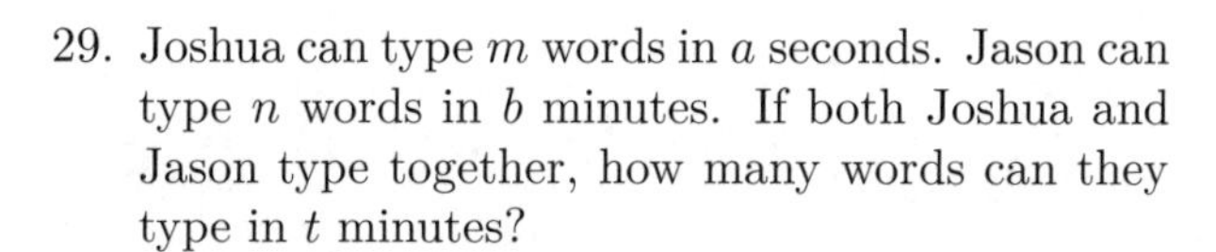

29. Joshua can type m words in a seconds. Jason can type n words in b minutes. If both Joshua and Jason type together, how many words can they type in t minutes?

(A) $t(60ma + bn)$

(B) $t(ma + bn)$

(C) $t\left(\dfrac{a}{60m} + \dfrac{b}{n}\right)$

(D) $t\left(\dfrac{m}{a} + \dfrac{n}{b}\right)$

(E) $t\left(\dfrac{60m}{a} + \dfrac{n}{b}\right)$

USE THIS SPACE FOR SCRATCH WORK

30. How long does it take for an initial amount to double in value if it is invested at 8% compounded annually? (Round your answer to the nearest integer.)

(A) 9 years

(B) 10 years

(C) 11 years

(D) 12 years

(E) 13 years

31. The region R is enclosed by the line $2x+3y=6$, the x-axis, and the y-axis. If the region R is rotated about the y-axis, what is the volume of the resulting solid?

(A) 4π

(B) 6π

(C) 8π

(D) 12π

(E) 15π

32. $\sec^2\left(\frac{11\pi}{12}\right) - \tan^2\left(\frac{11\pi}{12}\right) =$

(A) $\frac{\sqrt{3}}{3}$

(B) $\frac{\sqrt{2}}{2}$

(C) 1

(D) $\frac{1}{2}$

(E) $\frac{1}{3}$

USE THIS SPACE FOR SCRATCH WORK

33. Which of the following set of numbers has the smallest standard deviation?

    (A) $0, 5, 10, 15, 20$

    (B) $2, 6, 10, 14, 18$

    (C) $4, 7, 10, 13, 16$

    (D) $6, 8, 10, 12, 14$

    (E) $8, 9, 10, 11, 12$

34. Which of the following rectangular equation is equivalent to $r = 3\csc\theta$?

    (A) $y = 3$

    (B) $x = 3$

    (C) $x^2 + y^2 = 3$

    (D) $x^2 + y^2 = 3x$

    (E) $x^2 + y^2 = 3y$

35. The shaded region in Figure 7 represents the set S of points (x, y) in the triangular region. If $f : (x, y) \to (x - 1, y + 2)$ for every pair (x, y) in the shaded region, what is the area of the mapping of set S by f ?

    (A) 64

    (B) 48

    (C) 32

    (D) 24

    (E) 16

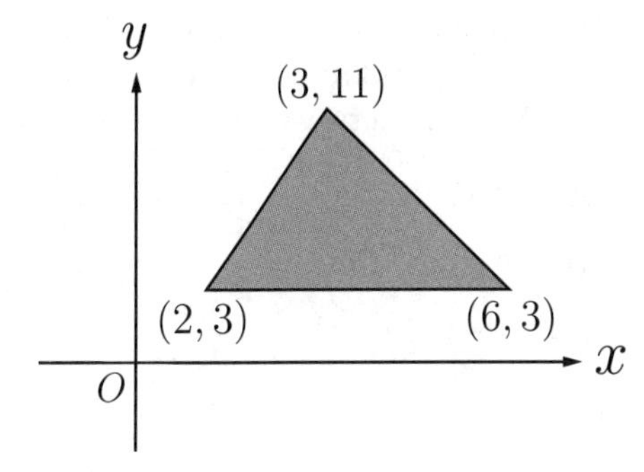

Figure 7

USE THIS SPACE FOR SCRATCH WORK

36. In the right triangle ABC, $m\angle C = 90°$. If $\sin B = 0.75$, what is the value of $\tan A$?

(A) 2.35

(B) 1.78

(C) 1.13

(D) 0.88

(E) 0.62

37. What is the sum of the infinite geometric series $\frac{2}{3} - \frac{1}{3} + \frac{1}{6} + \cdots$?

(A) 2

(B) $\frac{4}{3}$

(C) $\frac{4}{9}$

(D) $\frac{1}{4}$

(E) $\frac{1}{5}$

38. If the line $y = -x + 3$ intersects both lines $y = 2x$ and $y = 2x - 3$ at point A and point B, what is the distance between points A and B ?

(A) 2.45

(B) 2.23

(C) 1.73

(D) 1.41

(E) 1.21

USE THIS SPACE FOR SCRATCH WORK

39. In Figure 8, $\triangle ABC$ is inscribed in the semicircle with radius 5. If $\theta = 65°$, what is the area of $\triangle ABC$?

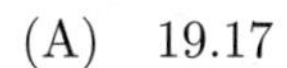

(A) 19.17

(B) 21.45

(C) 26.39

(D) 30.57

(E) 36.85

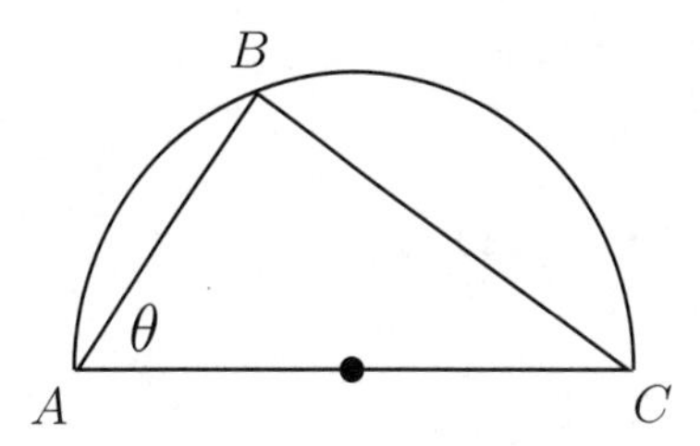

Figure 8

40. Solve: $x - 3 = \sqrt{2x - 3}$

(A) $x = -2$ only

(B) $x = 6$ only

(C) $x = -2$ and $x = 6$

(D) $x = 2$ and $x = -6$

(E) $x = 6$ and $x = -6$

41. If $\sin\theta = \frac{3}{5}$, $\frac{\pi}{2} < \theta < \pi$, what is the value of $\cos 2\theta$?

(A) $-\frac{24}{25}$

(B) $-\frac{16}{25}$

(C) $-\frac{7}{25}$

(D) $\frac{7}{25}$

(E) $\frac{24}{25}$

USE THIS SPACE FOR SCRATCH WORK

42. Simplify: $2\ln x - 3\ln y + \dfrac{1}{2}\ln z$

(A) $\ln \dfrac{x^2y^3}{\sqrt{z}}$

(B) $\ln \dfrac{x^2\sqrt{z}}{y^3}$

(C) $\ln \dfrac{x^2}{y^3\sqrt{z}}$

(D) $\dfrac{\ln(x^2y^3)}{\ln\sqrt{z}}$

(E) $\dfrac{\ln(x^2\sqrt{z})}{\ln y^3}$

43. If the ratio of the three sides of a triangle is $5:5:8$, what is the largest angle of the triangle? (Round your answer to the nearest integer)

(A) $122°$

(B) $114°$

(C) $106°$

(D) $92°$

(E) $80°$

44. What is the domain of $y = \dfrac{2}{\sqrt[3]{1-2x}}$?

(A) $x \geq \dfrac{1}{2}$

(B) $x > \dfrac{1}{2}$

(C) $x < \dfrac{1}{2}$

(D) All real numbers

(E) All real numbers except $x = \dfrac{1}{2}$

USE THIS SPACE FOR SCRATCH WORK

45. What is the period of $y = |\sin 2x|$?

(A) 4π

(B) 2π

(C) π

(D) $\dfrac{\pi}{2}$

(E) $\dfrac{\pi}{4}$

46. Suppose $f(x) = \sqrt{2x+4}$ and $g(x) = \dfrac{1}{2}x^2 - 2$.
If $x < 0$, what is $f(g(x))$?

(A) $|x|$

(B) x

(C) $-x$

(D) $\sqrt{x^2 - 4}$

(E) Undefined

47. What are the real zeros of $f(x) = x^3 - 3x^2 - 4x + 12$?

(A) $x = -2, 1, 2$

(B) $x = -2, 2, 3$

(C) $x = -1, 0, 2$

(D) $x = 0, 1, 3$

(E) $x = 1, 2, 3$

USE THIS SPACE FOR SCRATCH WORK

48. Jason rolls two dice to form a two-digit integer. If the number on the first die represents the tens digit and the number on the second die represents the units digit, what is the probability that the integer formed is divisible by 8?

(A) $\frac{2}{9}$

(B) $\frac{7}{36}$

(C) $\frac{1}{6}$

(D) $\frac{5}{36}$

(E) $\frac{1}{9}$

49. When the two circles $x^2+y^2=4$ and $(x-3)^2+y^2=4$ intersect in the xy-coordinate plane, there are two intersection points. What are the x and y-coordinates of the two intersection points?

(A) $(1.5, 1.32)$ and $(1.5, -1.32)$

(B) $(1.5, 1.75)$ and $(1.5, -1.75)$

(C) $(-1.5, 1.32)$ and $(-1.5, -1.32)$

(D) $(-1.5, 1.75)$ and $(-1.5, -1.75)$

(E) $(1.25, 1.32)$ and $(-1.25, 1.32)$

USE THIS SPACE FOR SCRATCH WORK

50. Figure 9 shows a right prism with triangular bases. The prism has the five faces. The height of the prism is 2 feet, and the length of the two sides of the triangular bases are 3 feet and 4 feet. What is the surface area of the prism?

    (A) 54 ft^2

    (B) 48 ft^2

    (C) 42 ft^2

    (D) 36 ft^2

    (E) 30 ft^2

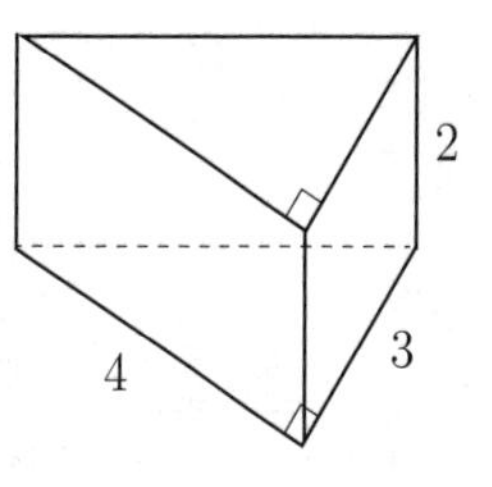

Figure 9

STOP

Mathematics Scoring Worksheet

Directions: In order to calculate your score correctly, fill out the table below. After calculating your raw score, round the raw score to the nearest whole number. The scaled score can be determined using the "Math Test Score Conversion Table".

Mathematics Score			
A. Number Correct		**B.** Number Incorrect ÷ 4	
Total Unrounded Raw Score $A - B$		**Total Rounded Raw Score** **Round to nearest whole number**	

Math Test Score Conversion Table					
Raw Score	Scaled Score	Raw Score	Scaled Score	Raw Score	Scaled Score
50	800	29	660	8	490
49	800	28	650	7	480
48	800	27	640	6	480
47	800	26	630	5	470
46	800	25	630	4	460
45	800	24	620	3	450
44	800	23	610	2	440
43	800	22	600	1	430
42	790	21	590	0	410
41	780	20	580	−1	390
40	770	19	570	−2	370
39	760	18	560	−3	360
38	750	17	560	−4	340
37	740	16	550	−5	340
36	730	15	540	−6	330
35	720	14	530	−7	320
34	710	13	530	−8	320
33	700	12	520	−9	320
32	690	11	510	−10	320
31	680	10	500	−11	310
30	670	9	500	−12	310

ANSWERS AND SOLUTIONS

Answers

1. A	11. B	21. E	31. B	41. D
2. D	12. D	22. C	32. C	42. B
3. B	13. E	23. B	33. E	43. C
4. D	14. C	24. B	34. A	44. E
5. E	15. B	25. C	35. E	45. D
6. E	16. C	26. E	36. D	46. C
7. C	17. C	27. C	37. C	47. B
8. A	18. A	28. B	38. D	48. D
9. D	19. A	29. E	39. A	49. A
10. C	20. D	30. A	40. B	50. D

Solutions

1. (A)

Multiply each side of the inequality $\frac{x}{y} < 1$ by y. Since y is the positive integer, the inequality symbol remains the same.

$$\frac{x}{y} < 1 \qquad \text{Multiply each side by } y$$
$$x < y \qquad \text{Subtract each side by } y$$
$$x - y < 0$$

Therefore, (A) is the correct answer.

2. (D)

In order to evaluate $f(x+h)$, substitute $x+h$ for x in $f(x) = -x^2+1$. We found that $f(x+h) = -(x+h)^2+1 = -x^2-2xh-h^2+1$. Thus,

$$\begin{aligned} f(x+h) - f(x) &= -(x+h)^2 + 1 - (-x^2+1) \\ &= -x^2 - 2xh - h^2 + 1 + x^2 - 1 \\ &= -2xh - h^2 \end{aligned}$$

Therefore, $f(x+h) - f(x) = -2xh - h^2$.

3. (B)

> Tip
> 1. $f(x)$ is an odd function when $f(-x) = -f(x)$ for all values of x. Any odd function is symmetric with respect to the origin. Whereas, $f(x)$ is an even function when $f(-x) = f(x)$ for all values of x. Any even function is symmetric with respect to the y-axis.
> 2. $\cos x$ is an even function and is symmetric with respect to the y-axis since $\cos(-x) = \cos x$.

$\sin x$, $\tan x$ and x^3 are odd functions. So eliminate answer choices (A), (C), and (D). Let $f(x) = x^2 + 2x + 3$.

$$f(-x) = (-x)^2 + 2(-x) + 3 = x^2 - 2x + 3$$
$$-f(x) = -(x^2 + 2x + 3) = -x^2 - 2x - 3$$

Since $f(-x) \neq -f(x)$ and $f(-x) \neq f(x)$, $f(x)$ does not satisfy the definition of an odd function or an even function. Thus, $f(x) = x^2 + 2x + 3$ is neither. So eliminate answer choice (E). Therefore, (B) is the correct answer.

4. (D)

> Tip In order to convert degrees to radians, multiply degrees by $\frac{\pi}{180°}$.

$$105° \times \frac{\pi}{180°} = \frac{105°}{180°}\pi = \frac{7\pi}{12}$$

5. (E)

$$\frac{(n+1)!}{(n-1)!} = \frac{(n+1) \cdot n \cdot (n-1)!}{(n-1)!} = n(n+1)$$

6. (E)

> Tip
> 1. The general equation of a circle: $(x-h)^2 + (y-k)^2 = r^2$.
> 2. In order to avoid a common mistake when finding the center of conics (a circle, an ellipse and a hyperbola), set $x - h = 0$ and $y - k = 0$ and solve for x and y. Thus, the x and y coordinates of the center of the conics are $x = h$ and $y = k$.

In order to write the general equation of the circle $x^2 + y^2 + 4x - 2y - 3 = 0$, complete the squares in x and y.

$x^2 + y^2 + 4x - 2y - 3 = 0$

$x^2 + y^2 + 4x - 2y = 3$	Add 3 to each side
$x^2 + 4x + y^2 - 2y = 3$	Rearrange the terms
$(x+2)^2 + y^2 - 2y = 7$	Add 4 to each side to complete squares in x
$(x+2)^2 + (y-1)^2 = 8$	Add 1 to each side to complete squares in y

In order to find the center of the circle, set $x + 2 = 0$ and $y - 1 = 0$ and solve for x and y. Thus, $x = -2$ and $y = 1$. Therefore, the center of the circle is $(-2, 1)$).

7. (C)

Since t is a time in hours, change 90 minutes to 1.5 hours. In order to find the number of bacteria after 90 minutes, substitute 100 for N_0 and 1.5 for t.

$$N(t) = N_0 e^{0.35t} \qquad \text{Substitute 100 for } N_0 \text{ and 1.5 for } t$$
$$N(1.5) = 100e^{0.35\times1.5} = 169$$

Therefore, the number of bacteria after 90 minutes is 169.

8. (A)

In order to find the function $g(x)$ for which $g(f(x)) = 3x + 1$, substitute $6x + 4$ for x in each answer choice. Let $g(x) = \dfrac{x}{2} - 1$ in answer choice (A). Then,

$$\begin{aligned} g(f(x)) &= g(6x+4) \\ &= \frac{6x+4}{2} - 1 \\ &= 3x + 1 \end{aligned}$$

Therefore, $g(x) = \dfrac{x}{2} - 1$.

9. (D)

Tip

1. The conjugate pairs theorem states that complex zeros and irrational zeros always occur in conjugate pairs.
2. Vieta's formulas relate the coefficients of a polynomial to the sum and product of its zeros and are described below. For a quadratic function $f(x) = x^2 + bx + c$, let z_1 and z_2 be the zeros of f.

$$z_1 + z_2 = -b \quad \text{Sum of zeros equals the opposite of the coefficient of } x$$
$$z_1 z_2 = c \quad \text{Product of zeros equals the constant term}$$

According to the conjugate pairs theorem, $2 + \sqrt{3}$ is also a zero of the quadratic function. Thus, $2 + \sqrt{3}$ and $2 - \sqrt{3}$ are zeros of the quadratic function. Since the irrational zeros are given, use Vieta's formulas to write a quadratic function with leading coefficient 1.

$$\text{Sum of zeros: } (2+\sqrt{3}) + (2-\sqrt{3}) = 4 \xrightarrow{\text{Opposite}} -4 \text{ (Coefficient of } x)$$
$$\text{Product of zeros: } (2+\sqrt{3})(2-\sqrt{3}) = 1 \xrightarrow{\text{Same}} 1 \text{ (Constant term)}$$

Therefore, the quadratic function whose zeros are $2 + \sqrt{3}$ and $2 - \sqrt{3}$ is $x^2 - 4x + 1$.

10. (C)

The graph in Figure 1 contains the point $(1, -2)$. Thus, substitute 1 for x and -2 for y in each answer choice to see if the equation holds true. As shown below, we found that only two functions in answer choices (A) and (C) contain the point $(1, -2)$.

$$y = \left(\frac{1}{2}\right)^{x-1} - 3 \xrightarrow{\text{When } x = 1 \text{ and } y = -2} -2 = \left(\frac{1}{2}\right)^{1-1} - 3 \ \checkmark(\text{true})$$

$$y = 2^{x-1} - 3 \xrightarrow{\text{When } x = 1 \text{ and } y = -2} -2 = 2^{1-1} - 3 \ \checkmark(\text{true})$$

Furthermore, the graph in Figure 1 represents the graph of the exponential growth function. This implies that the base of the exponential function must be greater than 1. Therefore, $y = 2^{x-1} - 3$ in answer (C) is the correct answer.

11. (B)

Tip

For the rational function

$$f(x) = \frac{p(x)}{q(x)} = \frac{ax^m + \cdots}{bx^n + \cdots}$$

where m is the degree of the numerator and n is the degree of the denominator, a horizontal asymptote can be determined by the following three cases.

- Case 1: If $n < m$, there is no horizontal asymptote. Whereas, there is a slant (or oblique) asymptote.
- Case 2: If $n = m$, f has a horizontal asymptote of $y = \frac{a}{b}$, where a and b are the leading coefficients of the numerator and denominator.
- Case 3: If $n > m$, f has a horizontal asymptote of $y = 0$.

$\lim_{x \to \infty} \frac{2x^2 - x - 1}{x^2 - 3}$ means finding the horizontal asymptote of $\frac{2x^2 - x - 1}{x^2 - 3}$. Since the numerator is a second degree polynomial $(m = 2)$ and the denominator is a second degree polynomial $(n = 2)$, the horizontal asymptote is $y = 2$.

12. (D)

Of 187 adults, 79 said "Yes". This implies that $\frac{79}{187}$ or about 42.2% of the sample of 187 adults have a tablet computer. Therefore, the total number of U.S. adults who own a tablet computer is 192 million $\times$ 0.422 $=$ 81 million.

13. (E)

The domain of a function is the set of the x-values. The graph in Figure 2 shows that the domain of f is all real numbers.

14. (C)

> **Tip** Cancelling out a common factor $x-c$ from both the numerator and the denominator of a rational function produces a hole in the graph of the rational function.

The rational function $f(x)=\dfrac{x-2}{x^2-4}$ has a common factor of $x-2$ in both the numerator and the denominator.

$$f(x)=\frac{x-2}{x^2-4}=\frac{\cancel{(x-2)}}{(x+2)\cancel{(x-2)}}=\frac{1}{x+2}$$

Thus, cancelling out $x-2$ produces a hole in the graph of f at $x=2$ and does not create a vertical asymptote at $x=2$. Since f simplifies to $\dfrac{1}{x+2}$, f has a vertical asymptote at $x=-2$. Furthermore, the numerator of $\dfrac{1}{x+2}$ is a constant $(m=0)$ and the denominator of $\dfrac{1}{x+2}$ is a first degree polynomial $(n=1)$, the horizontal asymptote of $\dfrac{1}{x+2}$ is $y=0$. Therefore, $f(x)=\dfrac{x-2}{x^2-4}$ has a vertical asymptote at $x=-2$ and a horizontal asymptote at $y=0$.

15. (B)

The slope of the line that passes through the points $(-1,1)$ and $(2,-5)$ is as follows:

$$\text{Slope}=\frac{-5-1}{2-(-1)}=-2$$

The slope of the perpendicular line is the negative reciprocal of -2 or $\dfrac{1}{2}$. Since the slope of the line $x-2y=6$ is $\dfrac{1}{2}$, (B) is the correct answer.

16. (C)

In order to find out where the particle is located at $t=\dfrac{5\pi}{6}$, substitute $t=\dfrac{5\pi}{6}$ into the parametric equations $x=\sin t$ and $y=\cos t$.

$$x=\sin t \quad\Longrightarrow\quad x=\sin\frac{5\pi}{6}=\sin\frac{\pi}{6}==0.5$$

$$y=\cos t \quad\Longrightarrow\quad y=\cos\frac{5\pi}{6}=-\cos\frac{\pi}{6}=-0.87$$

Therefore, the particle is located at $(0.5,-0.87)$ when $t=\dfrac{5\pi}{6}$.

17. (C)

Tip

The Law of Sines: If a, b, and c are the lengths of the sides of a triangle, and A, B, and C are the opposite angles, then

$$\frac{a}{\sin A} = \frac{b}{\sin B} = \frac{c}{\sin C}$$

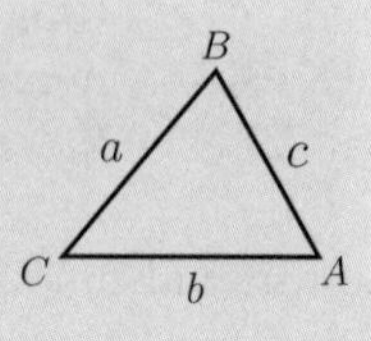

Since $AC = 1$, $m\angle A = 64°$, and $m\angle B = 32°$, $\triangle ABC$ is a SAA triangle as shown below.

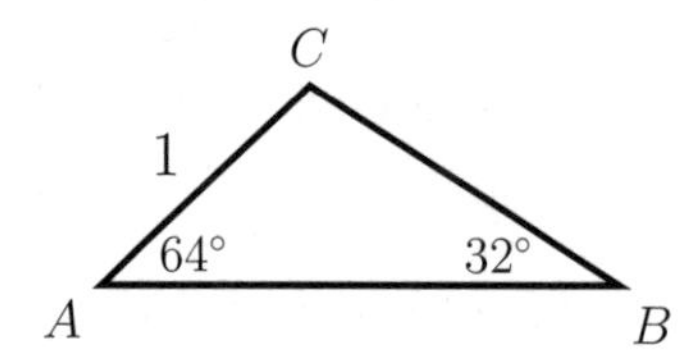

In order to find BC, use the Law of Sines.

$$\frac{BC}{\sin 64°} = \frac{1}{\sin 32°} \implies BC = \frac{\sin 64°}{\sin 32°} = 1.70$$

Therefore, $BC = 1.70$.

18. (A)

Tip $\cos^2 x + \sin^2 x = 1$

$$\begin{aligned} \tan x + \cot x &= \frac{\sin x}{\cos x} + \frac{\cos x}{\sin x} = \frac{\sin^2 x}{\sin x \cos x} + \frac{\cos^2 x}{\sin x \cos x} \\ &= \frac{1}{\sin x \cos x} = \frac{1}{\sin x} \cdot \frac{1}{\cos x} \\ &= \csc x \sec x \end{aligned}$$

19. (A)

In order to graph $|f(x)|$, start with the graph shown in Figure 4A.

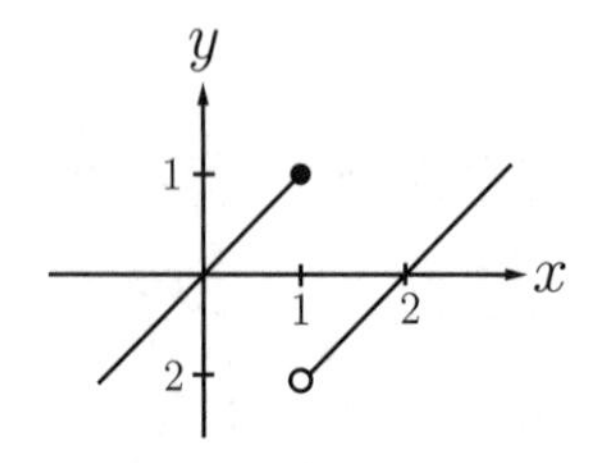

Fig 4A: $f(x)$

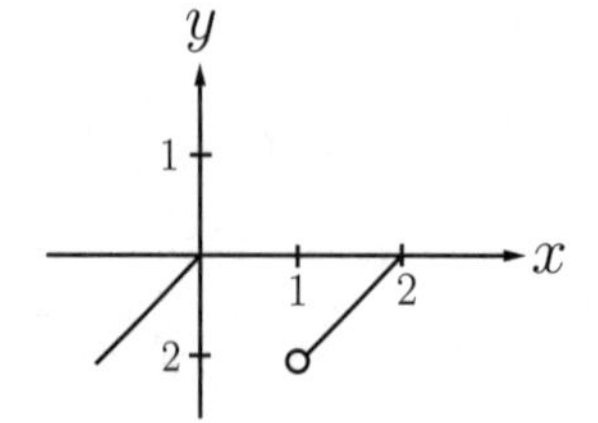

Fig 4B: Part below the x-axis

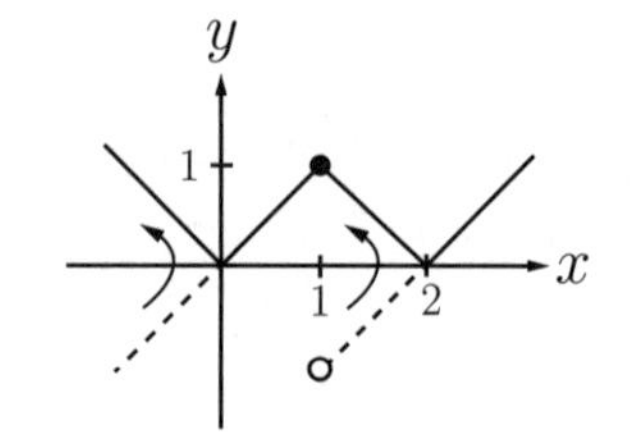

Fig 4C: $y = |f(x)|$

Determine the part of the graph that lies below the x-axis as shown in Figure 4B. Lastly, reflect the part of the graph that lies below the x-axis about the x-axis as shown in Figure 4C.

20. (D)

> **Tip** The quadratic formula is a general formula for solving quadratic equations. The solutions to the quadratic equation $ax^2 + bx + c = 0$ are as follows:
>
> $$x = \frac{-b \pm \sqrt{b^2 - 4ac}}{2a}$$

Since the quadratic equation $3x^2 + 2x - 4 = 0$ cannot be factored, use the quadratic formula to solve the equation.

$$\begin{aligned} x &= \frac{-b \pm \sqrt{b^2 - 4ac}}{2a} && \text{Substitute 3 for } a, \text{ 2 for } b, \text{ and } -4 \text{ for } c \\ &= \frac{-2 \pm \sqrt{(2)^2 - 4(3)(-4)}}{2(3)} \\ &= \frac{-2 \pm 2\sqrt{13}}{6} \end{aligned}$$

$$x = 0.87 \quad \text{or} \quad x = -1.54$$

Therefore, the solutions to $3x^2 + 2x = 4$ are $x = 0.87$ or $x = -1.54$.

21. (E)

The least common denominators of the rational expressions is $x^2 - 9$ or $(x + 3)(x - 3)$.

$$\begin{aligned} \frac{1}{x-3} - \frac{6}{x^2-9} &= \frac{x+3}{(x+3)(x-3)} - \frac{6}{(x+3)(x-3)} \\ &= \frac{x-3}{(x+3)(x-3)} \\ &= \frac{1}{x+3} \end{aligned}$$

Therefore, $\frac{1}{x-3} - \frac{6}{x^2-9} = \frac{1}{x+3}$.

22. (C)

$$\begin{aligned} \sum_{n=1}^{10}(-1)^n 2n &= -2 + 4 - 6 + 8 - 10 + 12 - 14 + 16 - 18 + 20 \\ &= (-2+4) + (-6+8) + (-10+12) + (-14+16) + (-18+20) \\ &= 2 + 2 + 2 + 2 + 2 \\ &= 10 \end{aligned}$$

Therefore, $\sum_{n=1}^{10}(-1)^n 2n = 10$.

23. (B)

Divide each side of the equation $16x^2 + 9y^2 = 144$ by 144. Thus, $\frac{x^2}{3^2} + \frac{y^2}{4^2} = 1$. Therefore, the area of the ellipse is $\pi(3)(4) = 12\pi$.

24. (B)

In order to find the inverse function, switch the x and y variables and solve for y.

$y = 7^{x-3}$	Switch the x and y variables
$7^{y-3} = x$	Convert the equation to a logarithmic equation
$y - 3 = \log_7 x$	Add 3 to each side
$y = \log_7 x + 3$	

Therefore, the inverse function of $y = 7^{x-3}$ is $\log_7 x + 3$.

25. (C)

> Tip: Two complex numbers are equal if and only if their real parts are equal and their imaginary parts are equal.

Since $z_1 = 2y + (x+6)i$ and $z_2 = (3-x) + yi$ are equal, $2y = 3 - x$, and $x + 6 = y$. Since $2y = 3 - x$ is equivalent to $x + 2y = 3$, and $x + 6 = y$ is equivalent to $x - y = -6$, use the linear combination method to find the values of x and y.

$$\begin{aligned} x + 2y &= 3 \\ x - y &= -6 \quad \text{Subtract the two equations} \\ \hline 3y &= 9 \\ y &= 3 \end{aligned}$$

Since $y = 3$ and $x - y = -6$, $x = -3$. Therefore, the value of $z_3 = x + yi = -3 + 3i$.

26. (E)

> Tip: Let $(x-c)^m$ be a factor of a polynomial function f.
>
> If $m =$ odd $\Longrightarrow$ graph of f crosses the x-axis at $x = c$.
>
> If $m =$ even $\Longrightarrow$ graph of f touches the x-axis at $x = c$.

The graph of the cubic function touches the x-axis at $x = 3$. This implies that 3 is a zero of multiplicity 2 and $(x-3)^2$ is a factor of the cubic function. Additionally, the graph of the cubic function crosses the x-axis at $x = -1$. This implies that -1 is a zero of multiplicity 1 and $(x+1)$ is a factor of the cubic function. Furthermore, the graph of the cubic function goes down as x increases and goes up as x decreases, which means that the leading coefficient is negative. Therefore, the cubic function $-(x+1)(x-3)^2$ best represents the graph in Figure 5.

27. (C)

> Tip
> 1. $\log_a a = 1$
> 2. $\log_a x^n = n \log_a x$
> 3. $\log_{a^n} x = \frac{1}{n} \log_a x$

Since $\log_2 \sqrt{2} = \frac{1}{2}$, $\log_4 2 = \frac{1}{2}$, and $\log_8 2 = \frac{1}{3}$ as shown below,

$$\log_2 \sqrt{2} = \log_2 2^{\frac{1}{2}} = \frac{1}{2} \log_2 2 = \frac{1}{2}$$
$$\log_4 2 = \log_{2^2} 2 = \frac{1}{2} \log_2 2 = \frac{1}{2}$$
$$\log_8 2 = \log_{2^3} 2 = \frac{1}{3} \log_2 2 = \frac{1}{3}$$

$\log_2 \sqrt{2} - \log_4 2 + \log_8 2 = \frac{1}{2} - \frac{1}{2} + \frac{1}{3} = \frac{1}{3}$. Therefore, (C) is the correct answer.

28. (B)

$\triangle ABC$ is a right triangle as shown below. In order to find AB, use the Pythagorean theorem: $b^2 = a^2 + AB^2$. Thus, $AB = \sqrt{b^2 - a^2}$.

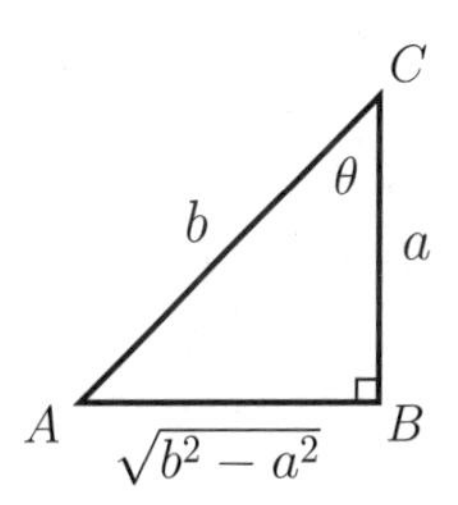

Since the definition of the sine function is $\frac{\text{opposite side}}{\text{hypotenuse}}$, $\sin\theta = \frac{\sqrt{b^2 - a^2}}{b}$.

29. (E)

Joshua can type m words in a seconds. That implies that Joshua can type $\frac{m}{a}$ words in 1 second, $60\left(\frac{m}{a}\right)$ words in 60 seconds, or $60t\left(\frac{m}{a}\right)$ words in t minutes. In addition, Jason can type n words in b minutes. This implies that Jason can type $\frac{n}{b}$ words in 1 minute, or $t\left(\frac{n}{b}\right)$ words in t minutes. Since both Joshua and Jason type together, the total number of words that they can type in t minutes is as follows:

$$60t\left(\frac{m}{a}\right) + t\left(\frac{n}{b}\right) = t\left(\frac{60m}{a} + \frac{n}{b}\right)$$

30. (A)

> **Tip**
>
> If an initial amount P is invested at an annual interest rate (expressed as a decimal) r compounded n times per year, the amount of money A accumulated in t years is as follows:
>
> $$A = P\left(1+\frac{r}{n}\right)^{nt}$$

The interest rate is 8% compounded annually. Thus, $r = 0.08$, and $n = 1$. For simplicity, let the initial amount P be \$100. Since the initial amount is doubled, the amount of money is $A = \$200$.

$A = 100(1+0.08)^t$	Substitute 200 for A
$100(1+0.08)^t = 200$	Divide each side by 100
$(1.08)^t = 2$	Convert the equation to a logarithmic equation
$t = \log_{1.08} 2$	
$t = \dfrac{\log_{10} 2}{\log_{10} 1.08} = 9$	

Therefore, it takes 9 years for an initial amount to double in value.

31. (B)

> **Tip**
>
> 1.
>
> To find the x-intercept of a line $\Longrightarrow$ Substitute 0 for y and solve for x
>
> To find the y-intercept of a line $\Longrightarrow$ Substitute 0 for x and solve for y
>
> 2. The volume V of a cone: $V = \frac{1}{3}\pi r^2 h$

The x-intercept and the y-intercept of the line $2x + 3y = 6$ are 3 and 2, respectively. The shaded area in Figure A shows the region R enclosed by the line $2x + 3y = 6$, the x-axis, and the y-axis. Figure B shows the resulting solid after the region R is rotated about the y-axis.

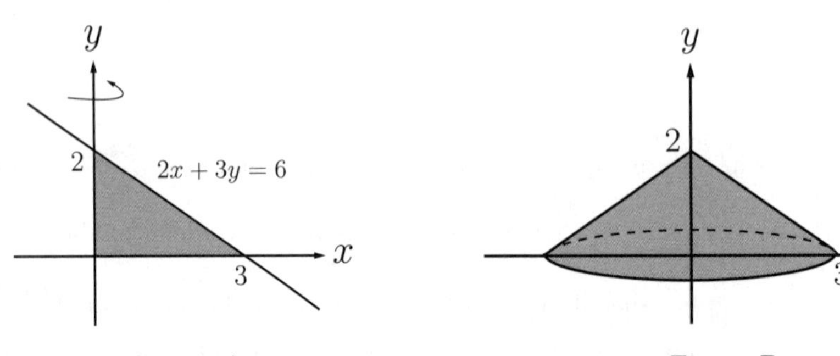

Figure A

Figure B

The resulting solid is a cone with a radius of 3 and a height of 2. Therefore, the volume of the cone is $V = \frac{1}{3}\pi r^2 h = \frac{1}{3}\pi(3)^2(2) = 6\pi$.

32. (C)

> Tip — Pythagorean Identity: $1 + \tan^2\theta = \sec^2\theta \implies \sec^2\theta - \tan^2\theta = 1.$

According to the Pythagorean identity, $\sec^2\left(\frac{11\pi}{12}\right) - \tan^2\left(\frac{11\pi}{12}\right) = 1.$

33. (E)

> Tip — The standard deviation measures the amount of variation or dispersion from the mean. In other words, it is a measure of how spread out the numbers are. A small standard deviation indicates that the numbers tend to be very close to the mean.

The mean of each set of numbers in all of the answer choices is 10. Since the numbers in answer choice (E) are closest to the mean, the set of numbers in answer choice (E) has the smallest standard deviation.

34. (A)

> Tip — To convert from a polar equation to a rectangular equation,
>
> $x = r\cos\theta, \qquad y = r\sin\theta$

Since $\csc\theta = \frac{1}{\sin\theta}$, $r = 3\csc\theta$ is equivalent to $r = \frac{3}{\sin\theta}$.

$$r = \frac{3}{\sin\theta} \qquad \text{Multiply each side by } \sin\theta$$
$$r\sin\theta = 3$$
$$y = 3$$

Therefore, the rectangular equation $y = 3$ is equivalent to $r = 3\csc\theta$.

35. (E)

$f : (x, y) \to (x - 1, y + 2)$ for every pair (x, y) in the triangular region means to move every pair (x, y) 1 unit to the left and 2 units up as shown below.

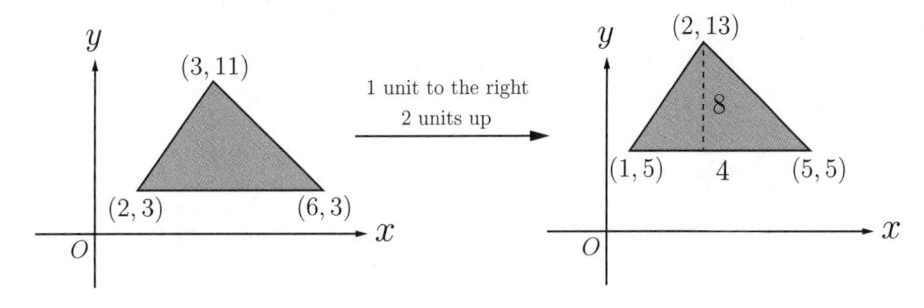

The mapping of set S by f is also a triangular region whose base is 3 and height is 4. Therefore, the area of the mapping of set S by f is $\frac{1}{2}(4)(8) = 16$.

36. (D)

The sum of the measures of the interior angles of a triangle is 180°. In $\triangle ABC$, $m\angle C = 90°$. Thus, $m\angle A + m\angle B = 90°$. $\sin B = 0.75$. Thus, $m\angle B = \sin^{-1} 0.75 = 48.59°$. Since $m\angle A = 90° - m\angle B = 90° - 48.59° = 41.41°$. Therefore, the value of $\tan A$ is $\tan 41.41° = 0.88$.

37. (C)

The common ratio is $-\frac{1}{2}$. Since $|r| < 1$,

$$S = \frac{a_1}{1-r} = \frac{\frac{2}{3}}{1-(-\frac{1}{2})} = \frac{\frac{2}{3}}{\frac{3}{2}} = \frac{4}{9}$$

Therefore, the sum of the infinite geometric series $\frac{2}{3} - \frac{1}{3} + \frac{1}{6} + \cdots$ is $\frac{4}{9}$.

38. (D)

Tip Distance Formula: $D = \sqrt{(x_2 - x_1)^2 + (y_2 - y_1)^2}$

The line $y = -x + 3$ intersects the line $y = 2x$ as shown in the figure below. In order to find the intersection point A, substitute $2x$ for y in $y = -x + 3$.

$$y = -x + 3 \qquad \text{Substitute } 2x \text{ for } y$$
$$2x = -x + 3 \qquad \text{Solve for } x$$
$$x = 1$$

Since $x = 1$, $y = 2x = 2$. Thus, the intersection point A is $(1, 2)$.

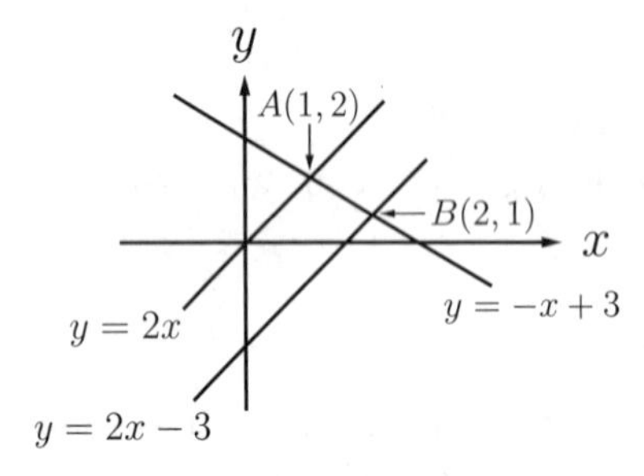

Additionally, the line $y = -x + 3$ intersects the line $y = 2x - 3$. In order to find the intersection point B, substitute $2x - 3$ for y in $y = -x + 3$.

$$y = -x + 3 \qquad \text{Substitute } 2x - 3 \text{ for } y$$
$$2x - 3 = -x + 3 \qquad \text{Solve for } x$$
$$x = 2$$

Since $x = 2$, $y = 2x - 3 = 1$. Thus, the intersection point B is $(2, 1)$. In order to find the distance between points $A(1, 2)$ and $B(2, 1)$, use the distance formula.

$$D = \sqrt{(x_2 - x_1)^2 + (y_2 - y_1)^2} = \sqrt{(2-1)^2 + (1-2)^2} = \sqrt{2}$$

Therefore, the distance between points $A(1, 2)$ and $B(2, 1)$ is $\sqrt{2}$ or 1.41.

39. (A)

> Tip — If triangle ABC shown at the right is a SAS triangle (a, b, and $m\angle C$ are known), the area of triangle ABC is as follows:
> $$A = \frac{1}{2}ab\sin C$$

Since $\triangle ABC$ is inscribed in a semicircle, $\triangle ABC$ is a right triangle with $AC = 10$ as shown below.

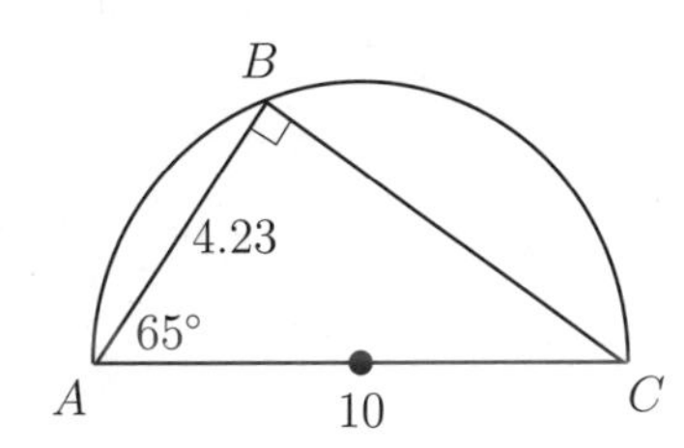

In order to find AB, use the definition of the cosine function.

$$\cos 65^\circ = \frac{\text{Adjacent side}}{\text{Hypotenuse}} = \frac{AB}{10} \quad \Longrightarrow \quad AB = 10\cos 65^\circ = 4.23$$

Therefore, the area of $\triangle ABC$ is $A = \frac{1}{2}(4.23)(10)\sin 65^\circ = 19.17$.

40. (B)

> Tip — Binomial expansion formula: $(x-y)^2 = x^2 - 2xy + y^2$

$x - 3 = \sqrt{2x-3}$	Square both sides
$(x-3)^2 = 2x - 3$	Use the binomial expansion formula
$x^2 - 6x + 9 = 2x - 3$	Subtract $2x - 3$ from each side
$x^2 - 8x + 12 = 0$	Factor the quadratic expression
$(x-2)(x-6) = 0$	Use the zero product property: If $ab = 0$, then $a = 0$ or $b = 0$.
$x = 2$ or $x = 6$	

Substitute 2 and 6 for x in the original equation to check the solutions.

$$2 - 3 = \sqrt{4-3} \qquad\qquad 6 - 3 = \sqrt{12-3}$$
$$-1 \neq 1 \quad \text{(Not a solution)} \qquad\qquad 3 = 3 \quad \checkmark\ \text{(Solution)}$$

Therefore, the only solution to $x - 3 = \sqrt{2x-3}$ is $x = 6$.

41. (D)

> Tip — $\cos 2\theta = 1 - 2\sin^2\theta$

Since $\sin\theta = \frac{3}{5}$, $\cos 2\theta = 1 - 2\sin^2\theta = 1 - 2\left(\frac{3}{5}\right)^2 = \frac{7}{25}$.

42. (B)

> **Tip**
> 1. $n \log_a x = \log_a x^n$
> 2. $\log_a x + \log_a y = \log_a xy$
> 3. $\log_a x - \log_a y = \log_a \frac{x}{y}$

$$\begin{aligned} 2\ln x - 3\ln y + \frac{1}{2}\ln z &= \ln x^2 + \ln z^{\frac{1}{2}} - \ln y^3 \\ &= (\ln x^2 + \ln\sqrt{z}) - \ln y^3 \\ &= \ln x^2\sqrt{z} - \ln y^3 \\ &= \ln\frac{x^2\sqrt{z}}{y^3} \end{aligned}$$

Therefore, $2\ln x - 3\ln y + \frac{1}{2}\ln z = \ln\frac{x^2\sqrt{z}}{y^3}$.

43. (C)

> **Tip**
> If triangle ABC shown at the right is a SSS triangle (a, b, and c are known), the measure of angle A can be calculated by the Law of Cosines.
>
> $$m\angle A = \cos^{-1}\left(\frac{a^2 - b^2 - c^2}{-2bc}\right)$$
>
> Note that side a is opposite angle A.

In the figure below, let the three sides b, c and a be 5, 5, and 8, respectively, since the ratio of the three sides is $5 : 5 : 8$. The Law of Sines implies that the largest angle is opposite the longest side. Thus, $\angle A$ is the largest angle since side a is the longest side.

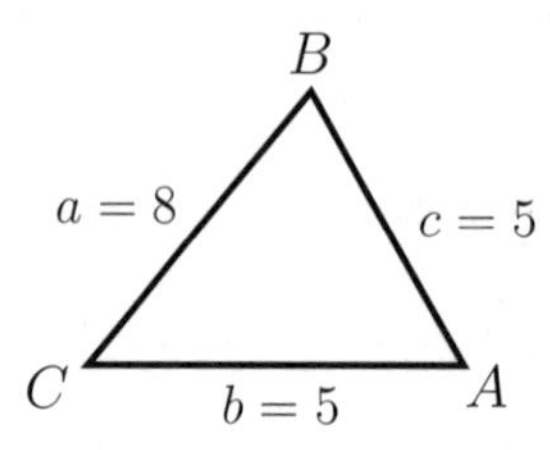

In order to find the measure of angle A, use the Law of Cosines. Since side a is opposite angle A,

$$m\angle A = \cos^{-1}\left(\frac{a^2 - b^2 - c^2}{-2bc}\right) = \cos^{-1}\left(\frac{8^2 - 5^2 - 5^2}{-2(5)(5)}\right) = 106.26°$$

Therefore, the largest angle of the triangle is 106°.

44. (E)

Tip The domain of a cube root function is the set of all real numbers.

The domain of the cube root function $\sqrt[3]{1-2x}$ is the set of all real numbers. However, since $\sqrt[3]{1-2x}$ is in the denominator, you must exclude $x = \dfrac{1}{2}$ from the domain because the function $y = \dfrac{2}{\sqrt[3]{1-2x}}$ is undefined when $x = \dfrac{1}{2}$. Therefore, (E) is the correct answer.

45. (D)

As shown in Figure A, the period of $\sin 2x$ is $\dfrac{2\pi}{2} = \pi$.

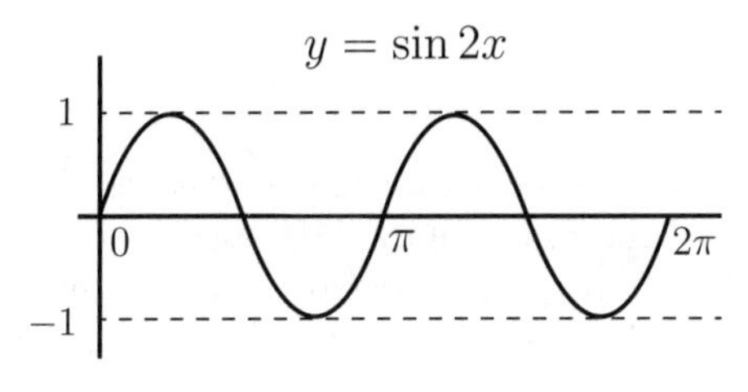

Figure A

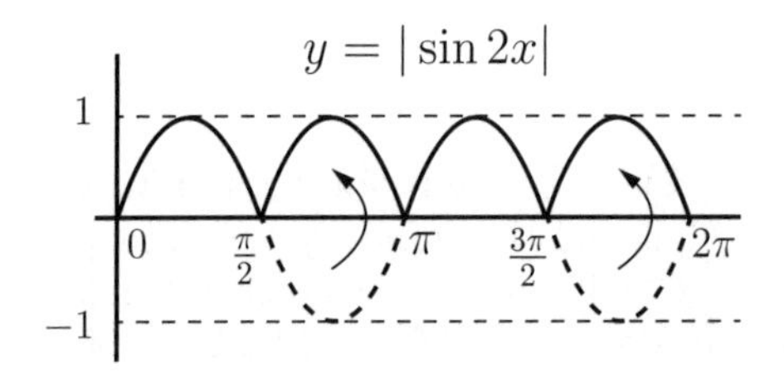

Figure B

However, the graph of $y = |\sin 2x|$ completes its cycle every $\dfrac{\pi}{2}$ as shown in Figure B, the period of $y = |\sin 2x|$ is $\dfrac{\pi}{2}$.

46. (C)

Tip
1. $\sqrt{x^2} = |x|$
2. When $x < 0$, $|x| = -x$

$$f(g(x)) = f\left(\frac{1}{2}x^2 - 2\right) = \sqrt{2\left(\frac{1}{2}x^2 - 2\right) + 4}$$
$$= \sqrt{x^2} = |x|$$

Since $x < 0$, $|x| = -x$. Therefore, when $x < 0$, $f(g(x)) = -x$.

47. (B)

> Tip
> 1. The x-intercepts, zeros, roots, or solutions of a function $f(x)$ are the values of x that make $f(x) = 0$.
> 2. $x^2 - y^2 = (x+y)(x-y)$

Substitute 0 for y and solve for x.

$$\begin{aligned} x^3 - 3x^2 - 4x + 12 &= 0 \\ x^2(x-3) - 4(x-3) &= 0 \\ (x-3)(x^2-4) &= 0 \\ (x-3)(x+2)(x-2) &= 0 \\ x = -2 \quad \text{or} \quad x = 2 \quad &\text{or} \quad x = 3 \end{aligned}$$

Therefore, the real zeros of $f(x) = x^3 - 3x^2 - 4x + 12$ are $x = -2, 2, 3$.

48. (D)

Jason rolls two dice to form a two-digit integer. Since the number on the first die represents the tens digit and the number on the second die represents the units digit, there are a total number of $6 \times 6 = 36$ possible two-digit integers. Out of these 36 integers, there are only 5 integers that are divisible by 8: 16, 24, 32, 56, 64. Note that 40 and 48 can not be formed because a die has numbers 1 through 6. Therefore, the probability that the integer formed is divisible by 8 is $\frac{5}{36}$.

49. (A)

Since both circles $x^2 + y^2 = 4$ and $(x-3)^2 + y^2 = 4$ intersect, substitute $(x-3)^2 + y^2$ for 4 in $x^2 + y^2 = 4$.

$$\begin{aligned} x^2 + y^2 &= (x-3)^2 + y^2 && \text{Substitute } (x-3)^2 + y^2 \text{ for } 4 \\ x^2 + y^2 &= x^2 - 6x + 9 + y^2 && \text{Expand} \\ 0 &= -6x + 9 && \text{Subtract } x^2 \text{ and } y^2 \text{ from each side} \\ 6x &= 9 \\ x &= 1.5 \end{aligned}$$

Thus, the x-coordinates of the two intersection points are both $x = 1.5$. In order to find the y-coordinate of the two intersection points, substitute $x = 1.5$ into $x^2 + y^2 = 4$ and solve for y.

$$\begin{aligned} (1.5)^2 + y^2 &= 4 \\ y &= \pm\sqrt{4 - (1.5)^2} = \pm 1.32 \end{aligned}$$

Therefore, the x and y-coordinates of the two intersection points are $(1.5, 1.32)$ and $(1.5, -1.32)$.

50. (D)

As shown in the figure below, the triangular base is a right triangle. Use the Pythagorean theorem to find the length of the hypotenuse: $c^2 = 3^2 + 4^2$. Thus, the length of the hypotenuse is 5.

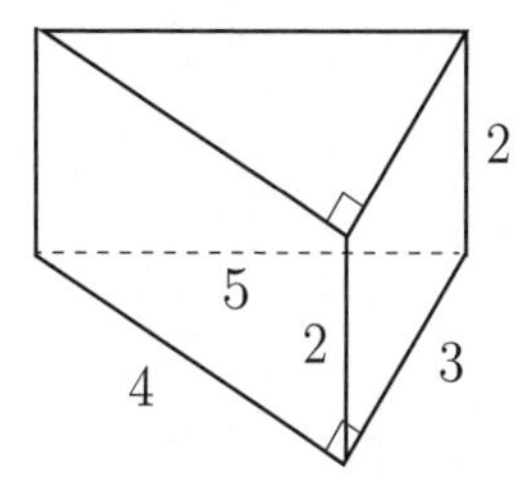

The area of the five faces of the right prism are as follows:

	Top	Bottom	Left	Right	Back
Area	$\frac{1}{2}(4)(3)$	$\frac{1}{2}(4)(3)$	4×2	3×2	5×2

Therefore, the surface area of the prism is $6 + 6 + 8 + 6 + 10 = 36$.

SAT II MATH LEVEL 2 TEST 3

Directions: Among the given answer choices, choose the BEST answer for each problem. If the exact numerical value is not within the given answer choices, select the answer that best approximates this value. Afterwards, fill in the corresponding oval on the answer sheet.

Notes:

1. A calculator may be required to answer some of the questions. Scientific and graphing calculators are allowed during this test.
2. For some questions, it is up to you to decide whether the calculator should be in degree mode or radian mode.
3. Provided figures for questions are drawn as accurately as possible UNLESS otherwise specified by the problem. Unless otherwise indicated, all figures are assumed to lie in a plane.
4. Unless otherwise specified, it can be assumed that the domain of any function f is to be the set of all real numbers x for which $f(x)$ is a real number.

Reference Information: Use the following information and formulas as a reference in answering questions on this test.

1. If the radius and height of a right circular cone are r and h respectively, then the Volume V of the cone is $V = \frac{1}{3}\pi r^2 h$.
2. If the circumference and slant height of a right circular cone are c and ℓ respectively, then the Lateral Area A of the cone is $A = \frac{1}{2}c\ell$.
3. At any given radius r, the Volume V of a sphere is $V = \frac{4}{3}\pi r^3$.
4. At any given radius r, the Surface Area A of a sphere is $A = 4\pi r^2$.
5. The Volume V of a pyramid is $V = \frac{1}{3}Bh$; given that B and h represent the base area and height of the pyramid respectively.

USE THIS SPACE FOR SCRATCH WORK

1. If the line $y = kx - 2$ passes only through quadrants III and IV, what must be the value of k ?

   (A) $k < 0$

   (B) $k \leq 0$

   (C) $k = 0$

   (D) $k > 0$

   (E) $k \geq 0$

2. Figure 1 shows the graph of f. In which of the following interval is f decreasing?

   (A) $(1, 3)$

   (B) $(2, 4)$

   (C) $(1, 4)$

   (D) $(-1, 1)$

   (E) $(-1, 2)$

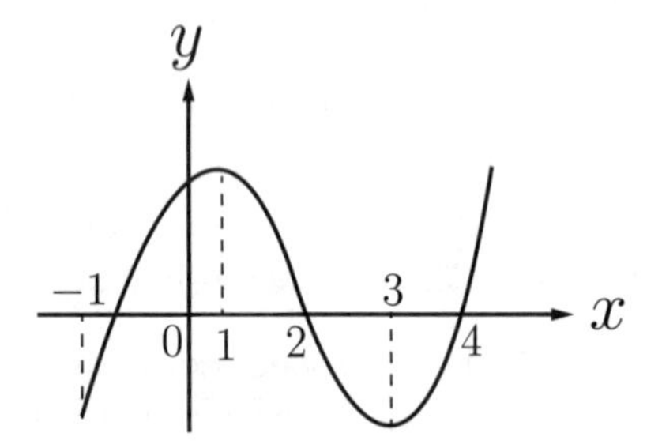

Figure 1

3. If $\sin^2 x = \dfrac{3}{4}$, what is the value of $\cos^2 x$?

   (A) $\dfrac{1}{8}$

   (B) $\dfrac{1}{4}$

   (C) $\dfrac{1}{2}$

   (D) $\dfrac{3}{4}$

   (E) $\dfrac{\sqrt{3}}{2}$

USE THIS SPACE FOR SCRATCH WORK

4. If $\sqrt{x} \cdot \sqrt[3]{x} = x^n$, what is the value of n ?

(A) $\frac{7}{2}$

(B) $\frac{7}{6}$

(C) $\frac{5}{6}$

(D) $\frac{2}{5}$

(E) $\frac{1}{6}$

5. If $\tan\theta = 0.8,\ 0 < \theta < \frac{\pi}{2}$, what is the measure of angle θ ?

(A) 0.39

(B) 0.48

(C) 0.67

(D) 1.21

(E) 2.08

6. If vector **V** has a terminal point of $(-2, 3)$ and an initial point of $(3, -4)$, what is the position vector **V** ?

(A) $\langle 5, -7 \rangle$

(B) $\langle 1, -1 \rangle$

(C) $\langle 0, -2 \rangle$

(D) $\langle -1, 1 \rangle$

(E) $\langle -5, 7 \rangle$

USE THIS SPACE FOR SCRATCH WORK

7. Joshua is on a Ferris Wheel and his position is modeled by the parametric equations $x(t) = 30\sin\left(\frac{\pi}{20}t\right)$, $y(t) = -30\cos\left(\frac{\pi}{20}t\right) + 36$, where $0 \leq t \leq 40$. If Joshua is 6 feet above the ground at $t = 0$, how high is Joshua above the ground at $t = 20$?

   (A) 66

   (B) 49

   (C) 31

   (D) 25

   (E) 6

8. According to the table shown below, there are 55 juniors and 45 seniors in a high school. The number of students who are taking Calculus is 60 and the number of students who are taking Chemistry is 40. If the number of juniors who are taking Calculus is 25, what is the number of seniors who are taking Chemistry?

	Calculus	Chemistry	
Junior			55
Senior			45
	60	40	100

   (A) 5

   (B) 10

   (C) 25

   (D) 30

   (E) 35

USE THIS SPACE FOR SCRATCH WORK

9. If the complex number z is equal to $2-2i$, what is the magnitude of z ?

(A) $4\sqrt{3}$

(B) $4\sqrt{2}$

(C) 4

(D) $2\sqrt{3}$

(E) $2\sqrt{2}$

10. If $f(x)=-(x+1)^2+2$ and $g(x)=f(x-2)+3$, what is the vertex of $g(x)$?

(A) $(-3,-1)$

(B) $(-3,5)$

(C) $(-1,5)$

(D) $(1,5)$

(E) $(1,-1)$

11. What is the range of $y=-2\sin\left(3t+\dfrac{3\pi}{4}\right)-1$?

(A) $-3\leq y\leq 1$

(B) $-2\leq y\leq 2$

(C) $-1\leq y\leq 0$

(D) $0\leq y\leq 3$

(E) $1\leq y\leq 3$

USE THIS SPACE FOR SCRATCH WORK

12. If $f(x) = \dfrac{x}{2} + 1$ and $g(x) = 2x^3 + 1$, what is $f(g(0.64))$?

(A) 6.49

(B) 4.92

(C) 2.14

(D) 1.76

(E) 1.39

13. $\sin\dfrac{\pi}{3} + \sin\dfrac{2\pi}{3} + \sin\dfrac{4\pi}{3} + \sin\dfrac{5\pi}{3} =$

(A) 0

(B) $\dfrac{\sqrt{3}}{2}$

(C) $\sqrt{3}$

(D) $2\sqrt{3}$

(E) $4\sqrt{3}$

14. A proportion of all U.S. retail sales that involves the internet is modeled by

$$P(t) = \frac{0.7}{1 + 3e^{-0.28t}}$$

where t represents years after 2010. For example, $t = 0$ represents 2010, $t = 1$ represents 2011, and so on. What proportion of U.S. retail sales involve the internet in 2020 ?

(A) 0.56

(B) 0.59

(C) 0.62

(D) 0.65

(E) 0.68

USE THIS SPACE FOR SCRATCH WORK

15. If $10^{\log_{10} x} < 10$, what is the largest possible positive integer value of x ?

(A) 11

(B) 10

(C) 9

(D) 8

(E) 7

16. If $\sin 2\theta = \frac{2}{3}$, what is the value of $(\sin\theta + \cos\theta)^2$?

(A) $\frac{1}{3}$

(B) $\frac{2}{3}$

(C) 1

(D) $\frac{4}{3}$

(E) $\frac{5}{3}$

17. Which of the following statements is NOT true about the inverse function?

(A) The inverse function of $f(x)$ is $\frac{1}{f(x)}$.

(B) The domain of the inverse function is the range of $f(x)$.

(C) The graph of the inverse function is obtained by reflecting the graph of $f(x)$ about the line $y = x$.

(D) If $f^{-1}(x)$ is the inverse function of $f(x)$, $f(f^{-1}(x)) = x$.

(E) The inverse function of $f(x) = \frac{1}{x}$ is $f^{-1}(x) = \frac{1}{x}$.

USE THIS SPACE FOR SCRATCH WORK

18. $\lim_{x \to 0} \dfrac{e^x - 1}{\sin x} =$

(A) 0

(B) 1

(C) 2

(D) 58

(E) 60

19. A lamp post casts a shadow of 25 feet when the sun makes a 35° angle of elevation. What is the height of the lamp post?

(A) 17.5 feet

(B) 15.6 feet

(C) 14.1 feet

(D) 12.3 feet

(E) 11.8 feet

20. A water tank has the shape of an inverted cone with a diameter of 8 feet and a height of 5 feet as shown in Figure 2. The water is being pumped into the tank so that the water level is rising. If the height of the water is 2 feet, what is the radius of the surface of the water?

(A) $\dfrac{1}{5}$ feet

(B) $\dfrac{2}{5}$ feet

(C) $\dfrac{3}{5}$ feet

(D) $\dfrac{6}{5}$ feet

(E) $\dfrac{8}{5}$ feet

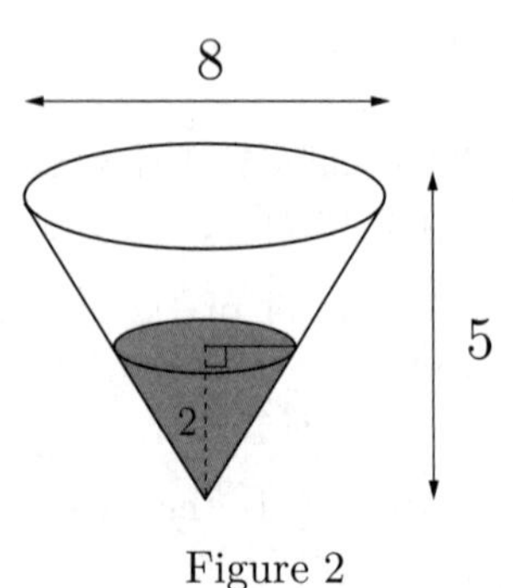

Figure 2

USE THIS SPACE FOR SCRATCH WORK

21. What is the value of x for which $\frac{1}{2}x^3 - 1 = 0.6$?

(A) 0.21

(B) 0.89

(C) 1.13

(D) 1.47

(E) 1.79

22. A is a $m \times 2$ matrix and B is a 2×4 matrix. In order for the product of the two matrices, AB, to be defined, what must be the value of m ?

(A) 2

(B) 3

(C) 4

(D) 6

(E) Any positive integers

23. y varies directly with the cube of x and inversely with the square root of z. When $x = 2$ and $z = 16$, $y = 6$. What is the value of y when $x = \frac{1}{3}$ and $z = \frac{1}{4}$?

(A) $\frac{9}{2}$

(B) $\frac{5}{9}$

(C) $\frac{2}{5}$

(D) $\frac{2}{9}$

(E) $\frac{1}{9}$

USE THIS SPACE FOR SCRATCH WORK

24. If a sequence is defined by $a_{n+2} = a_{n+1} \cdot a_n$ for $n \geq 1$, $a_1 = 2$, and $a_2 = 3$, which of the following is NOT a factor of the fifth term?

    (A) 6

    (B) 12

    (C) 18

    (D) 24

    (E) 27

25. Which of the following best represents the set of points equidistant from two points in the xy-plane?

    (A) A line

    (B) A triangle

    (C) A square

    (D) A circle

    (E) A ellipse

26. Figure 3 shows a part of a trigonometric function. Which of the following trigonometric function best represents the graph in Figure 3?

    (A) $y = \cot\left(x - \frac{\pi}{4}\right)$

    (B) $y = \tan\left(x - \frac{\pi}{4}\right)$

    (C) $y = \tan\left(x + \frac{\pi}{4}\right)$

    (D) $y = \csc\left(x - \frac{\pi}{4}\right)$

    (E) $y = \sec\left(x + \frac{\pi}{4}\right)$

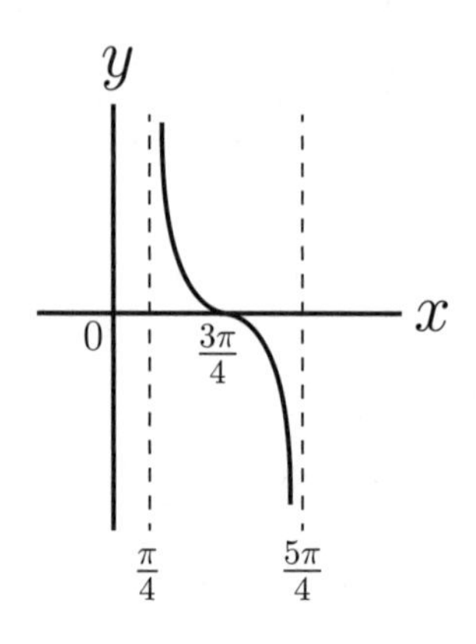

Figure 3

USE THIS SPACE FOR SCRATCH WORK

27. In $\triangle ABC$ shown in Figure 4, $BD = 10$, $m\angle ABD = 15°$, and $m\angle CBD = 25°$. What is AC ?

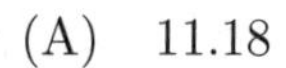

(A) 11.18

(B) 9.42

(C) 8.26

(D) 7.34

(E) 6.63

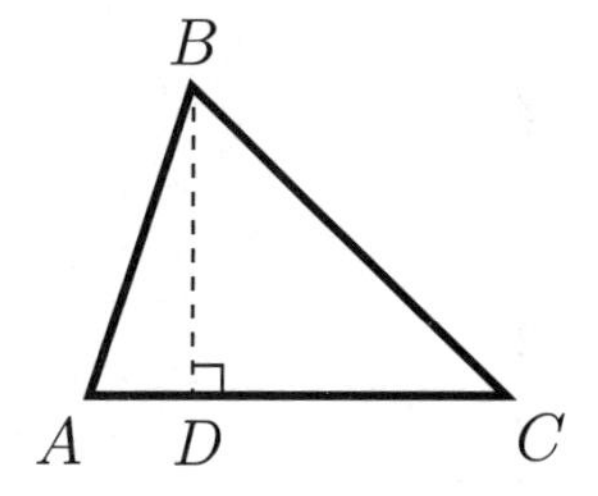

Figure 4

28. Which of the following expression is a factor of $f(x) = x^3 - 3x^2 + 4x - 4$?

(A) $x + 2$

(B) $x + 1$

(C) $x - 2$

(D) $x - 3$

(E) $x - 4$

29. If $f(x) = -|x - 1| - 3$, what is the x-intercept of f ?

(A) $x = 4$ only

(B) $x = 2$ only

(C) $x = -2$ only

(D) $x = 4$ or $x = -2$

(E) There is no x-intercept.

USE THIS SPACE FOR SCRATCH WORK

30. If $\dfrac{7p+2q}{p+q} = 4$, what is the value of $\dfrac{p}{q}$?

(A) $\dfrac{7}{2}$

(B) $\dfrac{5}{3}$

(C) $\dfrac{2}{3}$

(D) $\dfrac{3}{5}$

(E) $\dfrac{2}{7}$

31. $\csc^2 x(1-\cos^2 x) =$

(A) $\sin^2 x$

(B) $\sec^2 x$

(C) $\tan^2 x$

(D) $\cot^2 x$

(E) 1

32. If the equation of the ellipse $x^2+4y^2-16y=0$ is given, what is the length of the major axis of the ellipse?

(A) 16

(B) 12

(C) 10

(D) 8

(E) 4

USE THIS SPACE FOR SCRATCH WORK

33. Solve: $2(3)^x + 2 = 7$

(A) 0.69

(B) 0.75

(C) 0.83

(D) 0.92

(E) 1.04

34. An identity equation is an equation that is true no matter what value is substituted for the variable. For the following identity equation, what is the value of A ?

$$\frac{4x-1}{(x-1)(2x-1)} = \frac{A}{x-1} - \frac{2}{2x-1}$$

(A) 1

(B) 2

(C) 3

(D) 4

(E) 5

35. If the polynomial function $f(x) = x^3 - 3x^2 + 2x + 4$, which of the following function is the reflection of f about the y-axis?

(A) $-x^3 - 3x^2 - 2x + 4$

(B) $-x^3 - 3x^2 + 2x + 4$

(C) $-x^3 + 3x^2 + 2x - 4$

(D) $-x^3 + 3x^2 - 2x + 4$

(E) $-x^3 + 3x^2 - 2x - 4$

USE THIS SPACE FOR SCRATCH WORK

36. The equation $(x-x_0)^2+(y-y_0)^2+(z-z_0)^2 = r^2$ represents a sphere with center (x_0, y_0, z_0) and radius r. If the equation of the sphere is $x^2+y^2+(z+1)^2 = 11$, which of the following point (x, y, z) in three dimensional space lies inside or on the surface of the sphere?

(A) $(1, 2, 2)$

(B) $(2, 2, 1)$

(C) $(3, 1, 0)$

(D) $(3, 2, -1)$

(E) $(3, 3, 4)$

37. $\dfrac{1}{\sqrt{3}+\sqrt{2}} + \dfrac{1}{\sqrt{2}+1} =$

(A) $\sqrt{3}-1$

(B) $\sqrt{3}+1$

(C) $\sqrt{3}-\sqrt{2}$

(D) $\sqrt{3}+\sqrt{2}-2$

(E) $2\sqrt{3}-3$

38. How many arrangements can be formed using all the letters in the word SOLOMON?

(A) 720

(B) 840

(C) 1680

(D) 2520

(E) 5040

USE THIS SPACE FOR SCRATCH WORK

39. A pitcher had won 60% of the games he pitched. For the next five games, the pitcher won 2 games and lost 3, to finish the season having won 56% of his games. How many games did the pitcher play in all?

(A) 20

(B) 25

(C) 28

(D) 40

(E) 45

40. Which of the following rational function best represents the graph in Figure 5?

(A) $y = \dfrac{1}{x}$

(B) $y = \dfrac{1}{x(x-2)}$

(C) $y = \dfrac{x-2}{x(x-2)}$

(D) $y = \dfrac{x-2}{x(x+2)}$

(E) $y = \dfrac{(x+2)}{x(x+2)}$

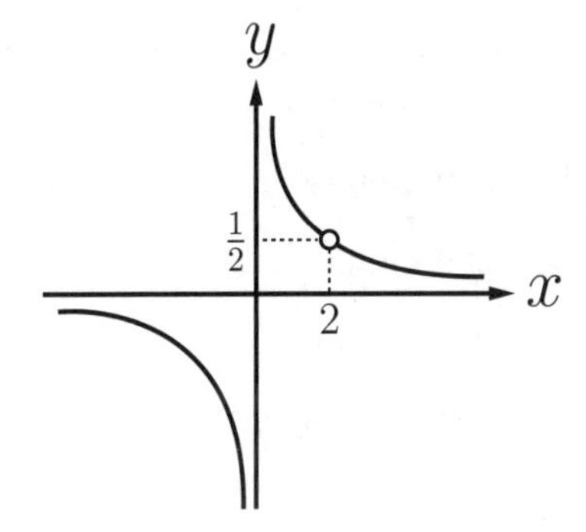

Figure 5

41. A manufacturing company that produces light bulbs found out that 3% of their production is defective. If you have purchased five light bulbs from the company, what is the probability that none of them are defective?

(A) 0.97

(B) 0.92

(C) 0.86

(D) 0.15

(E) 0.03

USE THIS SPACE FOR SCRATCH WORK

42. If $\log_2 3 = m$ and $\log_2 5 = n$, what is $\log_2 30$ in terms of m and n ?

(A) $m + n$

(B) $m + n + 1$

(C) mn

(D) $mn + 1$

(E) $m(n + 1)$

43. A local telephone company provides wireless communication services to residential customers for a monthly charge of $13.50 plus 1.5 cents per minute for the first 800 minutes used, and 2.5 cents thereafter in one month. If the company charged a customer $26.75 for October, how many minutes did the customer use in October?

(A) 850 minutes

(B) 800 minutes

(C) 750 minutes

(D) 700 minutes

(E) 650 minutes

44. Solve: $\tan^2\theta = 1$, where $0 \le \theta < 2\pi$.

(A) $\frac{\pi}{4}$

(B) $\frac{3\pi}{4}$

(C) $\frac{\pi}{4}, \frac{5\pi}{4}$

(D) $\frac{3\pi}{4}, \frac{7\pi}{4}$

(E) $\frac{\pi}{4}, \frac{3\pi}{4}, \frac{5\pi}{4}, \frac{7\pi}{4}$

USE THIS SPACE FOR SCRATCH WORK

45. $\dfrac{3-4i}{2-3i} =$

(A) $-\dfrac{6}{5} - \dfrac{i}{5}$

(B) $-\dfrac{18}{13} - \dfrac{i}{13}$

(C) $\dfrac{6}{5} - \dfrac{i}{5}$

(D) $\dfrac{6}{13} + \dfrac{i}{13}$

(E) $\dfrac{18}{13} + \dfrac{i}{13}$

46. Given two vectors $\mathbf{A} = \langle 3, 2 \rangle$ and $\mathbf{B} = \langle -1, k \rangle$, what is the value of k such that $3\mathbf{B} - 2\mathbf{A} = \langle -9, 11 \rangle$?

(A) 1

(B) 2

(C) 3

(D) 4

(E) 5

47. In the geometric sequence, the 3rd term is $\dfrac{1}{3}$ and the 7th term is 27. What is the 9th term?

(A) 36

(B) 81

(C) 108

(D) 243

(E) 360

USE THIS SPACE FOR SCRATCH WORK

48. What are the equations of the asymptotes of $16x^2 - 9y^2 = 144$?

(A) $y = \pm\frac{16}{9}x$

(B) $y = \pm\frac{4}{3}x$

(C) $y = \pm\frac{3}{4}x$

(D) $y = \pm\frac{4}{9}x$

(E) $y = \pm\frac{9}{16}x$

49. Which of the following point (x, y) satisfies the system of inequalities shown below?

$$y > x^2$$
$$y < x + 2$$

(A) $(-1, -2)$

(B) $(-1, -1)$

(C) $(0, 3)$

(D) $(1, 2)$

(E) $(2, 5)$

USE THIS SPACE FOR SCRATCH WORK

50. From city A to city B, Jason traveled at a rate of 60 miles per hour for 2 hours and 15 minutes. From city B to city C, Jason traveled 30 miles for 54 minutes. If Jason did not take any breaks from city A to city B to city C, what is his average speed for the entire trip?

(A) 68.75 miles per hour

(B) 63.27 miles per hour

(C) 59.14 miles per hour

(D) 52.38 miles per hour

(E) 45.23 miles per hour

Mathematics Scoring Worksheet

Directions: In order to calculate your score correctly, fill out the table below. After calculating your raw score, round the raw score to the nearest whole number. The scaled score can be determined using the "Math Test Score Conversion Table".

Mathematics Score			
A. Number Correct		**B.** Number Incorrect ÷ 4	
Total Unrounded Raw Score $A - B$		**Total Rounded Raw Score** **Round to nearest whole number**	

Math Test Score Conversion Table					
Raw Score	Scaled Score	Raw Score	Scaled Score	Raw Score	Scaled Score
50	800	29	660	8	490
49	800	28	650	7	480
48	800	27	640	6	480
47	800	26	630	5	470
46	800	25	630	4	460
45	800	24	620	3	450
44	800	23	610	2	440
43	800	22	600	1	430
42	790	21	590	0	410
41	780	20	580	−1	390
40	770	19	570	−2	370
39	760	18	560	−3	360
38	750	17	560	−4	340
37	740	16	550	−5	340
36	730	15	540	−6	330
35	720	14	530	−7	320
34	710	13	530	−8	320
33	700	12	520	−9	320
32	690	11	510	−10	320
31	680	10	500	−11	310
30	670	9	500	−12	310

ANSWERS AND SOLUTIONS

Answers

1. C	11. A	21. D	31. E	41. C
2. A	12. D	22. E	32. D	42. B
3. B	13. A	23. D	33. C	43. A
4. C	14. B	24. D	34. C	44. E
5. C	15. C	25. A	35. A	45. E
6. E	16. E	26. A	36. C	46. E
7. A	17. A	27. D	37. A	47. D
8. B	18. B	28. C	38. B	48. B
9. E	19. A	29. E	39. B	49. D
10. D	20. E	30. C	40. C	50. D

Solutions

1. (C)

When $k > 0$, the line $y = kx - 2$ passes through quadrants I, III and IV as shown below. When $k < 0$, the line $y = kx - 2$ passes through quadrants II, III and IV.

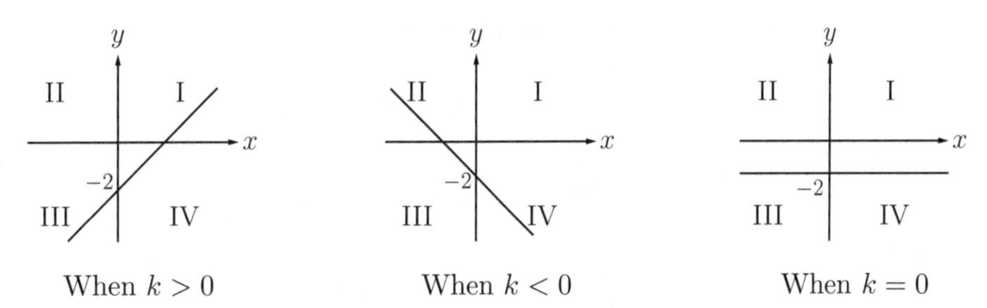

When $k = 0$, the line $y = kx - 2$ passes only through quadrants III and IV. Therefore, (C) is the correct answer.

2. (A)

When $1 < x < 3$, f is decreasing. Therefore, (A) is the correct answer.

3. (B)

Tip $\sin^2 x + \cos^2 x = 1 \implies \cos^2 x = 1 - \sin^2 x$

$$\cos^2 x = 1 - \sin^2 x = 1 - \frac{3}{4} = \frac{1}{4}$$

4. (C)

> **Tip**
> 1. $\sqrt[n]{a} = a^{\frac{1}{n}}$
> 2. $a^m \cdot a^n = a^{m+n}$

$$\sqrt{x} \cdot \sqrt[3]{x} = x^{\frac{1}{2}} \cdot x^{\frac{1}{3}} = x^{\frac{1}{2}+\frac{1}{3}} = x^{\frac{5}{6}}$$

Therefore, the value of n is $\frac{5}{6}$.

5. (C)

Set the angle mode to Radians in your calculator.

$$\tan\theta = 0.8 \quad \Longrightarrow \quad \theta = \tan^{-1}(0.8) = 0.67$$

Therefore, the measure of angle θ in radians is 0.67.

6. (E)

> **Tip**
> If point $A(x_1, y_1)$ is the initial point and point $B(x_2, y_2)$ is the terminal point, a vector, **AB**, is defined as
> $$\mathbf{AB} = \langle x_2 - x_1, y_2 - y_1 \rangle$$

Since the initial point is $(3, -4)$ and the terminal point is $(-2, 3)$, vector **V** is defined as follows:

$$\mathbf{V} = \langle x_2 - x_1, y_2 - y_1 \rangle = \langle -2 - 3, 3 - (-4) \rangle = \langle -5, 7 \rangle$$

Therefore, the position vector **V** is $\langle -5, 7 \rangle$.

7. (A)

Set the angle mode to Radians in your calculator. In order to determine Joshua's position at $t = 0$, substitute $t = 0$ into $x(t) = 30\sin\left(\frac{\pi}{20}t\right)$ and $y(t) = -30\cos\left(\frac{\pi}{20}t\right) + 36$. Since $x(0) = 0$ and $y(0) = 6$, the equation $y(t)$ determines Joshua's vertical position from the ground at time t. In order to determine how high Joshua is above the ground at $t = 20$, substitute $t = 20$ into $y(t)$.

$$y(t) = -30\cos\left(\frac{\pi}{20}t\right) + 36 \quad \Longrightarrow \quad y(20) = -30\cos\pi + 36 = 66$$

Therefore, Joshua is 66 feet above the ground at $t = 20$.

8. (B)

Since there are 55 total juniors and the number of juniors taking Calculus is 25, the number of juniors who are taking chemistry is $55 - 25 = 30$ as shown below.

	Calculus	Chemistry	
Junior	25	30	55
Senior		**10**	45
	60	40	100

Since there are 40 students who are taking Chemistry and the number of juniors who are taking Chemistry is 30, the number of seniors who are taking Chemistry is $40 - 30 = 10$. Therefore, (B) is the correct answer.

9. (E)

Tip: The absolute value or the magnitude of a complex number $a + bi$, denoted by $|a + bi|$, is the distance from the origin to the complex number in the complex plane. The formula for finding $|a + bi|$ is as follows:

$$|a + bi| = \sqrt{a^2 + b^2}$$

$$|2 - 2i| = \sqrt{(2^2 + (-2)^2} = 2\sqrt{2}$$

10. (D)

Tip: $f(x - 2) + 3$ means moving the graph of $f(x)$ 2 units to the right and 3 units up.

The quadratic function $f(x) = -(x + 1)^2 + 2$ is expressed in vertex form and the vertex of f is $(-1, 2)$. Since $g(x) = f(x - 2) + 3$, the vertex of g is obtained by moving the vertex of f 2 units to the right and 3 units up. Therefore, the vertex of g is $(1, 5)$.

11. (A)

For any angle t, the range of the sine function is $-1 \leq y \leq 1$.

$$-1 \leq \sin\left(3t + \frac{3\pi}{4}\right) \leq 1$$

$$-2 \leq -2\sin\left(3t + \frac{3\pi}{4}\right) \leq 2 \qquad \text{Multiply each side of the inequality by } -2$$

$$-3 \leq -2\sin\left(3t + \frac{3\pi}{4}\right) - 1 \leq 1 \qquad \text{Subtract 1 from each side of the inequality}$$

Therefore, the range of $y = -2\sin\left(3t + \frac{3\pi}{4}\right) - 1$ is $-3 \leq y \leq 1$.

12. (D)

$g(x) = 2x^3 + 1$. Thus, $g(0.64) = 2(0.64)^3 + 1 = 1.52$. Since $f(x) = \frac{x}{2} + 1$,

$$f(g(0.64)) = f(1.52) = \frac{1.52}{2} + 1 = 1.76$$

Therefore, $f(g(0.64)) = 1.76$

13. (A)

Since $\sin\frac{2\pi}{3} = \sin\frac{\pi}{3}$ and $\sin\frac{4\pi}{3} = \sin\frac{5\pi}{3} = -\sin\frac{\pi}{3}$,

$$\sin\frac{\pi}{3} + \sin\frac{2\pi}{3} + \sin\frac{4\pi}{3} + \sin\frac{5\pi}{3} = \sin\frac{\pi}{3} + \sin\frac{\pi}{3} - \sin\frac{\pi}{3} - \sin\frac{\pi}{3} = 0$$

Therefore, (A) is the correct answer.

14. (B)

$t = 10$ represents 2010. In order to find the proportion of U.S. retail sales involving the internet in 2020, substitute $t = 10$ into $P(t)$.

$$P(t) = \frac{0.7}{1 + 3e^{-0.28t}} \quad \Longrightarrow \quad P(10) = \frac{0.7}{1 + 3e^{-0.28(10)}} = 0.59$$

Therefore, the proportion of U.S. retail sales involving the internet in 2020 is 0.59.

15. (C)

Tip	$a^{\log_a x} = x^{\log_a a} = x$

Since $10^{\log_{10} x} = x$, $10^{\log_{10} x} < 10$ simplifies to $x < 10$. Therefore, the largest possible positive integer value of x for which $x < 10$ is 9.

16. (E)

Tip	
	1. $(a+b)^2 = a^2 + 2ab + b^2$
	2. $\cos^2\theta + \sin^2\theta = 1$
	3. $\sin 2\theta = 2\sin\theta\cos\theta$

$$\begin{aligned}(\sin\theta + \cos\theta)^2 &= \sin^2\theta + 2\sin\theta\cos\theta + \cos^2\theta \\ &= 1 + 2\sin\theta\cos\theta \\ &= 1 + \sin 2\theta \\ &= 1 + \frac{2}{3} \\ &= \frac{5}{3}\end{aligned}$$

Therefore, (E) is the correct answer.

17. (A)

The statements in answer choices (B), (C), (D), and (E) are correct statements about the inverse function of $f(x)$. Since the inverse function of $f(x)$ is not $\dfrac{1}{f(x)}$, (A) is the correct answer.

18. (B)

Set the angle mode to Radians in your calculator. Since $x \to 0$, substitute 0.01 for x.

$$\lim_{x\to 0} \frac{e^x - 1}{\sin x} = \frac{e^{0.01} - 1}{\sin(0.01)} = 1.005$$

Therefore, $\displaystyle\lim_{x\to 0} \frac{e^x - 1}{\sin x} = 1.$

19. (A)

Let x be the height of the lamp post as shown below.

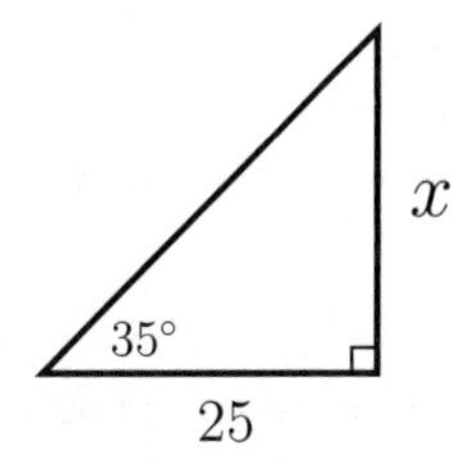

Since the lamp casts a shadow of 25 feet when the sun makes a 35° angle of elevation, use the definition of the tangent function to find the height of the lamp post.

$$\tan 35^\circ = \frac{x}{25} \quad \Longrightarrow \quad x = 25\tan 35^\circ = 17.5$$

Therefore, the height of the lamp post is 17.5.

20. (E)

$\triangle ABC$ and $\triangle ADE$ are similar triangle as shown below.

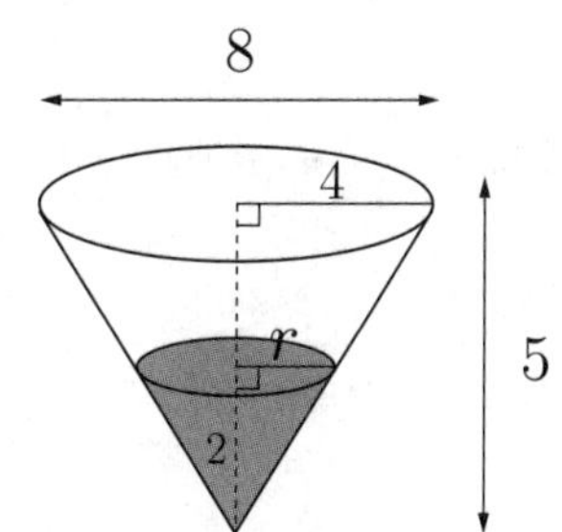

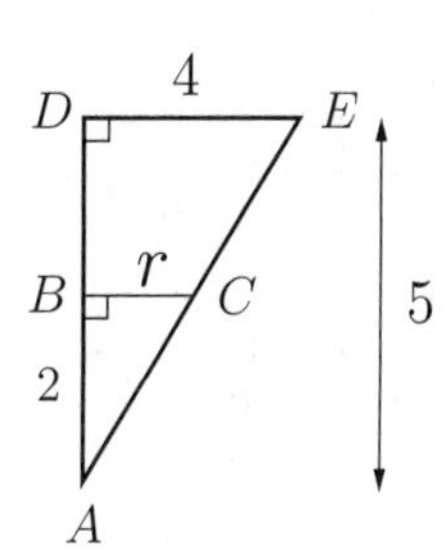

Set up a proportion to find the radius of the surface of the water.

$$\frac{r}{2} = \frac{4}{5} \quad \Longrightarrow \quad r = \frac{8}{5}$$

Therefore, the radius of the surface of the water is $\dfrac{8}{5}$.

21. (D)

$\frac{1}{2}x^3 - 1 = 0.6$	Add 1 to each side
$\frac{1}{2}x^3 = 1.6$	Multiply each side 2
$x^3 = 3.2$	Take the cube root of both sides
$x = \sqrt[3]{3.2} = 1.47$	

Therefore, the value of x for which $\frac{1}{2}x^3 - 1 = 0.6$ is 1.47.

22. (E)

> Tip Let A be a $m \times n$ matrix and B be a $n \times p$ matrix. If the number of columns in matrix A is equal to the number of rows in matrix B, the product of the two matrices, AB, is defined and is a $m \times p$ matrix.

A is a $m \times 2$ matrix and B is a 2×4 matrix. Since the number of columns in matrix A is equal to the number of rows in matrix B, the product of the two matrices, AB, is defined and is a $m \times 4$ matrix, where m is a positive integer. Therefore, (E) is the correct answer.

23. (D)

Since y varies directly with the cube of x and inversely with the square root of z, start with $y = \frac{kx^3}{\sqrt{z}}$. Substitute 2 for x, 16 for z, and 6 for y to find the value of k.

$y = \frac{kx^3}{\sqrt{z}}$	Substitute 2 for x, 16 for z, and 6 for y
$6 = \frac{8k}{4}$	Solve for k
$k = 3$	

Thus, the equation that relates x, y, and z is $y = \frac{3x^3}{\sqrt{z}}$. Substitute $\frac{1}{3}$ for x and $\frac{1}{4}$ for z to find the value of y.

$y = \frac{3x^3}{\sqrt{z}}$	Substitute $\frac{1}{3}$ for x and $\frac{1}{4}$ for z
$y = \frac{3\left(\frac{1}{3}\right)^3}{\sqrt{\frac{1}{4}}}$	Solve for y
$y = \frac{\frac{1}{9}}{\frac{1}{2}}$	
$y = \frac{2}{9}$	

Therefore, the value of y when $x = \frac{1}{3}$ and $z = \frac{1}{4}$ is $\frac{2}{9}$.

24. (D)

In order to evaluate the fifth term, we need to find the previous four terms as shown below.

$a_{n+2} = a_{n+1} \cdot a_n,\ a_1 = 2, a_2 = 3$	Recursive formula with $a_1 = 2$ and $a_2 = 3$
$a_3 = a_2 \cdot a_1 = 2(3) = 6$	Substitute 1 for n to find a_3
$a_4 = a_3 \cdot a_2 = 6(3) = 18$	Substitute 2 for n to find a_4
$a_5 = a_4 \cdot a_3 = 18(6) = 108$	Substitute 3 for n to find a_5

Since the fifth term of the sequence is 108, the factors of 108 are 1, 2, 3, 4, 6, 9, 12, 18, 27, 36, 54, 108. Therefore, (D) is the correct answer.

25. (A)

The perpendicular bisector of a line segment is the set of all points that are equidistant from its endpoints.

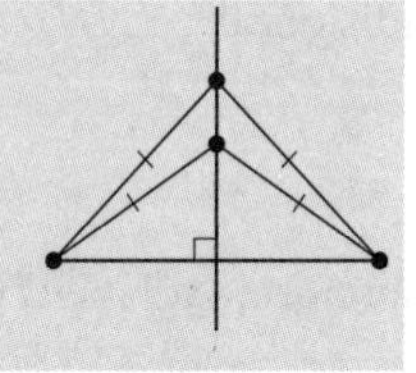

Draw a line segment that connects the two points in the xy-plane. The set of all points that are equidistant from the two points is the perpendicular bisector of the line segment. Since the perpendicular bisector of the line segment is a line, (A) is the correct answer.

26. (A)

Figure A shows the graph of $\cot x$ whose vertical asymptotes are at $x = 0$ and $x = \pi$.

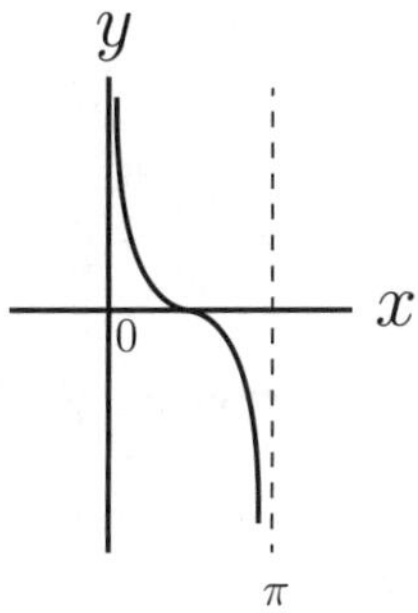

Fig. A: The graph of $\cot x$

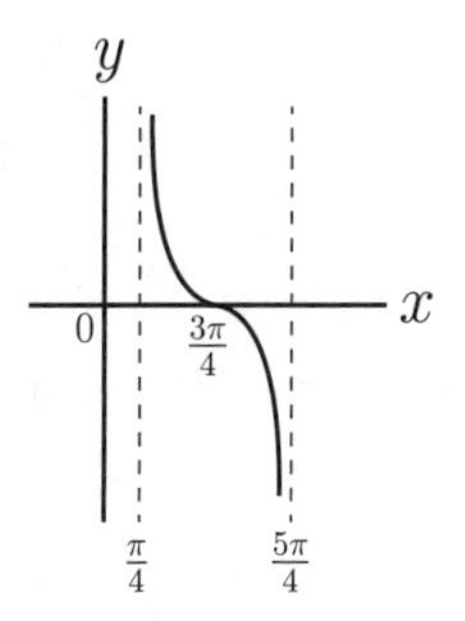

Fig. B: The graph of $\cot\left(x - \frac{\pi}{4}\right)$

Figure B shows the graph of $\cot\left(x - \frac{\pi}{4}\right)$, which is obtained by shifting the graph of $\cot x$ to the right $\frac{\pi}{4}$. The vertical asymptotes of $\cot\left(x - \frac{\pi}{4}\right)$ at $x = \frac{\pi}{4}$ and $x = \frac{5\pi}{4}$. Therefore, (A) is the correct answer.

27. (D)

In the figure below, $\triangle ABD$ is a right triangle.

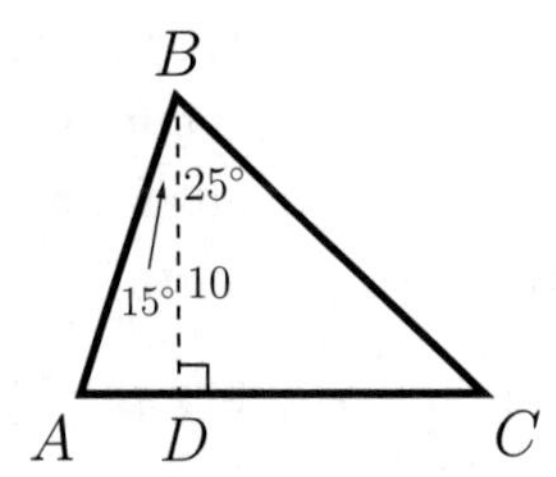

Use the definition of the tangent function to find AD.

$$\tan 15^\circ = \frac{AD}{10} \quad \Longrightarrow \quad AD = 10\tan 15^\circ = 2.68$$

Additionally, $\triangle BDC$ is a right triangle. Use the definition of the tangent function to find DC.

$$\tan 25^\circ = \frac{DC}{10} \quad \Longrightarrow \quad DC = 10\tan 25^\circ = 4.66$$

Therefore, $AC = AD + DC = 2.68 + 4.66 = 7.34$.

28. (C)

Tip: Factor Theorem: If $x - k$ is a factor of $f(x)$, then the remainder $r = f(k) = 0$.

Since $f(2) = (2)^3 - 3(2)^2 + 4(2) - 4 = 0$, $x - 2$ is a factor of $f(x) = x^3 - 3x^2 + 4x - 4$. Therefore, (C) is the correct answer.

29. (E)

The vertex of the absolute value function $f(x) = -|x-1| - 3$ is $(1, -3)$ as shown below.

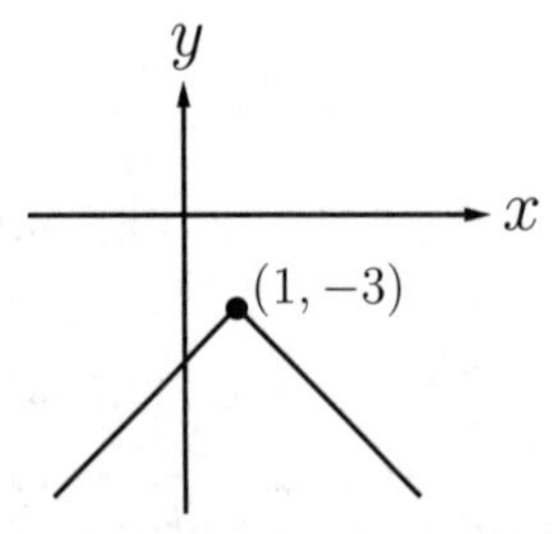

Since the graph of f never touches or crosses the x-axis, there is no x-intercept.

30. (C)

$$\frac{7p+2q}{p+q} = 4 \qquad \text{Multiply each side by } p+q$$
$$7p+2q = 4p+4q \qquad \text{Subtract } 2q \text{ and } 4p \text{ from each side}$$
$$3p = 2q$$
$$\frac{p}{q} = \frac{2}{3}$$

Therefore, the value of $\frac{p}{q}$ is $\frac{2}{3}$.

31. (E)

Tip
1. $\csc^2 x = \frac{1}{\sin^2 x}$
2. $\sin^2 x + \cos^2 x = 1 \implies \sin^2 x = 1 - \cos^2 x$

$$\csc^2 x(1-\cos^2 x) = \frac{1}{\sin^2 x} \cdot \sin^2 x = 1$$

32. (D)

In order to write a general equation for the ellipse $\frac{(x-h)^2}{a^2} + \frac{(y-k)^2}{b^2} = 1$, complete the squares in y.

$$x^2 + 4y^2 - 16y = 0$$
$$x^2 + 4(y^2 - 4y) = 0 \qquad \text{Factor out 4}$$
$$x^2 + 4(y^2 - 4y + 4) = 16 \qquad \text{Add 16 to each side}$$
$$x^2 + 4(y-2)^2 = 16 \qquad \text{Complete the squares in } y$$
$$\frac{x^2}{4^2} + \frac{(y-2)^2}{2^2} = 1 \qquad \text{Divide each side by 16}$$

The figure below shows the graph of the ellipse $\frac{x^2}{4^2} + \frac{(y-2)^2}{2^2} = 1$.

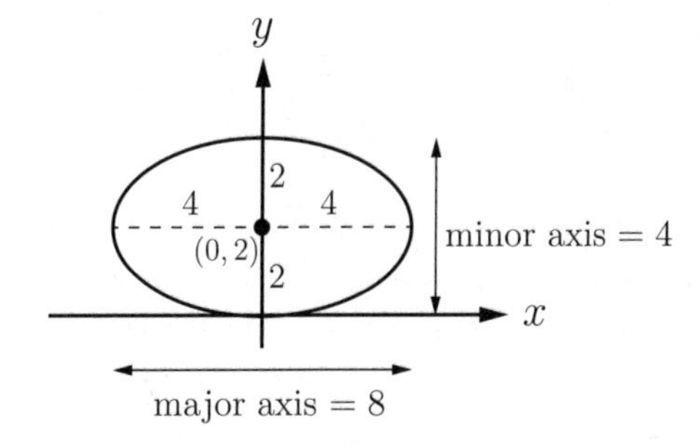

The center of the ellipse is $(0, 2)$ and the length of the major axis of the ellipse is 8. Therefore, (D) is the correct answer.

33. (C)

> Tip $a^x = y \iff x = \log_a y$

$$\begin{aligned} 2(3)^x + 2 &= 7 && \text{Subtract 2 from each side} \\ 2(3)^x &= 5 && \text{Divide each side by 2} \\ 3^x &= 2.5 && \text{Convert the equation to a logarithmic equation} \\ x &= \log_3 2.5 = 0.83 \end{aligned}$$

Therefore, the value of x for which $2(3)^x + 2 = 7$ is 0.83.

34. (C)

Multiply each side by $(x-1)(2x-1)$ and find the value for A.

$$\begin{aligned} (x-1)(2x-1) \cdot \frac{4x-1}{(x-1)(2x-1)} &= \left(\frac{A}{x-1} - \frac{2}{2x-1}\right)(x-1)(2x-1) \\ 4x-1 &= A(2x-1) - 2(x-1) \\ 4x-1 &= 2Ax - A - 2x + 2 \\ 4x-1 &= (2A-2)x - A + 2 \end{aligned}$$

Since the equation $4x - 1 = (2A-2)x - A + 2$ is an identity equation, the coefficients of x on each side of the equation must be the same; that is, $4 = 2A - 2$. Therefore, the value of A is 3.

35. (A)

> Tip $f(-x)$ involves reflecting the graph of $f(x)$ about the y-axis.

$f(x) = x^3 - 3x^2 + 2x + 4$. In order to find $f(-x)$, substitute $-x$ for x in $f(x)$.

$$f(-x) = (-x)^3 - 3(-x)^2 + 2(-x) + 4 = -x^3 - 3x^2 - 2x + 4$$

Therefore, (A) is the correct answer.

36. (C)

> Tip If the points (x_1, y_1, z_1) and (x_2, y_2, z_2) are given in three-dimensional space, the distance D between the two points is defined as follows:

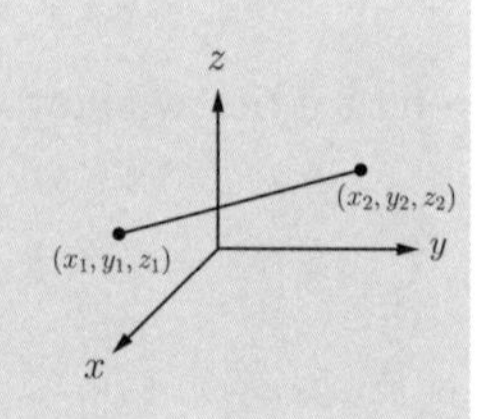

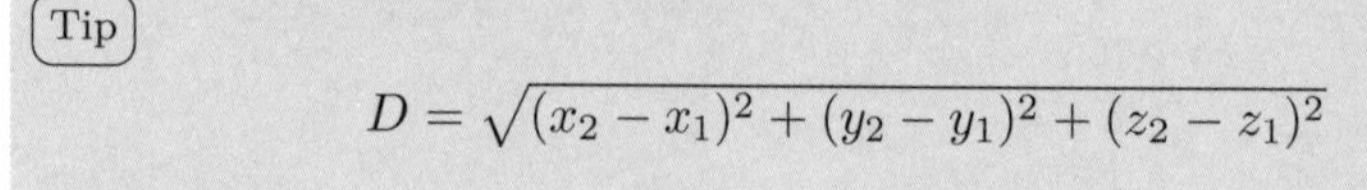

$$D = \sqrt{(x_2 - x_1)^2 + (y_2 - y_1)^2 + (z_2 - z_1)^2}$$

The center and the radius of the sphere $x^2 + y^2 + (z+1)^2 = 11$ are $(0, 0, -1)$ and $\sqrt{11}$, respectively. In order for a point to lie inside or on the surface of the sphere, the distance between the center of the sphere and the point must be less than or equal to the radius, $\sqrt{11}$. Since the distance between the center $(0, 0, -1)$ and $(3, 1, 0)$ is $\sqrt{11}$ as shown below,

$$D = \sqrt{(x_2 - x_1)^2 + (y_2 - y_1)^2 + (z_2 - z_1)^2} = \sqrt{(3-0)^2 + (1-0)^2 + (0-(-1))^2} = \sqrt{11}$$

the point $(3, 1, 0)$ is on the surface of the sphere. Therefore, (C) is the correct answer.

37. (A)

Tip

1. $(\sqrt{a}+\sqrt{b})(\sqrt{a}-\sqrt{b})=a-b$
2. In order to rationalize the denominator, multiply the numerator and the denominator by the denominator's conjugate. For instance,

$$\frac{1}{\sqrt{a}+\sqrt{b}}=\frac{1}{(\sqrt{a}+\sqrt{b})}\cdot\frac{\sqrt{a}-\sqrt{b}}{(\sqrt{a}-\sqrt{b})}=\frac{\sqrt{a}-\sqrt{b}}{a-b}$$

Rationalize both expressions $\frac{1}{\sqrt{3}+\sqrt{2}}$ and $\frac{1}{\sqrt{2}+1}$.

$$\frac{1}{\sqrt{3}+\sqrt{2}}=\frac{1}{(\sqrt{3}+\sqrt{2})}\cdot\frac{\sqrt{3}-\sqrt{2}}{(\sqrt{3}-\sqrt{2})}=\sqrt{3}-\sqrt{2},\qquad \frac{1}{\sqrt{2}+1}=\frac{1}{(\sqrt{2}+1)}\cdot\frac{\sqrt{2}-1}{(\sqrt{2}-1)}=\sqrt{2}-1$$

Therefore, $\frac{1}{\sqrt{3}+\sqrt{2}}+\frac{1}{\sqrt{2}+1}=\sqrt{3}-\sqrt{2}+\sqrt{2}-1=\sqrt{3}-1$

38. (B)

Tip

The number of permutations of n objects, where there are n_1 indistinguishable objects of one kind, and n_2 indistinguishable objects of a second kind, is given by

$$\text{Permutations with repetition}=\frac{n!}{n_1!\cdot n_2!}$$

The word SOLOMON has 7 letters. Since the letters S, O, L, M, and N are distinguishable, the order is important. However, there are 3 O's out of 7 letters. Thus,

$$\text{Permutations with repetition}=\frac{7!}{3!}=\frac{7\cdot 6\cdot 5\cdot 4\cdot \cancel{3!}}{\cancel{3!}}=840$$

Therefore, the number of different arrangements that can be formed using the letters in word SOLOMON is 840.

39. (B)

Define x as the number of games the pitcher played before the next five games. Since the pitcher has won 60% of the games he pitched, the number of games he has won can be expressed as $0.6x$. For the next five games, the pitcher won 2 games. Thus, the total number of games he has won can be expressed as $0.6x+2$. Furthermore, the pitcher won 2 games and lost 3 so that he finished the season having won 56% of his games. Thus, the total number of games he has won also can be expressed as $0.56(x+5)$. Set $0.6x+2$ and $0.56(x+5)$ equal to each other and solve for x.

$$\begin{aligned}0.6x+2&=0.56(x+5)\\0.6x+2&=0.56x+2.8\\0.04x&=0.8\\x&=20\end{aligned}$$

Therefore, the total number of games that the pitcher played in all is $x+5=25$.

40. (C)

> **Tip** Cancelling out a common factor $x - c$ from both the numerator and the denominator of a rational function produces a hole in the graph at $x = c$.

The figure below represents the graph of the rational function $y = \dfrac{1}{x}$.

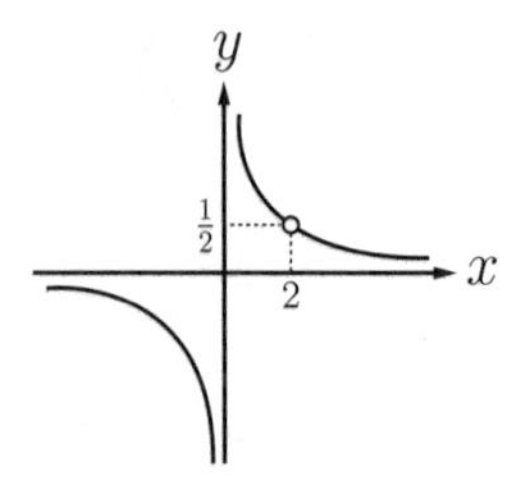

Since the graph has a hole at $x = 2$, the rational function $y = \dfrac{1}{x}$ must have the common factor $x-2$ from the numerator and the denominator; that is $y = \dfrac{x-2}{x(x-2)}$. Therefore, (C) is the correct answer.

41. (C)

A manufacturing company found out that 3% of their production was defective, which means that the probability that each light bulb you have purchased is NOT defective is $1 - 0.03$ or 0.97. Since you have purchased five light bulbs, the probability that none of them are defective is $(0.97)^5 = 0.86$.

42. (B)

> **Tip**
> 1. $\log_a a = 1$
> 2. $\log_a xy = \log_a x + \log_a y$

$$\begin{aligned}\log_2 30 &= \log_2(2 \cdot 3 \cdot 5)\\ &= \log_2 2 + \log_2 3 + \log_2 5\\ &= 1 + m + n\end{aligned}$$

Therefore, $\log_2 30$ in terms of m and n is $m + n + 1$.

43. (A)

The amount of money that the company charged a customer who used 800 minutes in October would be $\$13.5 + \$0.015(800) = \$25.5$. The company charged the customer \$26.75, which means that the customer used more than 800 minutes. Since the company charges 2.5 cents per minute after the first 800 minutes used, the customer used $\dfrac{26.75 - 25.5}{0.025} = 50$ minutes after 800 minutes. Therefore, the total number of minutes that the customer used was $800 + 50 = 850$ minutes.

44. (E)

$$\tan^2\theta = 1 \qquad \text{Take the square root of each side}$$
$$\tan\theta = \pm 1$$
$$\tan\theta = 1 \implies \theta = \frac{\pi}{4}, \frac{5\pi}{4}$$
$$\tan\theta = -1 \implies \theta = \frac{3\pi}{4}, \frac{7\pi}{4}$$

Therefore, (E) is the correct answer.

45. (E)

Tip

1. $i^2 = -1$
2. $(a+bi)(a-bi) = a^2 - bi^2 = a^2 + b^2$
3. In order to rationalize the denominator, multiply the numerator and the denominator by the denominator's conjugate. For instance,

$$\frac{1}{a+bi} = \frac{1}{(a+bi)} \cdot \frac{(a-bi)}{(a-bi)} = \frac{a-bi}{a^2+b^2}$$

$$\frac{3-4i}{2-3i} = \frac{(3-4i)}{(2-3i)} \cdot \frac{(2+3i)}{(2+3i)} = \frac{18+i}{13} = \frac{18}{13} + \frac{i}{13}$$

46. (E)

The two vectors are $\mathbf{A} = \langle 3, 2\rangle$ and $\mathbf{B} = \langle -1, k\rangle$. Thus,

$$\begin{aligned} 3\mathbf{B} - 2\mathbf{A} &= 3\langle -1, k\rangle - 2\langle 3, 2\rangle \\ &= \langle -3, 3k\rangle - \langle 6, 4\rangle \\ &= \langle -9, 3k-4\rangle \end{aligned}$$

Since $3\mathbf{B} - 2\mathbf{A} = \langle -9, 3k-4\rangle = \langle -9, 11\rangle$, $3k - 4 = 11$. Therefore, the value of k is 5.

47. (D)

Tip

1. The nth term of a geometric sequence: $a_n = a_1 \times r^{n-1}$
2. $a_9 = a_7 \times r^2$, where a_9 and a_7 are the 9th term and 7th term, respectively.

In the geometric sequence, the 3rd term is $\frac{1}{3}$ and the 7th term is 27. Use the nth term formula to find the common ratio, r.

$$\frac{a_7}{a_3} = \frac{a_1 \times r^6}{a_1 \times r^2} = \frac{27}{\frac{1}{3}}$$
$$r^4 = 81$$
$$r = \pm 3 \implies r^2 = 9$$

Since $a_9 = a_7 \times r^2$, the 9th term is $27 \times 9 = 243$.

48. (B)

In order to have a general equation of a hyperbola $\frac{(x-h)^2}{a^2} - \frac{(y-k)^2}{b^2} = 1$, divide each side of the equation $16x^2 - 9y^2 = 144$ by 144. Thus, $\frac{x^2}{3^2} - \frac{y^2}{4^2} = 1$. The graph of the hyperbola is shown below.

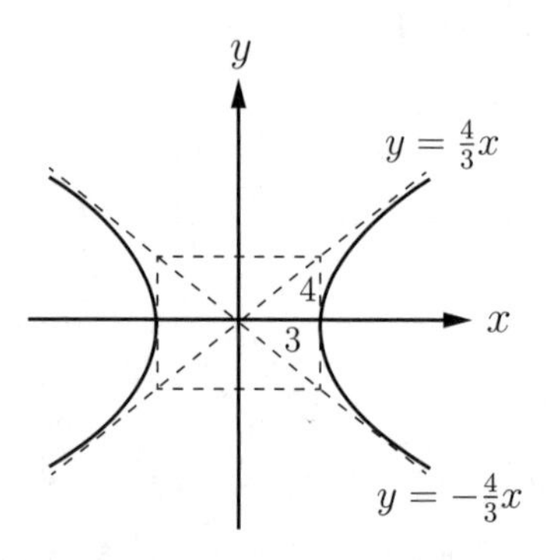

Since the slopes of the asymptotes are $\pm\frac{4}{3}$ and the asymptotes pass through the center $(0, 0)$, the equations of the asymptotes are $y = \pm\frac{4}{3}x$.

49. (D)

The point $(1, 2)$ satisfies the system of inequalities shown below.

$$\begin{aligned} y &> x^2 & \Longrightarrow \quad & 2 > 1^2 & \checkmark\text{(True)} \\ y &< x+1 & \Longrightarrow \quad & 2 < 1+2 & \checkmark\text{(True)} \end{aligned}$$

Therefore, (D) is the correct answer.

50. (D)

Tip
1. 15 minutes = 0.25 hour. 54 minutes = 0.9 hour.
2. distance = rate × time

Convert 15 minutes to $\frac{15\,\text{minutes}}{60\,\text{minutes}} = 0.25$ hour and 54 minutes to $\frac{54\,\text{minutes}}{60\,\text{minutes}} = 0.9$ hour. As shown in the table below, Jason traveled at a rate of 60 miles per hour for 2.25 hours from city A to city B, which means that Jason traveled $60 \times 2.25 = 135$ miles from city A to city B.

	From city A to city B	From city B to city C	Total
Distance	135 miles	30 miles	165 miles
Time	2.25 hours	0.9 hour	3.15 hours

Since Jason traveled 30 miles from city B to city C, the total distance of the entire trip is $135+30 = 165$ miles. Furthermore, it took Jason 2.25 hours to travel from city A to city B, and 0.9 hour from city B to city C. Thus, the total time for the entire trip is $2.25+0.9 = 3.15$ hours. Therefore, Jason's average speed for the entire trip is $\frac{165\,\text{miles}}{3.15\,\text{hour}} = 52.38$ miles per hour.

SAT II MATH LEVEL 2 TEST 4

Directions: Among the given answer choices, choose the BEST answer for each problem. If the exact numerical value is not within the given answer choices, select the answer that best approximates this value. Afterwards, fill in the corresponding oval on the answer sheet.

Notes:

1. A calculator may be required to answer some of the questions. Scientific and graphing calculators are allowed during this test.
2. For some questions, it is up to you to decide whether the calculator should be in degree mode or radian mode.
3. Provided figures for questions are drawn as accurately as possible UNLESS otherwise specified by the problem. Unless otherwise indicated, all figures are assumed to lie in a plane.
4. Unless otherwise specified, it can be assumed that the domain of any function f is to be the set of all real numbers x for which $f(x)$ is a real number.

Reference Information: Use the following information and formulas as a reference in answering questions on this test.

1. If the radius and height of a right circular cone are r and h respectively, then the Volume V of the cone is $V = \frac{1}{3}\pi r^2 h$.
2. If the circumference and slant height of a right circular cone are c and ℓ respectively, then the Lateral Area A of the cone is $A = \frac{1}{2}c\ell$.
3. At any given radius r, the Volume V of a sphere is $V = \frac{4}{3}\pi r^3$.
4. At any given radius r, the Surface Area A of a sphere is $A = 4\pi r^2$.
5. The Volume V of a pyramid is $V = \frac{1}{3}Bh$; given that B and h represent the base area and height of the pyramid respectively.

USE THIS SPACE FOR SCRATCH WORK

1. $\cos 35° \times \sec 35° =$

(A) $\sin 35°$

(B) $\csc 35°$

(C) $\tan 35°$

(D) $\cot 35°$

(E) 1

2. For positive integers x, y, and z, $x^3yz = 2xyz^2$. What is the value of $\frac{x^2}{z}$?

(A) 2

(B) 1

(C) $\frac{2}{3}$

(D) $\frac{1}{2}$

(E) $\frac{1}{3}$

3. Which of the following statement is NOT true about complex numbers?

(A) $i = \sqrt{-1}$

(B) $i^2 = -1$

(C) $i^5 = -i$

(D) $i^{11} = -i$

(E) $i(1+i) = -1+i$

USE THIS SPACE FOR SCRATCH WORK

4. For the following function g shown below, what is the value of $g(2) + g(3)$?

$$g(x) = \begin{cases} \frac{2}{3}x + 2, & x \geq 3 \\ -2x - 1, & x < 3 \end{cases}$$

(A) 3

(B) 2

(C) 1

(D) 0

(E) -1

5. If $\dfrac{n!}{(n-2)!} = 6$, what is the value of n ?

(A) 6

(B) 5

(C) 4

(D) 3

(E) 2

6. The knot is a unit of speed equal to one nautical mile or approximately 1.151 miles per hour. If a fishing boat is traveling at a rate of 30 miles per hour, how fast, in knots, is the boat traveling?

(A) 34.53 knots

(B) 32.35 knots

(C) 30.43 knots

(D) 28.96 knots

(E) 26.06 knots

USE THIS SPACE FOR SCRATCH WORK

7. $3\log_5 75 - 3\log_5 3 =$

(A) 4

(B) 5

(C) 6

(D) 8

(E) 9

8. Jason borrowed $20,000 from a bank to purchase a car. The bank charges him 0.5% interest per month on any unpaid balance and he will pay $300 toward the balance each month. Jason's balance each month after making a $300 payment is defined by $a_n = 1.005a_{n-1} - 300$, $a_0 = 20,000$, for $n \geq 1$. What is Jason's balance after he makes the third payment?

(A) $19,397

(B) $19,194

(C) $18,990

(D) $18,784

(E) $18,579

9. In the right triangle ABC, $m\angle C = 90°$. If $AB = 13$ and $AC = 12$, what is the exact value of $\cos B$?

(A) $\frac{1}{13}$

(B) $\frac{5}{13}$

(C) $\frac{5}{12}$

(D) $\frac{12}{13}$

(E) $\frac{13}{12}$

USE THIS SPACE FOR SCRATCH WORK

10. Figure 1 shows the graph of the logarithmic function $y = n\log_4 x$. If the graph contains the point $(2, 6)$, what is the value of n ?

(A) 12

(B) 8

(C) 6

(D) 4

(E) 2

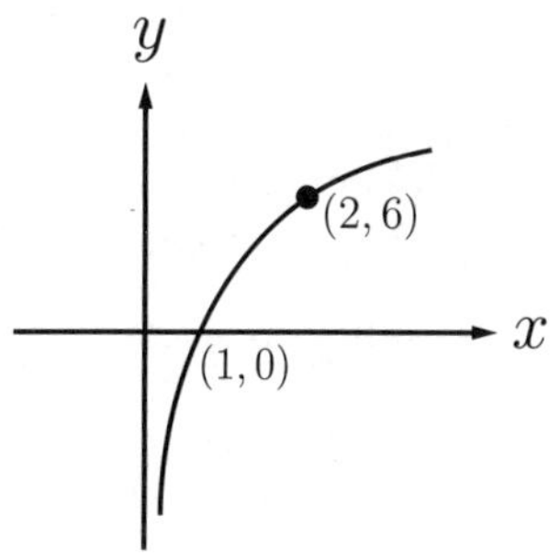

Figure 1

11. The two lines $y = -2x + k$ and $y = -x + k$ intersect at $(0, k)$, where $k > 0$. If the x-intercepts of the lines $y = -2x + k$ and $y = -x + k$ are p and q respectively, what is the value of $\dfrac{q}{p}$?

(A) 3

(B) 2.75

(C) 2.5

(D) 2.25

(E) 2

12. If $4^x + 4^x + 4^x + 4^x = 2^k$, which of the following expression is equivalent to k ?

(A) $x + 1$

(B) $2x + 2$

(C) $3x + 1$

(D) $4x + 1$

(E) $4x + 4$

USE THIS SPACE FOR SCRATCH WORK

13. The number of milligrams A of a certain drug in a patient's bloodstream t hours after the drug has been injected can be modeled by $A(t) = 10e^{-0.2t}$. What is the number of milligrams of the drug in a patient's blood stream after 10 hours?

(A) 1.35 milligrams

(B) 2.23 milligrams

(C) 3.68 milligrams

(D) 5.51 milligrams

(E) 6.98 milligrams

14. Figure 2 shows a scatterplot which displays the weight (in pounds) versus the height (in inches) of 18 students. The equation of a line of best fit is defined by $y = 5.25x - 192$, where y represents weight and x represents height. Using the line of best fit, what would be the weight of a person who is 5 feet 9 inches tall? (Round your answer to the nearest integer.)

(A) 160 pounds

(B) 165 pounds

(C) 170 pounds

(D) 175 pounds

(E) 180 pounds

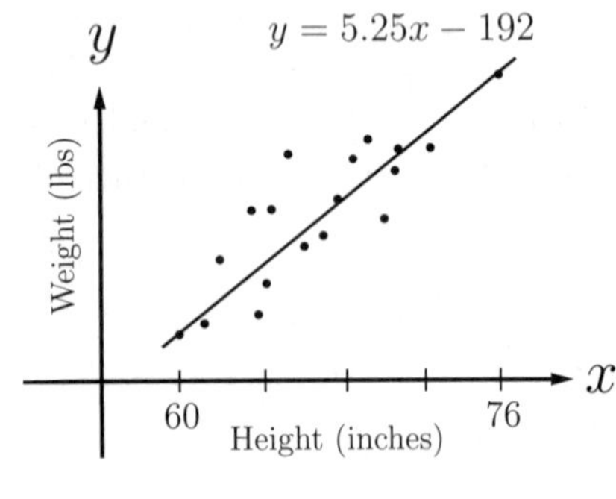

Figure 2

15. What is the value of x for which $x^2 = 3x$?

(A) $x = 3$ only

(B) $x = 1$ only

(C) $x = 3$ or $x = 0$

(D) $x = 3$ or $x = -3$

(E) No solution

USE THIS SPACE FOR SCRATCH WORK

16. What is the sum of the geometric series $\frac{1}{3}+\frac{1}{9}+\frac{1}{27}+\cdots$?

(A) $\frac{5}{2}$

(B) $\frac{4}{3}$

(C) $\frac{3}{4}$

(D) $\frac{2}{3}$

(E) $\frac{1}{2}$

17. If a cube with side length $2\sqrt{3}$ is inscribed in a sphere, what is the volume of the sphere?

(A) 32π

(B) 36π

(C) 42π

(D) 48π

(E) 64π

18. Set A has 5 numbers. The mean and the median of set A are 7 and 10, respectively. If 10 is added to set A to produce set B, which of the following best represents the mean and the median of set B ?

(A) Mean $=7$ and Median $=10$

(B) Mean $=7$ and Median $=13.5$

(C) Mean $=7.5$ and Median $=10$

(D) Mean $=7.5$ and Median $=13.5$

(E) Mean $=10$ and Median $=10$

USE THIS SPACE FOR SCRATCH WORK

19. Let $g(x) = x^2 - 1$ and $h(x) = x - 3$. If $f(x) = \dfrac{h(x)}{g(x)}$, what is the x-intercept of f ?

(A) 1

(B) 3

(C) 1 or -1

(D) 3 or 1

(E) 3 or 1 or -1

20. $\dfrac{x^2y + xy^2}{xy} =$

(A) $\dfrac{x}{y}$

(B) $\dfrac{y}{x}$

(C) $2xy$

(D) xy

(E) $x + y$

21. All human blood can be either A, B, O, or AB. The table below shows the distribution of blood types for a randomly chosen person in a small city. If the small city has a population of 250, how many people in the city have AB as their blood type?

Blood type	A	B	O	AB
Probability	0.35	0.15	0.38	

(A) 88

(B) 70

(C) 55

(D) 42

(E) 30

USE THIS SPACE FOR SCRATCH WORK

22. What is the period of $y = \sin\left(\frac{2\pi}{3}x - 1\right) + 2$?

(A) 3π

(B) 2π

(C) 3

(D) 2

(E) 1

23. If $\frac{ab}{a-b} < 0$, which of the following inequality could be true for a and b ?

(A) $0 < b < a$

(B) $b < 0 < a$

(C) $b < a < 0$

(D) $b < 0$ and $a < 0$

(E) $b > 0$ and $a > 0$

24. How many two-digit odd numbers are greater than 40?

(A) 45

(B) 40

(C) 35

(D) 30

(E) 25

USE THIS SPACE FOR SCRATCH WORK

25. In Figure 3, triangle ABC is a right triangle. If $m\angle ABC = 50°$ and $BC = 10$, what is the area of the triangle ABC ?

(A) 82.37

(B) 74.41

(C) 67.62

(D) 59.59

(E) 53.28

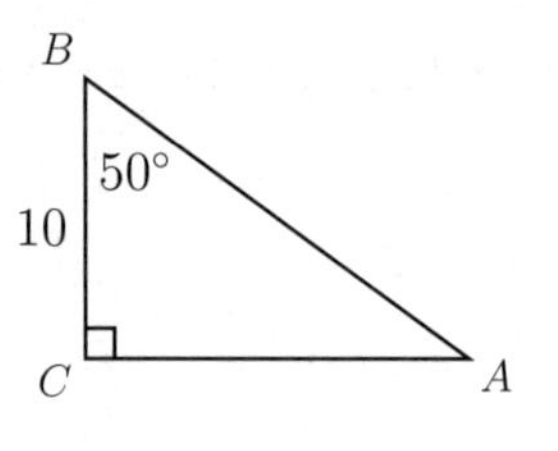

Figure 3

26. What is the inverse function of $y = \sqrt[3]{\dfrac{x-1}{2}} + 4$?

(A) $y = 2(x-4)^3 - 1$

(B) $y = 2(x-4)^3 + 1$

(C) $y = 2(x+4)^3 - 1$

(D) $y = 2(x+4)^3 + 1$

(E) $y = 4(x-2)^3 + 1$

27. Solve the equation: $\cos\theta = \sqrt{3}$, where $0 \le \theta < 2\pi$.

(A) $x = \dfrac{\pi}{6}$

(B) $x = \dfrac{\pi}{4}$

(C) $x = \dfrac{\pi}{6}$ or $x = \dfrac{11\pi}{6}$

(D) $x = \dfrac{\pi}{3}$ or $x = \dfrac{5\pi}{3}$

(E) No solution

USE THIS SPACE FOR SCRATCH WORK

28. Which of the following polar coordinates (r, θ) correspond with the rectangular coordinates $(-\sqrt{3}, -1)$?

    (A) $\left(2, \frac{\pi}{6}\right)$

    (B) $\left(2, \frac{5\pi}{6}\right)$

    (C) $\left(2, \frac{7\pi}{6}\right)$

    (D) $\left(2, \frac{4\pi}{3}\right)$

    (E) $\left(2, \frac{5\pi}{3}\right)$

29. Which of the following statement is logically equivalent to the conditional statement "If the weather is sunny, Joshua goes on a field trip." ?

    (A) "If Joshua does not go on a field trip, the weather is not sunny."

    (B) "If Joshua goes on a field trip, the weather is sunny."

    (C) "If Joshua goes on a field trip, the weather is sometimes sunny."

    (D) "If the weather is not sunny, Joshua does not go on a field trip."

    (E) "If the weather is sunny, Joshua sometimes goes on a field trip."

30. If $f(x) = 2x^3 - 3x + 2$ is divided by $x^2 + 1$, what is the remainder?

    (A) $-x - 2$

    (B) $-x + 2$

    (C) $-2x + 5$

    (D) $-5x - 5$

    (E) $-5x + 2$

USE THIS SPACE FOR SCRATCH WORK

31. Which of the following could be the graph of the parametric equations $y = \sin^2 t$, $x = \cos t$, where $\frac{\pi}{2} \le t \le \pi$?

(A)

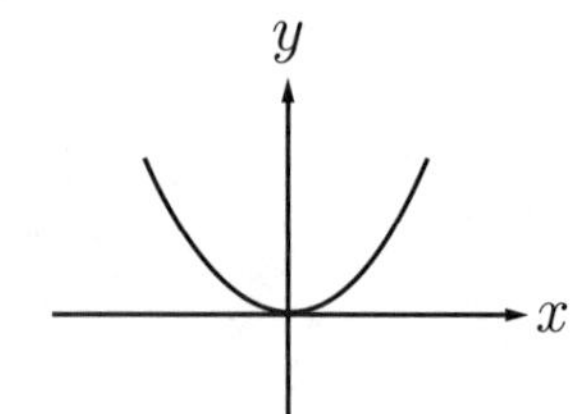

(B)

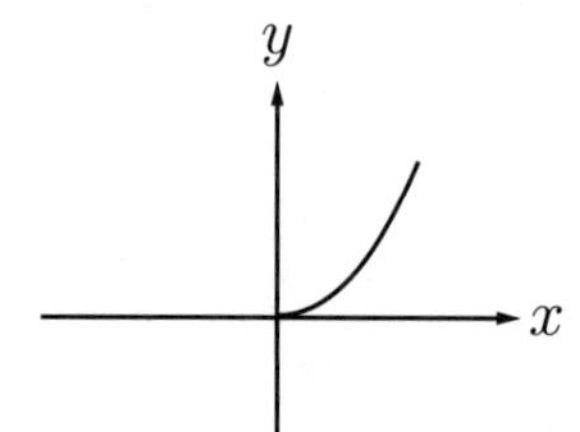

(C)

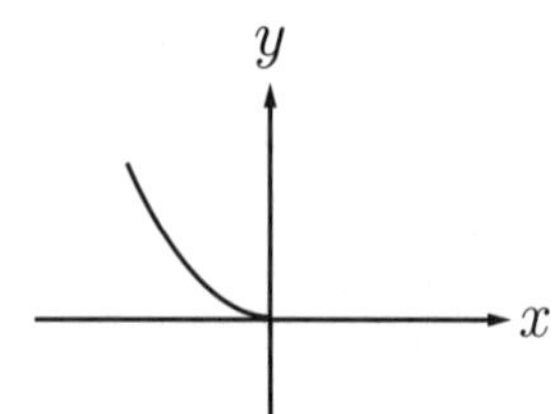

(D)

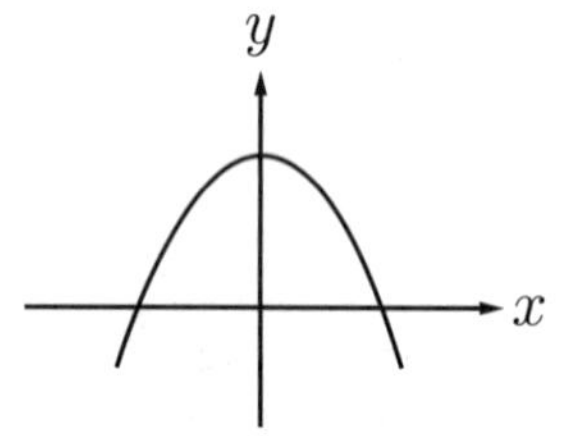

(E)

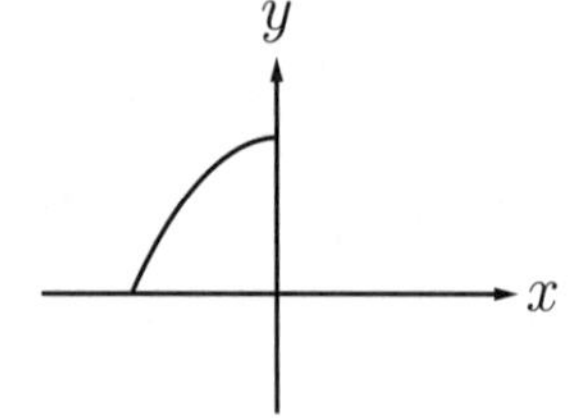

32. Solve the inequality: $e^x(x-2) > 0$

(A) $x > 0$

(B) $x < 0$

(C) $x > 2$

(D) $x < 2$

(E) $(-\infty, \infty)$

USE THIS SPACE FOR SCRATCH WORK

33. The points $A(3, 4)$, $B(3, -2)$, $C(7, -2)$ and $D(7, 4)$ form a rectangle. If the rectangle intersects the circle $(x-3)^2 + (y+2)^2 = 4$, what is the area of the common region that belongs to both the rectangle and the circle?

(A) 6π

(B) 4π

(C) 3π

(D) 2π

(E) π

34. If $f(x) = \log_5 x$ and $g(x) = 5^x$, what is the value of $f(g(3.78))$?

(A) 4.29

(B) 3.78

(C) 3.43

(D) 2.81

(E) 2.37

35. If $\cos(3\theta - 20) = \sin(\theta + 10)$, what is the degree measure of θ ?

(A) 15

(B) 20

(C) 25

(D) 30

(E) 35

USE THIS SPACE FOR SCRATCH WORK

36. If $10^m = 5^n$, what is the value of $\dfrac{n}{m}$?

(A) 1.43

(B) 1.18

(C) 0.96

(D) 0.85

(E) 0.70

37. If the line $y - x + 1 = 0$ intersects the parabola $2x - y^2 = 0$ for $x > 1$, what is the x-coordinate of the intersection point?

(A) 0.27

(B) 1.33

(C) 2.67

(D) 3.73

(E) 5.15

38. $\displaystyle\lim_{x \to 4} \frac{\sqrt{x+5}}{x-4} =$

(A) $\dfrac{1}{2}$

(B) 0

(C) 1

(D) 3

(E) The limit does not exist.

USE THIS SPACE FOR SCRATCH WORK

39. Two segments are drawn from point A to point B, and from point A to point C inside or on the surface of a cube as shown in Figure 4. If the length of the cube is 1, what is the degree measure of $\angle BAC$?

    (A) $35.26°$

    (B) $37.52°$

    (C) $40.71°$

    (D) $42.33°$

    (E) $44.79°$

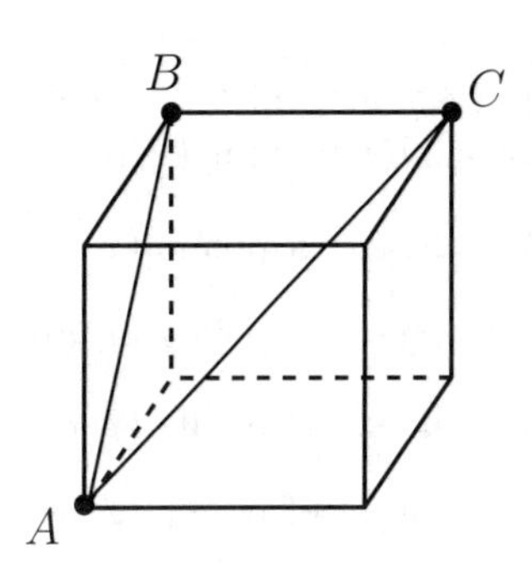

Figure 4

40. If $f(x)$ contains the point $(4, -2)$, which of the following point (x, y) must be on the graph of $-f(-x)$?

    (A) $(4, 2)$

    (B) $(4, -2)$

    (C) $(2, -4)$

    (D) $(-4, 2)$

    (E) $(-4, -2)$

41. The height H of a ball (in feet) thrown straight upward is given as a function of the time t (in seconds) by $H(t) = -16t^2 + 40t + 35$. At what time does the ball reach its maximum height?

    (A) 1.05 seconds

    (B) 1.25 seconds

    (C) 2.11 seconds

    (D) 2.50 seconds

    (E) 2.89 seconds

USE THIS SPACE FOR SCRATCH WORK

42. If $f:(x,y)\to(2xy, x+y)$ for every pair (x,y) in the plane, for which of the following point (x,y) must it be true that $(x,y)\to(2y, y+1)$?

    (A) The set of points (x,y) such that $x=1$

    (B) The set of points (x,y) such that $y=1$

    (C) The set of points (x,y) such that $y=0$

    (D) The set of points (x,y) such that $y=x$

    (E) $(0,0)$ only

43. Suppose $f(x)=ax^2+bx+c$. If $f(-2)=9$ and $f(-1)=-2$, what is the value of $3a-b$?

    (A) 4

    (B) 7

    (C) 8

    (D) 9

    (E) 11

44. If vectors $\mathbf{u}$ and $\mathbf{v}$ are given in Figure 5, what is $|2\mathbf{v}-\mathbf{u}|$?

    (A) 14.87

    (B) 10.24

    (C) 8.45

    (D) 6.08

    (E) 3.62

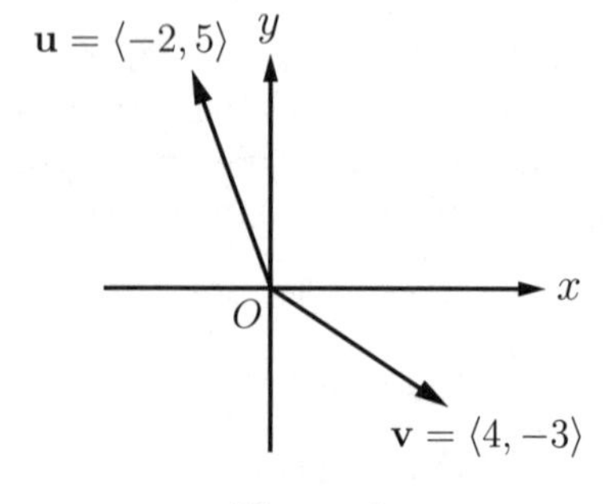

Figure 5

USE THIS SPACE FOR SCRATCH WORK

45. If a single pane of glass blocks 13% of the light passing through it, then the percent P of light that passes through n successive panes is given by $P(n) = 100e^{-0.13n}$. What is the smallest integer number of panes to block at least 80% of the light?

(A) 11

(B) 12

(C) 13

(D) 14

(E) 15

46. For which of the following radian measure of the angle θ, is the value of $\sin\theta$ the greatest?

(A) $\dfrac{\pi}{12}$

(B) $\dfrac{\pi}{3}$

(C) $\dfrac{7\pi}{12}$

(D) $\dfrac{3\pi}{4}$

(E) $\dfrac{5\pi}{6}$

47. Set A has 4 numbers. The standard deviation of set A is 4.44. If each number in set A is divided by 2, what is the new standard deviation of set A ?

(A) 1.11

(B) 2.22

(C) 4.44

(D) 6.44

(E) 8.88

USE THIS SPACE FOR SCRATCH WORK

48. If $\tan\theta = t$ for $0 < \theta < \frac{\pi}{2}$, what is $\sin\theta$ in terms of t ?

(A) $\frac{t}{\sqrt{1-t^2}}$

(B) $\frac{1}{\sqrt{1-t^2}}$

(C) $\frac{t}{t+1}$

(D) $\frac{t}{\sqrt{t^2+1}}$

(E) $\frac{1}{\sqrt{t^2+1}}$

49. What is the value of $\sin(\cos^{-1}(0.86))$?

(A) 0.35

(B) 0.51

(C) 0.67

(D) 0.83

(E) 0.96

50. Using the digits 1, 1, 2, and 2 exactly once, how many four-digit numbers can you make?

(A) 2

(B) 4

(C) 6

(D) 12

(E) 24

STOP

Mathematics Scoring Worksheet

Directions: In order to calculate your score correctly, fill out the table below. After calculating your raw score, round the raw score to the nearest whole number. The scaled score can be determined using the "Math Test Score Conversion Table".

Mathematics Score			
A. Number Correct		**B.** Number Incorrect ÷ 4	
Total Unrounded Raw Score $A - B$		**Total Rounded Raw Score** Round to nearest whole number	

Math Test Score Conversion Table					
Raw Score	Scaled Score	Raw Score	Scaled Score	Raw Score	Scaled Score
50	800	29	660	8	490
49	800	28	650	7	480
48	800	27	640	6	480
47	800	26	630	5	470
46	800	25	630	4	460
45	800	24	620	3	450
44	800	23	610	2	440
43	800	22	600	1	430
42	790	21	590	0	410
41	780	20	580	−1	390
40	770	19	570	−2	370
39	760	18	560	−3	360
38	750	17	560	−4	340
37	740	16	550	−5	340
36	730	15	540	−6	330
35	720	14	530	−7	320
34	710	13	530	−8	320
33	700	12	520	−9	320
32	690	11	510	−10	320
31	680	10	500	−11	310
30	670	9	500	−12	310

ANSWERS AND SOLUTIONS

Answers

1. E	11. E	21. E	31. E	41. B
2. A	12. B	22. C	32. C	42. A
3. C	13. A	23. B	33. E	43. E
4. E	14. C	24. D	34. B	44. A
5. D	15. C	25. D	35. C	45. C
6. E	16. E	26. B	36. A	46. C
7. C	17. B	27. E	37. D	47. B
8. A	18. C	28. C	38. E	48. D
9. B	19. B	29. A	39. A	49. B
10. A	20. E	30. E	40. D	50. C

Solutions

1. (E)

Tip $\sec\theta = \dfrac{1}{\cos\theta}$

$$\cos 35^\circ \times \sec 35^\circ = \cos 35^\circ \times \frac{1}{\cos 35^\circ} = 1$$

2. (A)

Since x, y and z are positive integers, divide each side of the equation by xyz^2.

$$x^3yz = 2xyz^2 \qquad \text{Divide each side by } xyz^2$$

$$\frac{x^3yz}{xyz^2} = 2 \qquad \text{Simplify}$$

$$\frac{x^2}{z} = 2$$

Therefore, the value of $\dfrac{x^2}{z}$ is 2.

3. (C)

Tip

1. The table below shows the powers of i, which repeat in a pattern: i, -1, $-i$, and 1.

Powers of i	i	i^2	i^3	i^4	i^5	i^6	i^7	i^8	$\cdots$	i^{12}	$\cdots$	i^{4n}
Value	i	-1	$-i$	1	i	-1	$-i$	1	$\cdots$	1	$\cdots$	1

2. If the power of i is a multiple of 4, the value is always equal to 1; that is $i^{4n} = 1$.

Note that $i^{11} = i^8 \cdot i^3 = 1 \cdot -i = -i$, and $i(1+i) = i + i^2 = -1 + i$. Since $i^5 = i^4 \cdot i = i$, (C) is the correct answer.

4. (E)

In order to evaluate the piecewise function at $x = 2$ and $x = 3$, check the conditions on the right side of the piecewise function to see where $x = 2$ and $x = 3$ belong. Since $x = 2$ satisfies the condition $x < 3$, the equation for g is $-2x - 1$. Thus, $g(2) = -2(2) - 1 = -5$. Likewise, $x = 3$ satisfies the condition $x \geq 3$, the equation for g is $\frac{2}{3}x + 2$. Thus, $g(3) = \frac{2}{3}(3) + 2 = 4$. Therefore, the value of $g(2) + g(3)$ is $-5 + 4 = -1$.

5. (D)

When $n = 3$, $n! = 3! = 6$, and $(n-2)! = (3-2)! = 1$. Thus, $\dfrac{n!}{(n-2)!} = \dfrac{3!}{1!} = 6$. Therefore, the value of n is 3.

6. (E)

Set up a proportion in terms of knots and miles per hour.

$$\begin{aligned} 1_{\text{knot}} : 1.151_{\text{mph}} &= x_{\text{knots}} : 30_{\text{mph}} \\ \frac{1}{1.151} &= \frac{x}{30} && \text{Use the cross product property} \\ 1.151x &= 30 \\ x &= 26.06 \end{aligned}$$

Therefore, the boat is traveling at 26.06 knots.

7. (C)

Tip

1. $\log_a x - \log_a y = \log_a \frac{x}{y}$
2. $\log_a x^n = n \log_a x$
3. $\log_a a = 1$

$$\begin{aligned} 3\log_5 75 - 3\log_5 3 &= 3(\log_5 75 - \log_5 3) = 3\log_5 \frac{75}{3} \\ &= 3\log_5 25 = 3\log_5 5^2 = 3 \cdot 2 \cdot \log_5 5 \\ &= 6 \end{aligned}$$

8. (A)

Jason's balance each month after making a \$300 payment is defined by $a_n = 1.005a_{n-1} - 300$, $a_0 = 20,000$, for $n \geq 1$. Thus,

$$\begin{aligned}
&\text{Balance after 1st payment:} \quad a_1 = 1.005a_0 - 300 = 1.005(\$20,000) - 300 = \$19,800 \\
&\text{Balance after 2nd payment:} \quad a_2 = 1.005a_1 - 300 = 1.005(\$19,800) - 300 = \$19,599 \\
&\text{Balance after 3rd payment:} \quad a_3 = 1.005a_2 - 300 = 1.005(\$19,599) - 300 = \$19,397
\end{aligned}$$

Therefore, Jason's balance after he makes the third payment is \$19, 397.

9. (B)

Tip $\cos\theta = \dfrac{\text{adjacent side}}{\text{hypotenuse}}$

In order to find BC, use the Pythagorean theorem: $13^2 = 12^2 + BC^2$. Thus, $BC = 5$ as shown in the figure below.

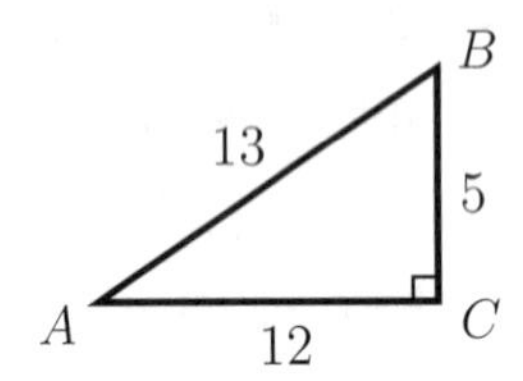

Therefore, $\cos B = \dfrac{\text{adjacent side}}{\text{hypotenuse}} = \dfrac{5}{13}$.

10. (A)

Tip $\log_{a^n} x = \frac{1}{n}\log_a x \quad \Longrightarrow \quad \log_4 2 = \log_{2^2} 2 = \frac{1}{2}\log_2 2 = \frac{1}{2}$

Since the graph of the logarithmic function $y = n\log_4 x$ contains the point $(2, 6)$ as shown in the figure below, substitute 2 for x and 6 for y to solve for n.

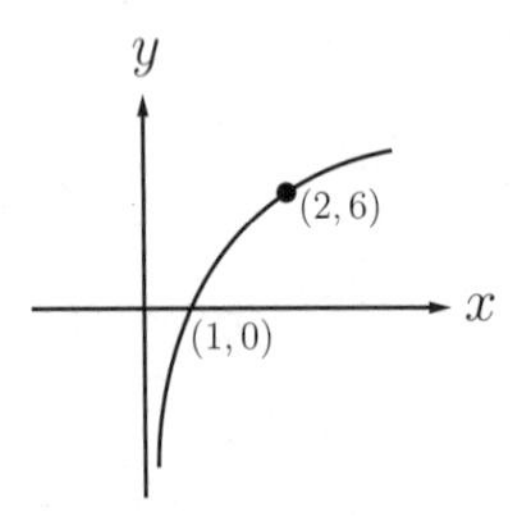

$$\begin{aligned}
y &= n\log_4 x && \text{Substitute 2 for } x \text{ and 6 for } y \\
6 &= n\log_4 2 && \text{Note that } \log_4 2 = \frac{1}{2} \\
6 &= \frac{1}{2}n \\
n &= 12
\end{aligned}$$

Therefore, the value of n is 12.

11. (E)

As shown in the figure below, the slope of the line that passes through the points $(0, k)$ and $(q, 0)$ is -1.

$$\text{slope} = \frac{0-k}{q-0} = -1 \implies q = k$$

Thus, q in terms of k is k.

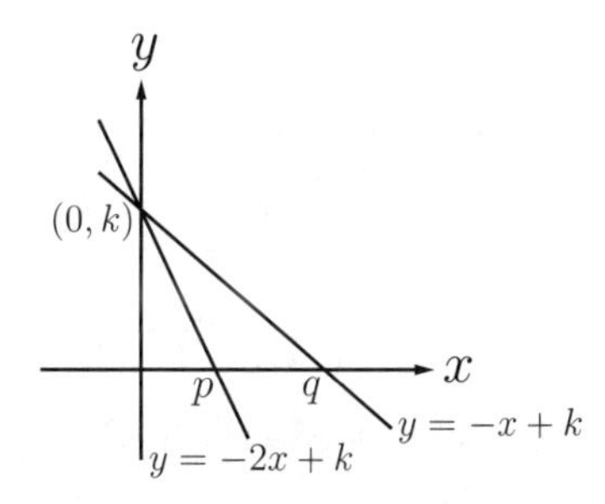

Likewise, since the slope of the line that passes through the points $(0, k)$ and $(p, 0)$ is -2, p in terms of k is $\frac{k}{2}$ as shown below.

$$\text{slope} = \frac{0-k}{p-0} = -2 \implies p = \frac{k}{2}$$

Therefore, the value of $\frac{q}{p}$ is $\frac{k}{\frac{k}{2}} = 2$.

12. (B)

$$\begin{aligned} 4^x + 4^x + 4^x + 4^x &= 2^k \\ 4 \cdot 4^x &= 2^k \\ 4^{x+1} &= 2^k \\ (2^2)^{x+1} &= 2^k \\ 2^{2x+2} &= 2^k \end{aligned}$$

Therefore, the expression that is equivalent to k is $2x + 2$.

13. (A)

In order to find the number of milligrams of the drug in a patient's blood stream after 10 hours, substitute 10 for t in $A(t) = 10e^{-0.2t}$. Thus, $A(10) = 10e^{-0.2(10)} = 1.35$. Therefore, the number of milligrams of the drug in a patient's blood stream after 10 hours is 1.35 milligrams.

14. (C)

Tip 1 foot = 12 inches

Convert 5 feet and 9 inches to 69 inches. In order to find the weight of a person who is 5 feet 9 inches tall using the line of best fit, substitute 69 for x in $y = 5.25x - 192$. Thus, $y = 5.25(69) - 192 = 170.25$. Therefore, (C) is the correct answer.

15. (C)

> Tip When solving an equation, do not divide each side of the equation by a variable, x, because it will eliminate some of the solutions.

$x^2 = 3x$	Subtract $3x$ from each side
$x^2 - 3x = 0$	Factor
$x(x-3) = 0$	Solve for x
$x = 0 \quad \text{or} \quad x = 3$	

Therefore, solutions to $x^2 = 3x$ are $x = 0$ or $x = 3$.

16. (E)

> Tip The infinite sum S of a geometric series is $S = \dfrac{a_1}{1-r}$ if $|r| < 1$, where r is the common ratio of the geometric sequence.

In the geometric series $\frac{1}{3} + \frac{1}{9} + \frac{1}{27} + \cdots$, $a_1 = \frac{1}{3}$ and r is $\frac{1}{3}$. Thus,

$$S = \frac{a_1}{1-r} = \frac{\frac{1}{3}}{1-\frac{1}{3}} = \frac{\frac{1}{3}}{\frac{2}{3}} = \frac{1}{2}$$

Therefore, the sum of the geometric series is $\frac{1}{2}$.

17. (B)

> Tip
> 1. The length of the longest diagonal of a cube with side length x is $x\sqrt{3}$.
> 2. When a cube with side length x is inscribed in a sphere, the diameter of the sphere is the same as longest diagonal of the cube. Thus, the diameter of the sphere is $x\sqrt{3}$ and the radius of the sphere is $\frac{x}{2}\sqrt{3}$.
> 3. The volume V of a sphere: $V = \frac{4}{3}\pi r^3$

The length of the cube is $2\sqrt{3}$. Thus, the longest diagonal of the cube and the diameter of the sphere are both $2\sqrt{3} \times \sqrt{3} = 6$. Since the radius of the sphere is 3, the volume of the sphere is $V = \frac{4}{3}\pi r^3 = \frac{4}{3}\pi(3)^3 = 36\pi$.

18. (C)

For simplicity, let set A be $\{1, 1, 10, 11, 12\}$ so that the mean and the median of the set are 7 and 10, respectively. Since 10 is added to set A to produce set B, set $B = \{1, 1, 10, 10, 11, 12\}$. Therefore, the mean and the median of set B are 7.5 and 10, respectively.

19. (B)

> Tip
> In order to find the x-intercept of a rational function $f(x) = \frac{h(x)}{g(x)}$, set the numerator $h(x)$ equal to 0 and solve for x.

In order to find the x-intercept of $f(x) = \dfrac{x-3}{x^2-1}$, set the numerator $x - 3$ equal to 0 and solve for x. Thus, $x = 3$. Therefore, the x-intercept of f is 3.

20. (E)

$$\frac{x^2y + xy^2}{xy} = \frac{xy(x+y)}{xy} = x + y$$

21. (E)

The probability of selecting a blood type AB is $1 - (0.35 + 0.15 + 0.38) = 0.12$ as shown in the table below.

Blood type	A	B	O	AB
Probability	0.35	0.15	0.38	0.12

Since the small city has a population of 250, the number of people who have AB as their blood type is $250 \times 0.12 = 30$.

22. (C)

> Tip
> The general forms of the sine function and cosine function are as follows:
>
> $$y = A\sin\big(B(x-C)\big) + D \qquad \text{or} \qquad y = A\cos\big(B(x-C)\big) + D$$
>
> where B affects the period. The period, P, is the horizontal length of one complete cycle and is obtained by $P = \dfrac{2\pi}{B}$.

Comparing $y = \sin\left(\dfrac{2\pi}{3}x - 1\right) + 2$ to the general form of $y = A\sin\big(B(x-C)\big) + D$, we found that $B = \dfrac{2\pi}{3}$. Therefore, the period is $P = \dfrac{2\pi}{\frac{2\pi}{3}} = 3$.

23. (B)

The inequality $b < 0 < a$ means that a is a positive number and b is a negative number. When $b < 0 < a$, $ab < 0$ and $a - b > 0$. Thus, $\dfrac{ab}{a-b} < 0$. Therefore, (B) is the correct answer.

24. (D)

Let xy be a two-digit number, where x and y represent the tens digit and units digit, respectively. The two-digit number xy must be greater than 40. Thus, there are 6 digits for x: 4, 5, 6, 7, 8, and 9. Since the two-digit number xy is an odd number, there are 5 digits for y: 1, 3, 5, 7, and 9. Therefore, the total number of two-digit odd numbers greater than 40 is $6 \times 5 = 30$.

25. (D)

Since triangle ABC is a right triangle as shown in the figure below, use the definition of the tangent function to find AC.

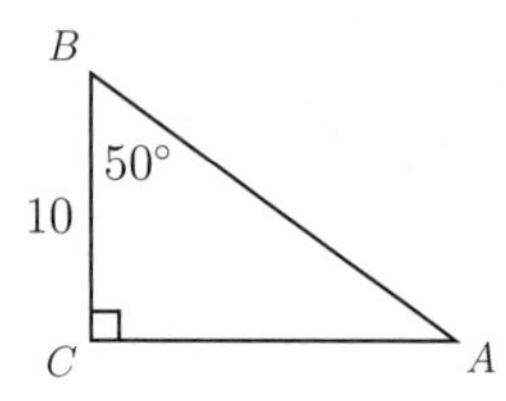

$$\tan 50^\circ = \frac{AC}{10} \quad \Longrightarrow \quad AC = 10\tan 50^\circ = 11.92$$

Therefore, the area of the triangle ABC is $\frac{1}{2}AC \cdot BC = \frac{1}{2}(11.92)(10) = 59.59$.

26. (B)

Switch the x and y variables and solve for y.

$y = \sqrt[3]{\frac{x-1}{2}} + 4$	Switch the x and y variables
$x = \sqrt[3]{\frac{y-1}{2}} + 4$	Subtract 4 from each side
$x - 4 = \sqrt[3]{\frac{y-1}{2}}$	Raise each side to the power of 3
$(x-4)^3 = \frac{y-1}{2}$	Multiply each side by 2
$y - 1 = 2(x-4)^3$	Add 1 to each side
$y = 2(x-4)^3 + 1$	

Therefore, the inverse function of $y = \sqrt[3]{\frac{x-1}{2}} + 4$ is $y = 2(x-4)^3 + 1$.

27. (E)

Solving the equation $\cos\theta = \sqrt{3}$ means to find the x-coordinates of the intersection points when two graphs $y = \cos\theta$ and $y = \sqrt{3}$ intersect.

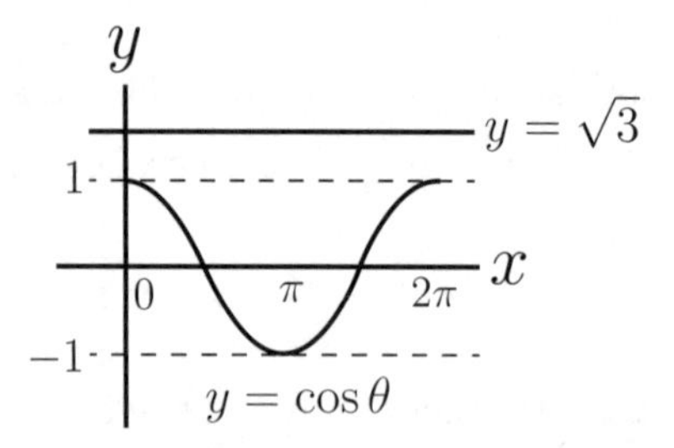

As shown in the figure above, the two graphs $y = \cos\theta$ and $y = \sqrt{3}$ never intersect. Therefore, there are no solutions to $\cos\theta = \sqrt{3}$.

28. (C)

To convert from the rectangular coordinates $(-\sqrt{3}, -1)$ to the polar coordinates (r, θ), do the following three steps:

- Step 1: Plot the point $(-\sqrt{3}, -1)$ as shown in Figure A. Since the point $(-\sqrt{3}, -1)$ is in the third quadrant, $\pi < \theta < \dfrac{3\pi}{2}$.
- Step 2: Find the distance between the point $(-\sqrt{3}, -1)$ and the origin, r, and the reference angle, β, formed by the positive x-axis and the terminal side as shown in Figure B.

$$r = \sqrt{x^2 + y^2} = \sqrt{(-\sqrt{3})^2 + (-1))^2} = 2$$
$$\beta = \left|\tan^{-1}\frac{y}{x}\right| = \left|\tan^{-1}\left(\frac{-1}{-\sqrt{3}}\right)\right| = \frac{\pi}{6}$$

- Step 3: Find θ using the reference angle based on the quadrant that θ lies in as shown in Figure C. Since $\beta = \dfrac{\pi}{6}$ and θ lies in the third quadrant, $\theta = \pi + \beta = \pi + \dfrac{\pi}{6} = \dfrac{7\pi}{6}$.

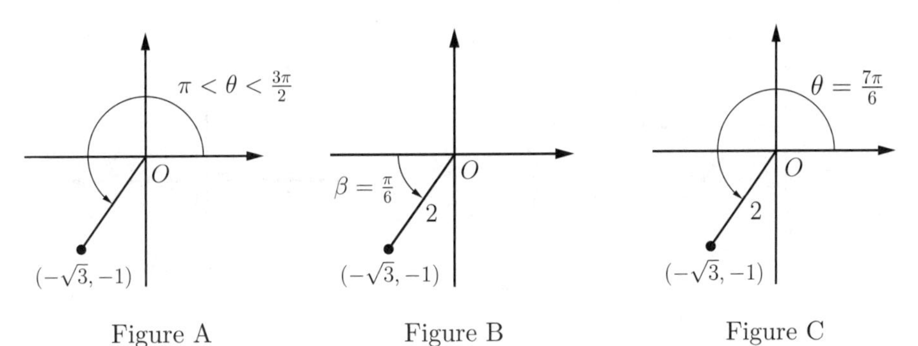

Figure A Figure B Figure C

Therefore, the polar coordinates that correspond with the rectangular coordinates $(-\sqrt{3}, -1)$ are $\left(2, \dfrac{7\pi}{6}\right)$.

29. (A)

Tip	**Conditional statement:**	If P, then Q
	Contrapositive:	If not Q, then not P

The conditional statement and its contrapositive are logically equivalent.

The contrapositive of the conditional statement "If the weather is sunny, Joshua goes on a field trip." is "If Joshua does not go on a field trip, the weather is not sunny." Therefore, (A) is the correct answer.

30. (E)

$$\begin{array}{r|l} & \quad\quad\quad 2x \\ x^2+1) & 2x^3 - 3x + 2 \\ & -\ 2x^3 - 2x \\ \hline & \quad\quad -\ 5x + 2 \end{array}$$

When $2x^3 - 3x + 2$ is divided by $x^2 + 1$, the remainder is $-5x + 2$.

31. (E)

Tip $\sin^2 t + \cos^2 t = 1 \quad \Longrightarrow \quad \sin^2 t = 1 - \cos^2 t$

Figure A shows the graph of $x = \cos t$, where $\frac{\pi}{2} \le t \le \pi$. The range of $\cos t$, $\frac{\pi}{2} \le t \le \pi$ is $-1 \le x \le 0$.

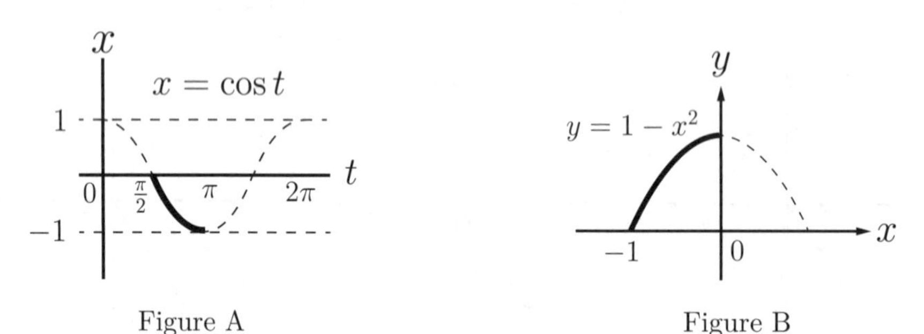

Figure A Figure B

Since $y = \sin^2 t = 1 - \cos^2 t$ and $x = \cos t$, $y = 1 - x^2$, where $-1 \le x \le 0$. Figure B shows the graph of $y = 1 - x^2$, $-1 \le x \le 0$. Therefore, (E) is the correct answer.

32. (C)

As shown in the figure below, the graph of $y = e^x$ lies above the x-axis, which implies that $e^x > 0$ for all real numbers x.

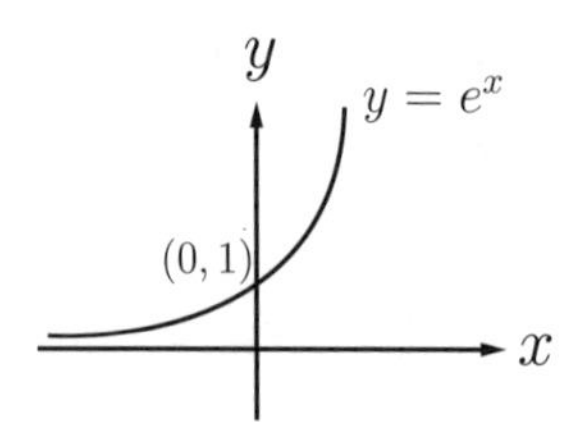

Since $e^x > 0$ for all real numbers x, divide each side of the inequality by e^x and solve for x.

$$\begin{aligned} e^x(x-2) &> 0 \qquad \text{divide each side of the inequality by } e^x \\ x - 2 &> 0 \\ x &> 2 \end{aligned}$$

Therefore, the solution for which $e^x(x-2) > 0$ is $x > 2$.

33. (E)

Tip The general equation of a circle is $(x-h)^2+(y-k)^2=r^2$, where (h,k) is the center of the circle, and r is the radius of the circle.

The center of the circle $(x-3)^2+(y+2)^2=2^2$ is $(3,-2)$, and the radius of the circle is 2 as shown in the figure below.

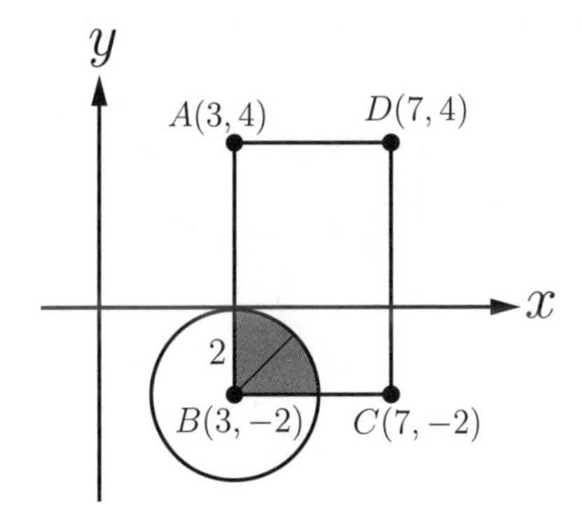

The rectangle $ABCD$ whose vertices are $A(3,4)$, $B(3,-2)$, $C(7,-2)$, and $D(7,4)$ intersects the circle. The shaded area in the figure above represents the common region that belongs to both the rectangle and the circle, which is equal to $\frac{1}{4}$ of the area of the circle. Therefore, the area of the common region is $\frac{1}{4}\pi(2)^2=\pi$.

34. (B)

Tip
1. $y=5^x$ is the inverse function of $y=\log_5 x$.
2. $f\left(f^{-1}(x)\right)=x$

Since $g(x)=5^x$ is the inverse function of $f(x)=\log_5 x$, $g(x)=f^{-1}(x)$. Thus,

$$f(g(3.78))=f\left(f^{-1}(3.78)\right)=3.78$$

Therefore, the value of $f(g(3.78))$ is 3.78.

35. (C)

The value of the sine function is equal to the value of the cosine function when two angles are complementary (their sum is 90°).

$$\begin{aligned}3\theta-20+\theta+10&=90\\4\theta-10&=90\\\theta&=25\end{aligned}$$

Therefore, the degree measure of θ is 25.

36. (A)

Tip | When each side of an exponential equation has a different base, convert the exponential equation to a logarithmic equation.

If $a^x = b \implies x = \log_a b$ e.g. If $2^x = 7$, then $x = \log_2 7$

Since each side of the exponential equation has a different base (the left side has a base of 10 and the right side has a base of 5), convert the exponential equation to a logarithmic equation.

$10^m = 5^n$ Convert the equation to a logarithmic equation

$m = \log_{10} 5^n$ Use $\log_a x^n = n \log_a x$

$m = n \log_{10} 5$ Divide each side by n

$\frac{m}{n} = \log_{10} 5 = 0.70$ Take the reciprocal of each side

$\frac{n}{m} = 1.43$

Therefore, the value of $\frac{n}{m}$ is 1.43.

37. (D)

Tip | The quadratic formula is a general formula for solving quadratic equations. The solutions to the quadratic equation $ax^2 + bx + c = 0$ are as follows:

$$x = \frac{-b \pm \sqrt{b^2 - 4ac}}{2a}$$

The line $y - x + 1 = 0$ intersects the parabola $2x - y^2 = 0$ for $x > 1$. In order to find the x-coordinate of the intersection point, change $y - x + 1 = 0$ to $y = x - 1$. Then, substitute $x - 1$ for y in $2x - y^2 = 0$ and solve for x.

$2x - y^2 = 0$ Substitute $x - 1$ for y

$2x - (x - 1)^2 = 0$ Simplify

$x^2 - 4x + 1 = 0$

Since the quadratic equation $x^2 - 4x + 1 = 0$ cannot be factored, use the quadratic formula to solve the equation.

$x = \frac{-b \pm \sqrt{b^2 - 4ac}}{2a}$ Substitute 1 for a, -4 for b, and 1 for c

$= \frac{-(-4) \pm \sqrt{(-4)^2 - 4(1)(1)}}{2(1)}$

$= \frac{4 \pm 2\sqrt{3}}{2}$

$= 2 \pm \sqrt{3}$

$x = 3.73$ or $x = 0.27$

The only solution to $x^2 - 4x + 1 = 0$ for $x > 1$ is 3.73. Therefore, the x-coordinate of the intersection point is 3.73.

38. (E)

In order to find the limit of the function $\frac{\sqrt{x+5}}{x-4}$ at $x=4$, plug-in $x=4$ to both the numerator and denominator.

$$\lim_{x\to 4}\frac{\sqrt{x+5}}{x-4}=\frac{\sqrt{4+5}}{4-4}=\frac{3}{0}=\text{undefined}$$

Therefore, the limit does not exist.

39. (A)

Tip

1. The length of the longest diagonal of a cube with side lengths x is $x\sqrt{3}$.

2. If triangle ABC shown at the right is a SSS triangle (a, b, and c are known), the measure of angle A can be calculated by the Law of Cosines.

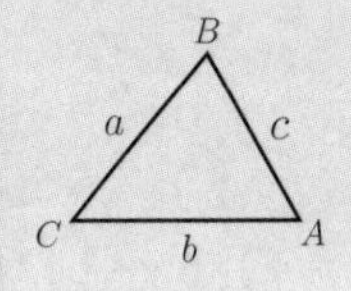

$$m\angle A=\cos^{-1}\left(\frac{a^2-b^2-c^2}{-2bc}\right)$$

Note that side a is opposite angle A.

Since the length of the cube is 1, $BC=1$. $\overline{AB}$ is a diagonal with $AB=\sqrt{2}$, and $\overline{AC}$ is the longest diagonal with $AC=\sqrt{3}$ as shown in Figure A.

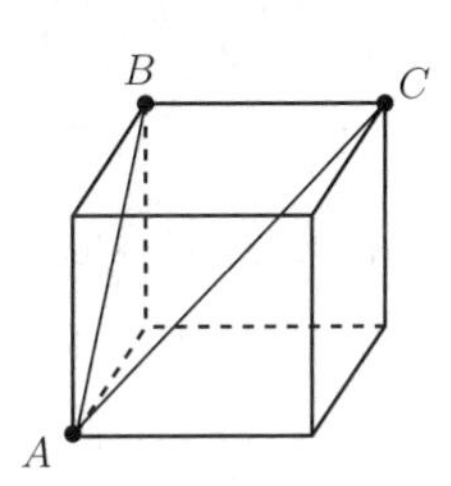

Figure A

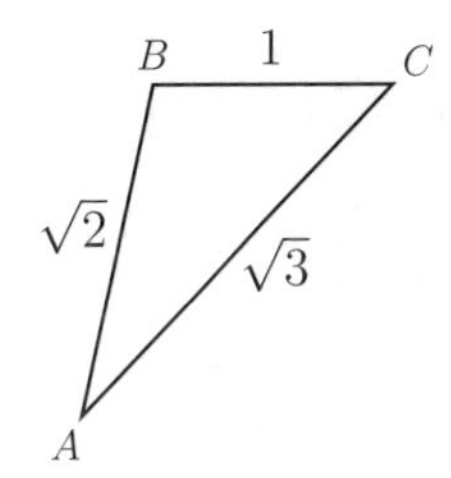

Figure B

$\triangle ABC$ is a SSS triangle (three sides are known) as shown in Figure B. Thus, the degree measure of $\angle BAC$ can be calculated by the Law of Cosines.

$$\begin{aligned}m\angle BAC&=\cos^{-1}\left(\frac{BC^2-AB^2-AC^2}{-2\cdot AB\cdot AC}\right)\\&=\cos^{-1}\left(\frac{1^2-(\sqrt{2})^2-(\sqrt{3})^2}{-2\cdot(\sqrt{2})\cdot(\sqrt{3})}\right)\\&=35.26^\circ\end{aligned}$$

Therefore, the degree measure of $\angle BAC$ is 35.26°.

40. (D)

Tip $-f(-x)$ means to reflect the graph of $f(x)$ about the origin.

If the point $(4,-2)$ is reflected about the origin, the new x and y coordinates of the point are $(-4,2)$. Therefore, the point $(-4,2)$ must be on the graph of $-f(-x)$.

41. (B)

Tip The x-coordinate of the vertex of a quadratic function $y = ax^2 + bx + c$ is $x = -\dfrac{b}{2a}$.

The quadratic function $H(t) = -16t^2 + 40t + 35$ reaches its maximum height at the vertex. In order to find the time at which the ball reaches its maximum height, find the x-coordinate of the vertex. Thus,

$$t = -\frac{b}{2a} = -\frac{40}{2(-16)} = 1.25$$

Therefore, at $t = 1.25$ seconds, the ball reaches its maximum height.

42. (A)

$f : (x, y) \to (2xy, x + y)$ for every pair (x, y) in the plane. In order for $(x, y) \to (2y, y + 1)$, $(2xy, x + y)$ must be the same as $(2y, y + 1)$. Thus, $x = 1$. Therefore, the set of points (x, y) such that $x = 1$ satisfies $(x, y) \to (2y, y + 1)$

43. (E)

Since $f(-2) = 9$ and $f(-1) = -2$, substitute -2 and -1 for x in $f(x) = ax^2 + bx + c$.

$$f(-2) = 4a - 2b + c = 9, \qquad f(-1) = a - b + c = -2$$

In order to find the value of $3a - b$, subtract $a - b + c = -2$ from $4a - 2b + c = 9$ as shown below.

$$\begin{array}{r} 4a - 2b + c = 9 \\ a - b + c = -2 \\ \hline 3a - b = 11 \end{array} \qquad \text{Subtract the two equations}$$

Therefore, the value of $3a - b$ is 11.

44. (A)

Tip The magnitude of a vector $\mathbf{V} = \langle a, b \rangle$, denoted by $|\mathbf{V}|$, is $|\mathbf{V}| = \sqrt{a^2 + b^2}$.

Since $\mathbf{v} = \langle 4, -3 \rangle$ and $\mathbf{u} = \langle -2, 5 \rangle$,

$$\begin{aligned} 2\mathbf{v} - \mathbf{u} &= 2\langle 4, -3 \rangle - \langle -2, 5 \rangle \\ &= \langle 8, -6 \rangle - \langle -2, 5 \rangle \\ &= \langle 10, -11 \rangle \end{aligned}$$

Thus, $2\mathbf{v} - \mathbf{u} = \langle 10, -11 \rangle$. Therefore, $|2\mathbf{v} - \mathbf{u}|$ is $\sqrt{10^2 + (-11)^2} = 14.87$.

45. (C)

Blocking 80% of the light means that 20% of the light passes through. In order to find the number of panes of glass so that 20% of the light passes through, substitute 20 for $P(n)$ in $P(n) = 100e^{-0.13n}$.

$P(n) = 100e^{-0.13n}$	substitute 20 for $P(n)$
$20 = 100e^{-0.13n}$	Divide each side by 100
$e^{-0.13n} = 0.2$	Covert the equation to a logarithmic equation
$-0.13n = \log_e 0.2 = -1.61$	Divide each side by -0.13
$n = 12.38$	

Thus, in order to block exactly 80% of the light, 12.38 of panes of glass are needed. However, the number of panes of glass should be an integer. Therefore, the smallest integer number of panes to block at least 80% of the light is 13.

46. (C)

The figure below shows the graph of $\sin\theta$, where $0 \le \theta \le \pi$.

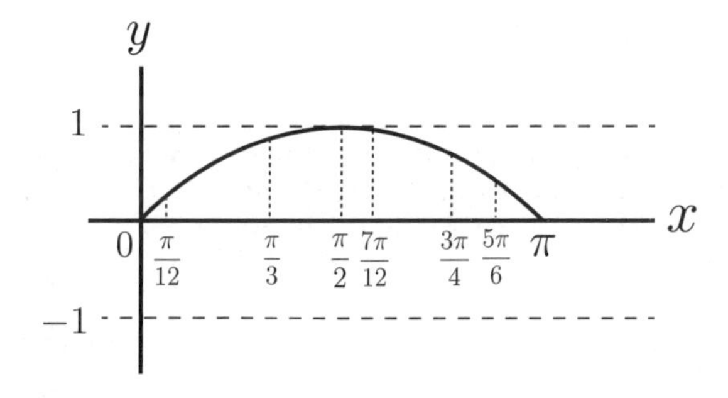

Since the value of the sine function reaches a maximum at $\theta = \frac{\pi}{2}$, the value of the sine function is greatest at $\theta = \frac{7\pi}{12}$, which is closest to $\frac{\pi}{2}$ as shown in the figure above. Therefore, (C) is the correct answer.

47. (B)

> Tip If all numbers in the data set are divided by the same number, k, the standard deviation is divided by k.

Since each number in set A is divided by 2, the new standard deviation of set A is $\frac{4.44}{2} = 2.22$.

48. (D)

> Tip $\tan\theta = \dfrac{\text{opposite side}}{\text{adjacent}}, \qquad \sin\theta = \dfrac{\text{opposite side}}{\text{hypotenuse}}$

Since $\tan\theta = \dfrac{\text{opposite side}}{\text{adjacent}} = t$, $BC = t$, and $AC = 1$ as shown in the figure below.

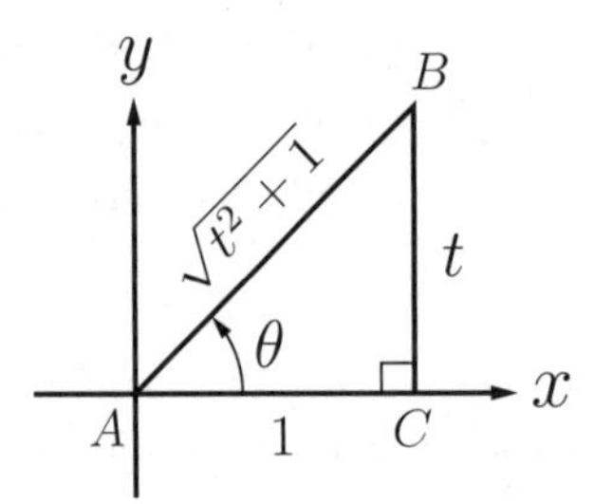

In order to find AB, use the Pythagorean theorem: $AB^2 = t^2 + 1^2$. Thus, $AB = \sqrt{t^2+1}$. Therefore, $\sin\theta = \dfrac{\text{opposite side}}{\text{hypotenuse}} = \dfrac{t}{\sqrt{t^2+1}}$.

49. (B)

Set the angle mode to Radians in your calculator. Since $\cos^{-1}(0.86) = 0.54$,

$$\sin(\cos^{-1}(0.86)) = \sin(0.54) = 0.51$$

Therefore, the value of $\sin(\cos^{-1}(0.86)) = 0.51$.

50. (C)

> Tip Permutations with repetition: The number of permutations of n objects, where there are n_1 indistinguishable objects of one kind, and n_2 indistinguishable objects of a second kind, is given by
>
> $$\text{Permutations with repetition} = \frac{n!}{n_1! \cdot n_2!}$$

Since 1 and 2 are distinguishable, the order is important. However, there are 2 1's and 2 2's out of 4 numbers.

$$\text{Permutations with repetition} = \frac{4!}{2! \cdot 2!} = \frac{4 \cdot 3 \cdot \cancel{2!}}{2! \cdot \cancel{2!}} = 6$$

Therefore, using the digits 1, 1, 2, and 2 exactly once, the number of four-digit numbers you can make is 6.

SAT II MATH LEVEL 2 TEST 5

Directions: Among the given answer choices, choose the BEST answer for each problem. If the exact numerical value is not within the given answer choices, select the answer that best approximates this value. Afterwards, fill in the corresponding oval on the answer sheet.

Notes:

1. A calculator may be required to answer some of the questions. Scientific and graphing calculators are allowed during this test.
2. For some questions, it is up to you to decide whether the calculator should be in degree mode or radian mode.
3. Provided figures for questions are drawn as accurately as possible UNLESS otherwise specified by the problem. Unless otherwise indicated, all figures are assumed to lie in a plane.
4. Unless otherwise specified, it can be assumed that the domain of any function f is to be the set of all real numbers x for which $f(x)$ is a real number.

Reference Information: Use the following information and formulas as a reference in answering questions on this test.

1. If the radius and height of a right circular cone are r and h respectively, then the Volume V of the cone is $V = \frac{1}{3}\pi r^2 h$.
2. If the circumference and slant height of a right circular cone are c and ℓ respectively, then the Lateral Area A of the cone is $A = \frac{1}{2}c\ell$.
3. At any given radius r, the Volume V of a sphere is $V = \frac{4}{3}\pi r^3$.
4. At any given radius r, the Surface Area A of a sphere is $A = 4\pi r^2$.
5. The Volume V of a pyramid is $V = \frac{1}{3}Bh$; given that B and h represent the base area and height of the pyramid respectively.

USE THIS SPACE FOR SCRATCH WORK

1. What is the inverse function of $y = \frac{1}{2}x + 3$?

(A) $y = -2x - 3$

(B) $y = -2x + 3$

(C) $y = 2x - 6$

(D) $y = 2x + 6$

(E) $y = 3x - 2$

2. Which of the following statement must be true?

(A) $0! = 0$

(B) $9! = 9 \times 8!$

(C) $4! + 5! = 9!$

(D) $3! \times 4! = 7!$

(E) $\frac{10!}{5!} = 2!$

3. $\mathbf{i}$ and $\mathbf{j}$ are unit vectors such that $\mathbf{i} = \langle 1, 0 \rangle$, and $\mathbf{j} = \langle 0, 1 \rangle$. Which of the following vector is equivalent to $3\mathbf{i} - 4\mathbf{j}$?

(A) $\langle 0, 0 \rangle$

(B) $\langle 4, 3 \rangle$

(C) $\langle 3, 4 \rangle$

(D) $\langle 4, -3 \rangle$

(E) $\langle 3, -4 \rangle$

USE THIS SPACE FOR SCRATCH WORK

4. Suppose $f(x) = x^2 + 1$. If $f(g(x)) = 4x^2 - 4x + 2$, which of following function must be $g(x)$?

(A) $\sqrt{4x}$

(B) $\dfrac{2}{x}$

(C) $4x^2 - 4x + 1$

(D) $|x - 4|$

(E) $2x - 1$

5. If f is an odd function, which of the following graph best represents f ?

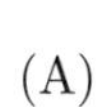

(A)

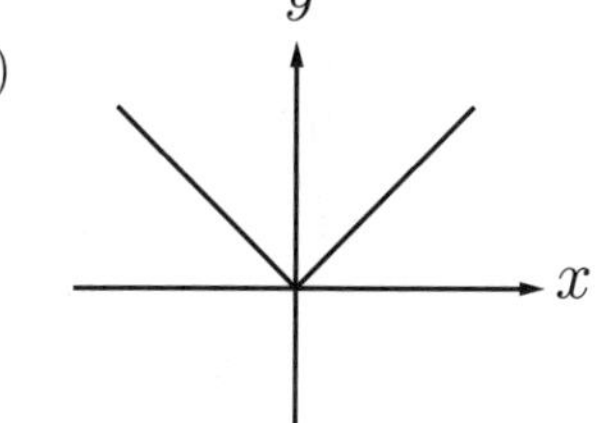

(B)

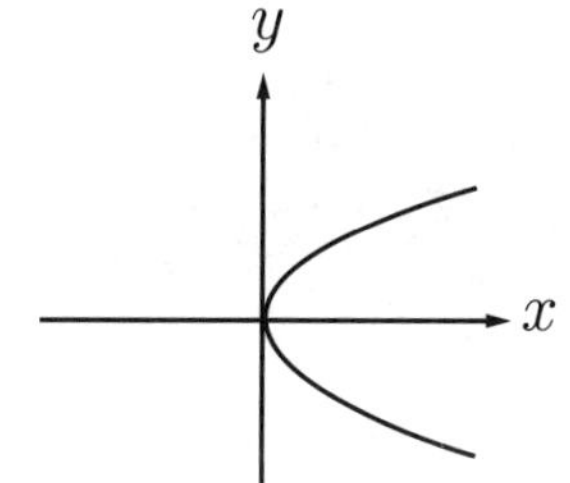

(C)

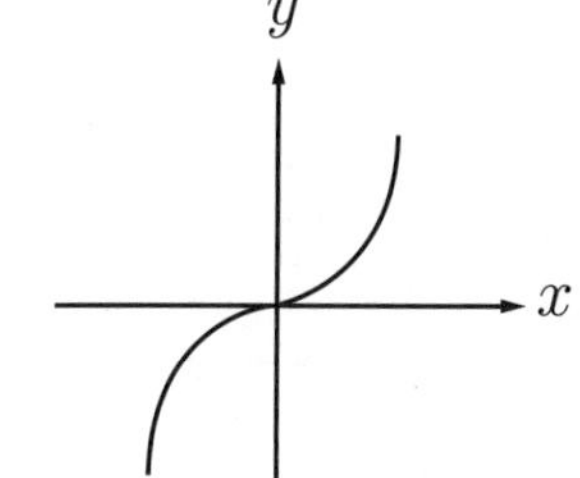

(D)

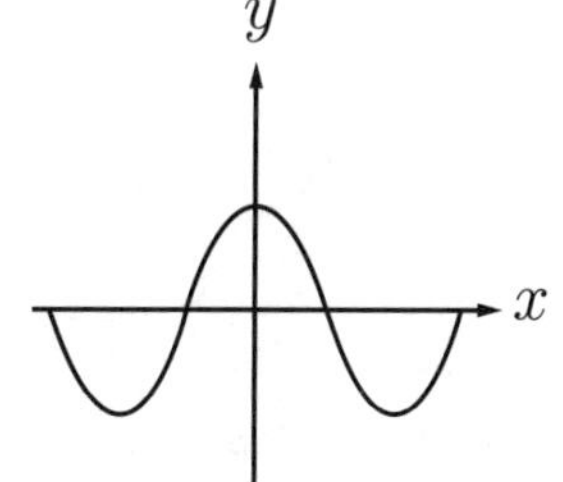

(E)

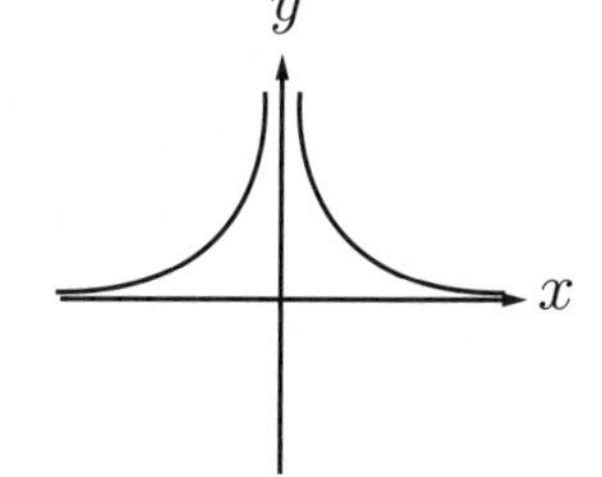

USE THIS SPACE FOR SCRATCH WORK

6. At Solomon Academy, 40% of the students are boys. 60% of the boy students and 40% of the girl students chose Mozzarella cheese as their favorite cheese. What percent of the students who chose Mozzarella cheese as their favorite cheese are boys?

   (A) 60%

   (B) 55%

   (C) 50%

   (D) 45%

   (E) 40%

7. Figure 1 shows the graph of a quadratic function $y = ax^2 + bx + c$, which is symmetric with respect to the y-axis. Which of the following statement is true about a, b, and c ?

   (A) $a < 0,\ b < 0,\ c < 0$

   (B) $a < 0,\ b = 0,\ c > 0$

   (C) $a > 0,\ b > 0,\ c < 0$

   (D) $a > 0,\ b = 0,\ c > 0$

   (E) $a > 0,\ b < 0,\ c = 0$

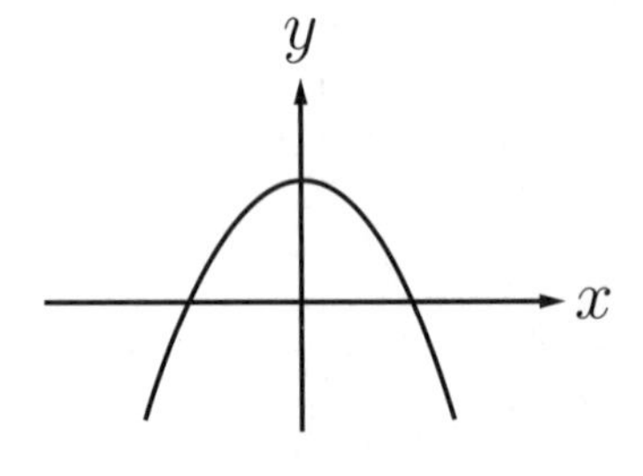

Figure 1

8. If $a < 0$, solve the inequality: $ax - 2a > 5a$

   (A) $x > -3$

   (B) $x < 3$

   (C) $x > 7$

   (D) $x < 7$

   (E) $x < -7$

USE THIS SPACE FOR SCRATCH WORK

9. A number is selected at random from the first 20 nonnegative integers. What is the probability that the number selected is an even number?

(A) $\frac{2}{5}$

(B) $\frac{9}{20}$

(C) $\frac{1}{2}$

(D) $\frac{11}{20}$

(E) $\frac{3}{5}$

10. Jason bought a stock for \$312. If the value of the stock increases 20% every year, how much is the stock worth in three years?

(A) \$476.78

(B) \$493.62

(C) \$504.79

(D) \$517.32

(E) \$539.14

11. If two solutions to $ax^4+bx^3+cx^2+dx+e=0$ are $2+\sqrt{2}$ and $1-3i$, what are the other solutions?

(A) $2+\sqrt{2}$ and $3-i$

(B) $2+\sqrt{2}$ and $1+3i$

(C) $2-\sqrt{2}$ and $3-i$

(D) $2-\sqrt{2}$ and $1+3i$

(E) $-2\sqrt{2}$ and $-3i$

USE THIS SPACE FOR SCRATCH WORK

12. If $\sin x = \dfrac{1}{\sqrt{5}}$ and $\cos x = \dfrac{2}{\sqrt{5}}$, what is the value of $\sin 2x$?

(A) $\dfrac{1}{5}$

(B) $\dfrac{3}{5}$

(C) $\dfrac{4}{5}$

(D) $\dfrac{5}{4}$

(E) $\dfrac{5}{3}$

13. Let $\begin{vmatrix} a & b \\ c & d \end{vmatrix} = ad - bc.$ $\begin{vmatrix} 1 & -1 \\ -4 & 2 \end{vmatrix} =$

(A) -2

(B) -1

(C) 2

(D) 4

(E) 6

14. If a car factory produces x cars in y months, how many cars does the car factory produce in z years?

(A) $\dfrac{xz}{y}$

(B) $\dfrac{xy}{z}$

(C) $\dfrac{12xy}{z}$

(D) $\dfrac{12yz}{x}$

(E) $\dfrac{12xz}{y}$

USE THIS SPACE FOR SCRATCH WORK

15. If $27^{3-x} = 9^{x+2}$, what is the value of x ?

(A) 2

(B) 1

(C) $\frac{1}{2}$

(D) $\frac{1}{3}$

(E) $\frac{1}{5}$

16. If $\sin\theta = -\frac{12}{13}$ and $\cos\theta > 0$, what is the value of $\tan\theta$?

(A) $-\frac{13}{5}$

(B) $-\frac{13}{12}$

(C) $-\frac{12}{5}$

(D) $-\frac{5}{12}$

(E) $-\frac{5}{13}$

17. Which of the following function is represented by the parametric equations $x = e^t$ and $y = e^{2t}$?

(A) $y = \sqrt{x}$

(B) $y = \frac{2}{x}$

(C) $y = \frac{1}{x^2}$

(D) $y = 2x$

(E) $y = x^2$

USE THIS SPACE FOR SCRATCH WORK

18. Figure 2 shows the graph of the line m which crosses the x-axis at -2 and crosses the y-axis at 4. If the line m is reflected over the y-axis, which of the following must be the equation of the new line?

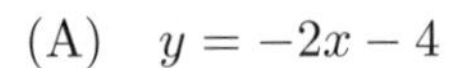

(A) $y = -2x - 4$

(B) $y = -2x + 4$

(C) $y = -\frac{1}{2}x + 4$

(D) $y = \frac{1}{2}x + 2$

(E) $y = \frac{1}{4}x - 2$

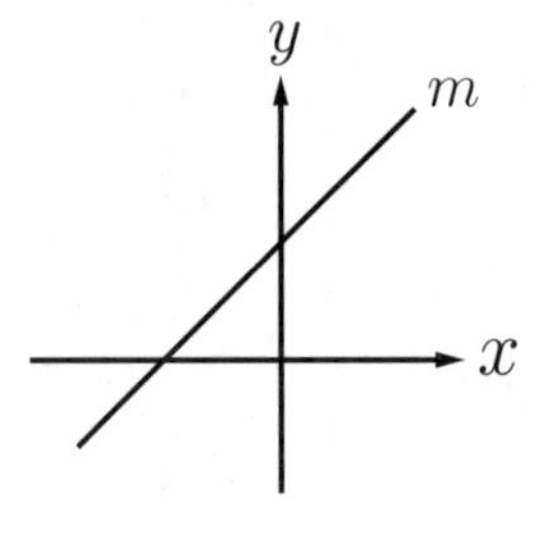

Figure 2

19. Solve: $\sqrt[3]{1-x} + 0.4 = 1$

(A) 0.95

(B) 0.91

(C) 0.86

(D) 0.82

(E) 0.78

20. Which of the following expression has the same value as $\cos 150°$?

(A) $\tan 90°$

(B) $\sin 120°$

(C) $\cos 180°$

(D) $\sin 240°$

(E) $\cos 300°$

USE THIS SPACE FOR SCRATCH WORK

21. Which of the following complex number is farthest away from the origin?

    (A) $3i$

    (B) $\sqrt{3}+i$

    (C) $-\sqrt{3}-2i$

    (D) $-2+\sqrt{2}i$

    (E) $-2-2i$

22. Which of the following rectangular equation is equivalent to the polar equation $r = 2\sec\theta$?

    (A) $y = x$

    (B) $y = 2$

    (C) $x = 2$

    (D) $x = y^2$

    (E) $x^2 + y^2 = 1$

23. If $f(x) = \sqrt{x+4} - 1$, what is the domain of the inverse function of f ?

    (A) $x \leq -1$

    (B) $x \geq -1$

    (C) $x \leq -4$

    (D) $x \geq -4$

    (E) $x \leq -5$

USE THIS SPACE FOR SCRATCH WORK

24. Vector **V** has a magnitude of 10 and vector **W** has a magnitude of 12. If the angle between the two vectors is $65°$, what is the magnitude of the resultant vector?

(A) 11.94

(B) 12.47

(C) 13.14

(D) 13.82

(E) 14.73

25. Which of the following logarithmic expression is equivalent to $\ln(x^2 + x - 2) - \ln(x - 1)$?

(A) $\ln x + 2$

(B) $\ln(x + 2)$

(C) $\ln x^2 - 1$

(D) $\ln(x^2 - 1)$

(E) $\ln(x^3 - 3x + 2)$

26. The eccentricity of an ellipse is a measure of how circular the ellipse is. At $e \approx 0$, an ellipse is nearly a circle. From the equation of the ellipse $\frac{x^2}{a^2} + \frac{y^2}{b^2} = 1$, the eccentricity e is defined by $e = \sqrt{1 - \frac{b^2}{a^2}}$. What is the eccentricity of the ellipse $x^2 + 4y^2 = 4$?

(A) 0.39

(B) 0.52

(C) 0.75

(D) 0.87

(E) 1.15

USE THIS SPACE FOR SCRATCH WORK

27. $\dfrac{x}{x-1} + \dfrac{x}{2x+4} =$

(A) $\dfrac{3x(x+1)}{2(x-1)(x+2)}$

(B) $\dfrac{x^2+x+4}{2(x-1)(x+2)}$

(C) $\dfrac{x^2+3x+3}{2(x-1)(x+2)}$

(D) $\dfrac{3(x+1)}{(x-1)(x+2)}$

(E) $\dfrac{2x}{(x-1)(x+2)}$

28. Solve the inequality: $x^2 + 7x < 18$

(A) $0 < x < 2$

(B) $2 < x < 5$

(C) $-2 < x < 7$

(D) $-9 < x < 2$

(E) $-9 < x < -2$

29. If $f(x) = \dfrac{-x+3}{x-1}$ and $g(x) = f(x-1) + 2$, what are the vertical and horizontal asymptotes of $g(x)$?

(A) $x = 2$ and $y = 1$

(B) $x = 2$ and $y = -1$

(C) $x = 1$ and $y = 2$

(D) $x = 1$ and $y = -1$

(E) $x = 0$ and $y = -3$

USE THIS SPACE FOR SCRATCH WORK

30. Figure 3 shows a half full cylindrical tank with a radius of 1 foot and a height of 20 feet. Water is being pumped into the cylindrical tank at a rate of $\frac{\pi}{6}$ ft^3 per minute so that the water level is rising. How many minutes does it take for the water level in the cylindrical tank to reach 15 feet?

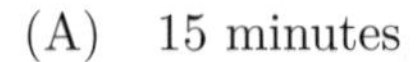

(A) 15 minutes

(B) 24 minutes

(C) 30 minutes

(D) 45 minutes

(E) 90 minutes

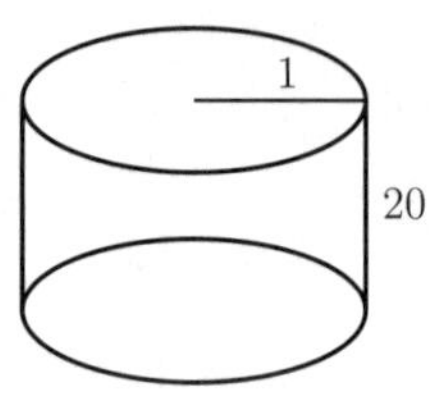

Figure 3

31. A set T has five numbers. The five numbers in set T form an arithmetic sequence with the third term being 10. If 5 is added to each number in set T, what is the sum of the new mean and median of set T ?

(A) 25

(B) 30

(C) 35

(D) 40

(E) 45

32. For $0 < \theta < \frac{\pi}{2}$, solve the trigonometric equation: $2\cos\theta + 0.36 = 1$.

(A) 1.25

(B) 1.48

(C) 1.73

(D) 1.96

(E) 2.27

USE THIS SPACE FOR SCRATCH WORK

33. $\sum_{n=10}^{21}(3n+1) =$

(A) 460

(B) 488.5

(C) 506

(D) 522.5

(E) 570

34. What are the solutions to the following system of nonlinear equations?

$$(x-2)^2 + (y-2)^2 = 4$$
$$y = -x - 1$$

(A) $(0,1)$ only

(B) $(2,2)$ only

(C) $(1,1)$ and $(3,2)$

(D) $(2,0)$ and $(-3,2)$

(E) No solutions

35. A function f is said to have a maximum point at $x = c$ if $f(c) \geq f(x)$ for all x on a given interval. Which of the following function has a maximum point in the interval $-2 < x < 2$?

(A) $y = \frac{3}{2}x - 1$

(B) $y = -x^2 + 2$

(C) $y = |x - 1|$

(D) $y = \sqrt{x+2}$

(E) $y = x^3 + 1$

USE THIS SPACE FOR SCRATCH WORK

36. $(\sin\theta + \cos\theta)^2 - (\sin\theta - \cos\theta)^2 =$

(A) 0

(B) 2

(C) $4(\sin\theta - \cos\theta)$

(D) $4\sin\theta\cos\theta$

(E) $2(\sin^2\theta - \cos^2\theta)$

37. Figure 4 shows a rectangular box with a square base that has a volume of 1000 cubic inches. What is the surface area of the box in terms of x ?

(A) $2x^2 + \dfrac{8000}{x}$

(B) $2x^2 + \dfrac{4000}{x}$

(C) $2x^2 + \dfrac{2000}{x}$

(D) $2x^2 + 2000x$

(E) $2x^2 + 1000x$

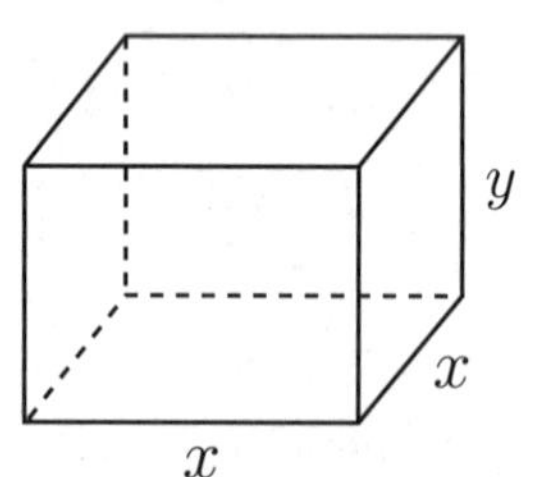

Figure 4

38. If $\ln y = -2x + \ln 3$, what is y as a function of x ?

(A) $y = 3e^{-2x}$

(B) $y = 3e^{2x}$

(C) $y = \dfrac{1}{3}e^{-2x}$

(D) $y = \dfrac{1}{3}e^{2x}$

(E) $y = e^{-2x} - 3$

USE THIS SPACE FOR SCRATCH WORK

39. Vector **V** has an initial point of $(-2, -3)$. Which of the following ordered pair represents the terminal point of vector **V** such that $|\mathbf{V}| = 10$?

(A) $(-8, -9)$

(B) $(-2, 11)$

(C) $(4, 6)$

(D) $(6, 3)$

(E) $(8, 5)$

40. Which of the following function is undefined at π and has a period of π ?

(A) $y = \tan x$

(B) $y = \tan\left(x - \frac{\pi}{2}\right)$

(C) $y = \sec x$

(D) $y = \cot\left(x - \frac{\pi}{2}\right)$

(E) $y = \csc x$

41. Solve the inequality: $\cos x < \tan x,\ \frac{\pi}{2} < x < \frac{3\pi}{2}$

(A) $\frac{\pi}{2} < x < \frac{3\pi}{4}$

(B) $\frac{\pi}{2} < x < 2.48$

(C) $\frac{3\pi}{4} < x < 3.83$

(D) $3.83 < x < 4.29$

(E) $2.48 < x < \frac{3\pi}{2}$

USE THIS SPACE FOR SCRATCH WORK

42. A polynomial function f contains a set of ordered pairs as shown in the table to the right. Which of the following function best represents f ?

(A) $y = 10(x-1)(x+2)$

(B) $y = -10(x+1)(x-2)$

(C) $y = 2(x-1)(x+2)(x+5)$

(D) $y = -2(x+1)(x-2)(x-5)$

(E) $y = \frac{1}{2}(x+1)(x-2)(x-5)$

x	-2	-1	0	2	3	5
y	56	0	-20	0	16	0

43. If $R = \dfrac{2}{1+\sqrt{5}}$, what is the value of $R^2 + \dfrac{1}{R}$?

(A) 2

(B) $2+\sqrt{5}$

(C) $2-2\sqrt{5}$

(D) $\dfrac{4-3\sqrt{5}}{2}$

(E) $\dfrac{4+2\sqrt{5}}{3}$

44. If $f(x) = \frac{1}{2}x^2 - 2x$ and $g(x) = e^x$, what is the domain of $g(f(x))$?

(A) $x \le 0$

(B) $x > 2$

(C) $0 < x < 4$

(D) $x \ge 4$

(E) All real numbers

USE THIS SPACE FOR SCRATCH WORK

45. If $y = 2x + 8$ intersects $y = -3x$ as shown in Figure 5, what is the area of the shaded region?

(A) 4.6

(B) 6.4

(C) 8.2

(D) 10.6

(E) 12.8

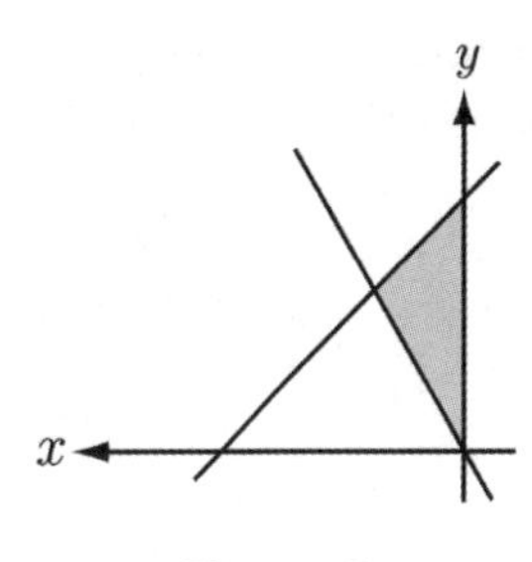

Figure 5

46. A pizza baked at $450°F$ is removed from the oven and is allowed to cool in a room with a temperature of $72°F$. The temperature of the pizza after t minutes is given by $P(t) = 72+(450-72)e^{-0.4t}$. At what time does the temperature of the pizza reach $85°F$?

(A) 9.18 minutes

(B) 8.42 minutes

(C) 7.31 minutes

(D) 6.56 minutes

(E) 4.73 minutes

47. Joshua is three times as old as his younger brother, Jason. In six years, Joshua will be twice as old as Jason. How many years from now will Jason be two-thirds as old as Joshua?

(A) 18

(B) 14

(C) 12

(D) 10

(E) 8

USE THIS SPACE FOR SCRATCH WORK

48. Which of the following piecewise function best represents the graph in Figure 6?

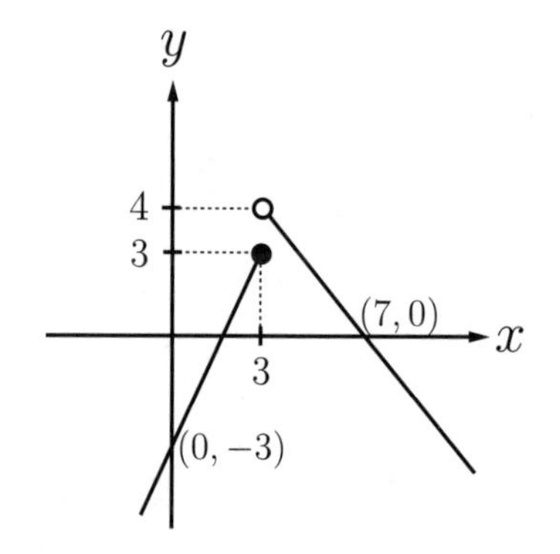

Figure 6

(A)

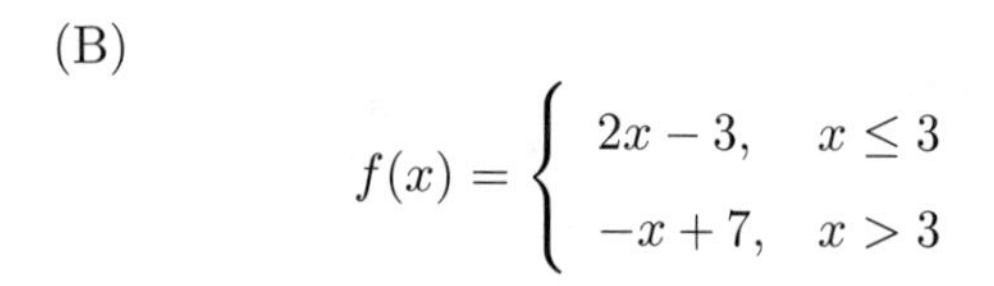

$$f(x) = \begin{cases} 2x - 3, & x \leq 3 \\ -2x + 5, & x > 3 \end{cases}$$

(B)

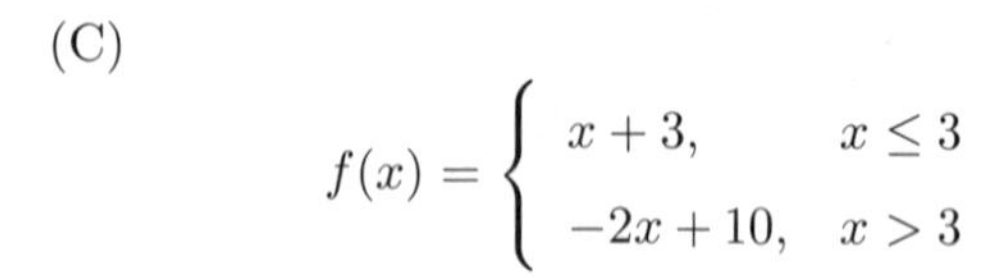

$$f(x) = \begin{cases} 2x - 3, & x \leq 3 \\ -x + 7, & x > 3 \end{cases}$$

(C)

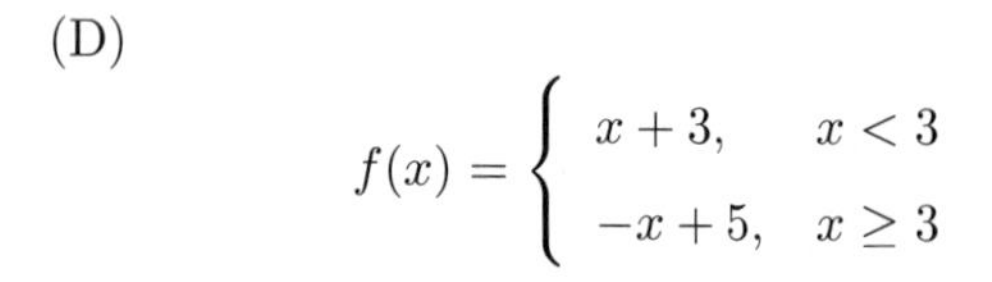

$$f(x) = \begin{cases} x + 3, & x \leq 3 \\ -2x + 10, & x > 3 \end{cases}$$

(D)

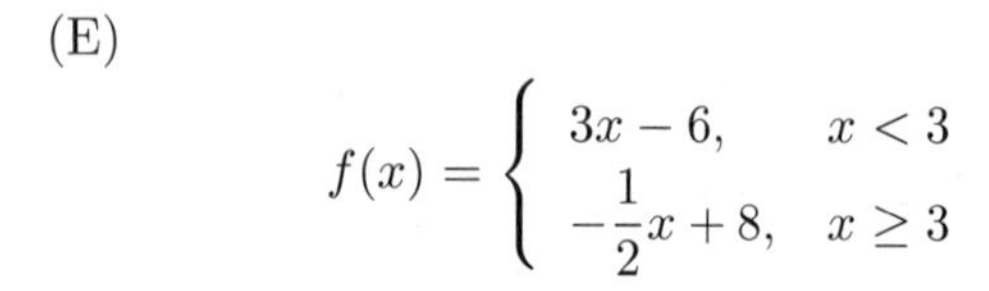

$$f(x) = \begin{cases} x + 3, & x < 3 \\ -x + 5, & x \geq 3 \end{cases}$$

(E)

$$f(x) = \begin{cases} 3x - 6, & x < 3 \\ -\frac{1}{2}x + 8, & x \geq 3 \end{cases}$$

49. Solve: $x - 1 = \sqrt{2x + 22}$

(A) $x = 2$ or $x = -1$

(B) $x = 3$ or $x = -7$

(C) $x = 7$ or $x = -3$

(D) $x = 7$ only

(E) $x = -3$ only

USE THIS SPACE FOR SCRATCH WORK

50. The frequency f of a vibrating guitar string is given by

$$f = \frac{1}{2L}\sqrt{\frac{T}{\mu}}$$

where L is the length of the string, T is the tension, and μ is the linear density. Which of the following statement CANNOT be true?

(A) If the length of the string is doubled, the frequency is reduced by half.

(B) The shorter the length of the sting, the higher the frequency.

(C) The higher the tension of the string, the higher the frequency.

(D) The larger the linear density, the higher the frequency.

(E) If the tension is quadrupled, the frequency is doubled.

STOP

Mathematics Scoring Worksheet

Directions: In order to calculate your score correctly, fill out the table below. After calculating your raw score, round the raw score to the nearest whole number. The scaled score can be determined using the "Math Test Score Conversion Table".

Mathematics Score			
A. Number Correct		**B.** Number Incorrect ÷ 4	
Total Unrounded Raw Score $A - B$		**Total Rounded Raw Score** **Round to nearest whole number**	

Math Test Score Conversion Table					
Raw Score	Scaled Score	Raw Score	Scaled Score	Raw Score	Scaled Score
50	800	29	660	8	490
49	800	28	650	7	480
48	800	27	640	6	480
47	800	26	630	5	470
46	800	25	630	4	460
45	800	24	620	3	450
44	800	23	610	2	440
43	800	22	600	1	430
42	790	21	590	0	410
41	780	20	580	−1	390
40	770	19	570	−2	370
39	760	18	560	−3	360
38	750	17	560	−4	340
37	740	16	550	−5	340
36	730	15	540	−6	330
35	720	14	530	−7	320
34	710	13	530	−8	320
33	700	12	520	−9	320
32	690	11	510	−10	320
31	680	10	500	−11	310
30	670	9	500	−12	310

ANSWERS AND SOLUTIONS

1. C	11. D	21. A	31. B	41. E
2. B	12. C	22. C	32. A	42. D
3. E	13. A	23. B	33. E	43. A
4. E	14. E	24. A	34. E	44. E
5. C	15. B	25. B	35. B	45. B
6. C	16. C	26. D	36. D	46. B
7. B	17. E	27. A	37. B	47. A
8. D	18. B	28. D	38. A	48. B
9. C	19. E	29. A	39. D	49. D
10. E	20. D	30. C	40. B	50. D

Solutions

1. (C)

Switch the x and y variables and solve for y.

$$y = \frac{1}{2}x + 3 \qquad \text{Switch the } x \text{ and } y \text{ variables}$$
$$x = \frac{1}{2}y + 3 \qquad \text{subtract 3 from each side}$$
$$x - 3 = \frac{1}{2}y \qquad \text{Multiply each side by 2}$$
$$y = 2x - 6$$

Therefore, the inverse function of $y = \frac{1}{2}x + 3$ is $y = 2x - 6$.

2. (B)

Tip
1. $0! = 1$
2. $n! = n \times (n-1)!$

Since $n! = n \times (n-1)!$, $9! = 9 \times 8!$. Therefore, (B) is the correct answer.

3. (E)

Since $\mathbf{i} = \langle 1, 0 \rangle$ and $\mathbf{j} = \langle 0, 1 \rangle$,

$$\begin{aligned} 3\mathbf{i} - 4\mathbf{j} &= 3\langle 1, 0 \rangle - 4\langle 0, 1 \rangle \\ &= \langle 3, 0 \rangle - \langle 0, 4 \rangle \\ &= \langle 3, -4 \rangle \end{aligned}$$

Therefore, $3\mathbf{i} - 4\mathbf{j}$ is $\langle 3, -4 \rangle$.

4. (E)

Tip $(a-b)^2 = a^2 - 2ab + b^2$

Suppose $f(x) = x^2 + 1$ and $g(x) = 2x - 1$. In order to evaluate $f(g(x))$, substitute $(2x - 1)$ for x in $f(x)$.

$$\begin{aligned} f(g(x)) &= f(2x-1) \\ &= (2x-1)^2 + 1 \\ &= 4x^2 - 4x + 1 + 1 \\ &= 4x^2 - 4x + 2 \end{aligned}$$

Therefore, $g(x)$ must be $2x - 1$.

5. (C)

Tip

	Odd functions	Even functions
Definition	$f(-x) = -f(x)$	$f(-x) = f(x)$
Graph	Symmetric with respect to the origin	Symmetric with respect to the y-axis
Example	x^3, $\sin x$, $\tan x$	x^2, $\cos x$

The graphs in answer choices (A), (D), and (E) are graphs of even functions because they are symmetric with respect to the y-axis. So eliminate (A), (D), and (E). Since the graph in answer choice (B) does not pass the vertical line test, it is not the graph of a function. So eliminate (B). Therefore, (C) is the correct answer.

6. (C)

Let x be the total number of students at Solomon Academy. Since 40% of the students are boys and 60% of the students are girls, the number of boys and girls can be expressed as $0.4x$ and $0.6x$, respectively. 60% of the boy students who chose Mozzarella cheese as their favorite cheese can be expressed as $0.6(0.4x) = 0.24x$. Likewise, 40% of the girl students who chose Mozzarella cheese as their favorite cheese can be expressed as $0.4(0.6x) = 0.24x$. Thus, the total number of students who chose Mozzarella cheese as their favorite cheese is $0.24x + 0.24x = 0.48x$. Therefore, the percent of boy students who chose Mozzarella cheese as their favorite cheese is $\dfrac{0.24x}{0.48x} = 0.5 = 50\%$.

7. (B)

Tip The axis of symmetry: $x = -\dfrac{b}{2a}$

The graph of a quadratic function $y = ax^2 + bx + c$ opens down as shown below. Thus, the leading coefficient is negative; that is $a < 0$.

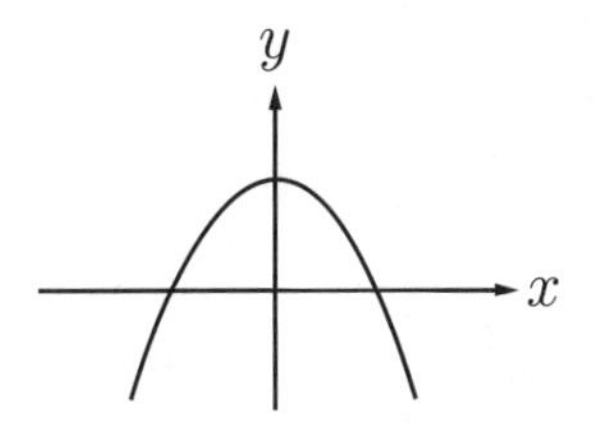

Since the graph is symmetric with respect to the y-axis, the axis of symmetry is zero; that is, $x = -\dfrac{b}{2a} = 0$, which implies that $b = 0$. The y-intercept of the quadratic function, c, lies above the x-axis. Thus, $c > 0$. Therefore, $a < 0$, $b = 0$, and $c > 0$.

8. (D)

Tip The inequality symbol must be reversed when you multiply or divide each side of the inequality by a negative number.

$ax - 2a > 5a$	Add $2a$ to each side
$ax > 7a$	Divide each side by a
$x < 7$	Reverse the inequality symbol since $a < 0$

Therefore, (D) is the correct answer.

9. (C)

Of the first 20 nonnegative integers: $0, 1, 2, \cdots, 18, 19$, there are 10 even numbers: $0, 2, 4, 6, 8, 10, 12, 14, 16$, and 18. Therefore, the probability that the number selected is an even number is $\dfrac{10}{20} = \dfrac{1}{2}$.

10. (E)

The value of the stock increases 20% per year. The value of the stock for the first three years is shown below.

The value of the stock after 1 year: $\$312(1 + 0.2) = \$374.4$

The value of the stock after 2 years: $\$312(1 + 0.2)^2 = \$449.28$

The value of the stock after 3 years: $\$312(1 + 0.2)^3 = \$539.14$

Therefore, the value of the stock after 3 years is \$539.14.

11. (D)

> Tip
>
> The conjugate pairs theorem states that complex zeros and irrational zeros always occur in conjugate pairs.
>
> If $a + bi$ is a zero of f, $\Longrightarrow$ $a - bi$ is also a zero of f.
>
> If $a + \sqrt{b}$ is a zero of f, $\Longrightarrow$ $a - \sqrt{b}$ is also a zero of f.

According to the conjugate pairs theorem, $2 + \sqrt{2}$ and $2 - \sqrt{2}$, and $1 - 3i$ and $1 + 3i$ are the solutions to $ax^4 + bx^3 + cx^2 + dx + e = 0$. Therefore, (D) is the correct answer.

12. (C)

> Tip
>
> $\sin 2x = 2 \sin x \cos x$

$$\sin 2x = 2 \sin x \cos x = 2\left(\frac{1}{\sqrt{5}}\right)\left(\frac{2}{\sqrt{5}}\right) = \frac{4}{5}$$

Therefore, the value of $\sin 2x$ is $\frac{4}{5}$.

13. (A)

Since $\begin{vmatrix} a & b \\ c & d \end{vmatrix} = ad - bc$, $\begin{vmatrix} 1 & -1 \\ -4 & 2 \end{vmatrix} = 1 \cdot 2 - (-1)(-4) = 2 - 4 = -2$.

14. (E)

There are 12 months in 1 year. Convert z years to $12z$ months. Define p as the number of cars that the car factory produce in $12z$ months. Set up a proportion in terms of cars and months.

$$x_{\text{cars}} : y_{\text{months}} = p_{\text{cars}} : 12z_{\text{months}}$$

$$\frac{x}{y} = \frac{p}{12z} \qquad \text{Use cross product property}$$

$$py = 12xz$$

$$p = \frac{12xz}{y}$$

Therefore, the number of cars that the car factory produces in z years is $\frac{12xz}{y}$.

15. (B)

The expressions on the left and right have different bases of 27 and 9, respectively. Change both bases to $27 = 3^3$ and $9 = 3^2$ and solve the equation.

$$27^{3-x} = 9^{x+2}$$
$$(3^3)^{3-x} = (3^2)^{x+2} \quad \text{Use the exponent property: } (a^m)^n = a^{mn}$$
$$3^{9-3x} = 3^{2x+4} \quad \text{Since both sides have the same base}$$
$$9 - 3x = 2x + 4$$
$$-5x = -5$$
$$x = 1$$

Therefore, the value of x is 1.

16. (C)

$\sin\theta$ is negative in the 3rd and 4th quadrants, and $\cos\theta$ is positive in the 1st and 4th quadrants. Thus, θ must lie in the 4th quadrant. Since $\sin\theta = -\frac{12}{13}$, the length of side opposite to θ is 12 and the length of the hypotenuse is 13. Using the Pythagorean theorem: $13^2 = a^2 + 12^2$, the length of the side adjacent to θ is 5 as shown in the figure below.

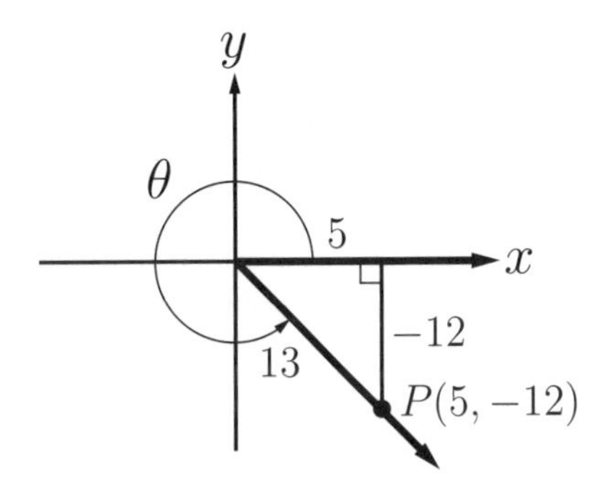

Suppose point $P(x, y)$ is on the terminal side of θ. Since θ lies in the 4th quadrant, the x and y coordinates of point P is $(5, -12)$. Thus,

$$\tan\theta = \frac{\text{opposite side}}{\text{adjacent side}} = -\frac{12}{5}$$

Therefore, the value of $\tan\theta = -\dfrac{12}{5}$.

17. (E)

Tip
1. Eliminate the parameter t so that the parametric equation become a rectangular equation $y = f(x)$.
2. $e^{2t} = (e^t)^2$

Since $x = e^t$ and $y = e^{2t} = (e^t)^2$, $y = x^2$. Therefore, the quadratic function $y = x^2$ is represented by the parametric equations $x = e^t$ and $y = e^{2t}$.

18. (B)

If the line m is reflected over the y-axis, the y-intercept remains the same $(0,4)$ and the x-intercept changes to $(2,0)$. Let's find the slope of the new line using the points, $(0,4)$ and $(2,0)$.

$$\text{Slope of the new line} = \frac{0-4}{2-0} = \frac{-4}{2} = -2$$

Thus, the slope of the new line is -2 and the y-intercept is 4. Therefore, the equation of the new line is $y=-2x+4$.

19. (E)

$\sqrt[3]{1-x}+0.4=1$	Subtract 0.4 from each side
$\sqrt[3]{1-x}=0.6$	Raise each side to a power of 3
$1-x=0.216$	Solve for x
$x=0.78$	

Therefore, the value of x for which $\sqrt[3]{1-x}+0.4=1$ is 0.78.

20. (D)

$\cos 150^\circ = -\cos 30^\circ = -\dfrac{\sqrt{3}}{2}$.

(A) $\tan 90^\circ = \text{undefined}$

(B) $\sin 120^\circ = \sin 60^\circ = \dfrac{\sqrt{3}}{2}$

(C) $\cos 180^\circ = -1$

(D) $\sin 240^\circ = -\sin 60^\circ = -\dfrac{\sqrt{3}}{2}$

(E) $\cos 300^\circ = \cos 60^\circ = \dfrac{1}{2}$

Therefore, (D) is the correct answer.

21. (A)

Tip: The absolute value of a complex number $z=a+bi$, denoted by $|z|$, is the distance from the origin to the complex number in the complex plane. The formula for finding the absolute value of a complex number $a+bi$ is as follows:

$$\text{If } z=a+bi, \qquad |z|=\sqrt{a^2+b^2}$$

(A) If $z=3i$, $\quad |z|=\sqrt{3^2}=3$

(B) If $z=\sqrt{3}+i$, $\quad |z|=\sqrt{(\sqrt{3})^2+1^2}=2$

(C) If $z=-\sqrt{3}-2i$, $\quad |z|=\sqrt{(-\sqrt{3})^2+(-2)^2}=\sqrt{7}$

(D) If $z=-2+\sqrt{2}i$, $\quad |z|=\sqrt{(-2)^2+(\sqrt{2})^2}=\sqrt{6}$

(E) If $z=-2-2i$, $\quad |z|=\sqrt{(-2)^2+(-2)^2}=2\sqrt{2}$

Therefore, the complex number farthest away from the origin is $3i$.

22. (C)

Tip $x = r\cos\theta$, $y = r\sin\theta$

$r = 2\sec\theta$	Since $\sec\theta = \dfrac{1}{\cos\theta}$
$r = \dfrac{2}{\cos\theta}$	Multiply each side by $\cos\theta$
$r\cos\theta = 2$	Since $x = r\cos\theta$
$x = 2$	

Therefore, the rectangular equation $x = 2$ is equivalent to the polar equation $r = 2\sec\theta$.

23. (B)

Tip
1. The domain of the inverse function is the range of the original function.
2. The range of the inverse function is the domain of the original function.

If $f(x) = \sqrt{x}$, $f(x+4) - 1$ in function notation represents $\sqrt{x+4} - 1$. The function notation $f(x+4) - 1$ suggest that the graph of $\sqrt{x+4} - 1$ involves a horizontal shift and a vertical shift from the graph of the parent function $y = \sqrt{x}$. Thus, perform the horizontal shift first and then perform the vertical shift as shown below.

$f(x) = \sqrt{x}$	The graph of the parent function as shown in Fig. A
$f(x+4) = \sqrt{x+4}$	Move the graph of $\sqrt{x}$ left 4 units as shown in Fig. B
$f(x+4) - 1 = \sqrt{x+4} - 1$	Move the graph of $\sqrt{x+4}$ down 1 unit as shown in Fig. C

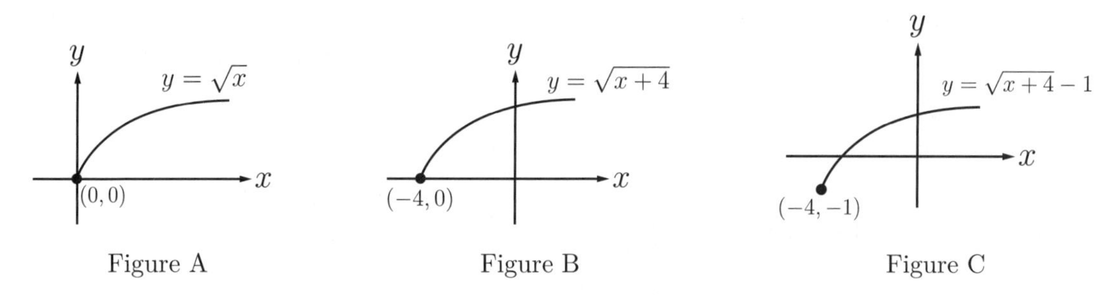

Figure A Figure B Figure C

As shown in Figure C, the range of the function $y = \sqrt{x+4} - 1$ is $y \geq -1$. Since the domain of the inverse function is the range of the original function, the domain of the inverse function is $x \geq -1$.

24. (A)

Tip If triangle ABC shown to the right is a SAS triangle (a, b, and $m\angle C$ are known), side c can be calculated by the Law of Cosines.

$$c^2 = a^2 + b^2 - 2ab\cos C$$

Note that side c is opposite angle C.

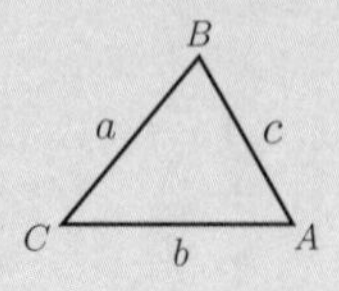

Define the magnitude of **V** as a, the magnitude of **W** as b, and the magnitude of the resultant vector as c as shown in the figure below. **V** has a magnitude of 10, **W** has a magnitude of 12, and the angle between the two vectors is 65°. Thus, $a = 10$, $b = 12$, and $m\angle C = 65°$.

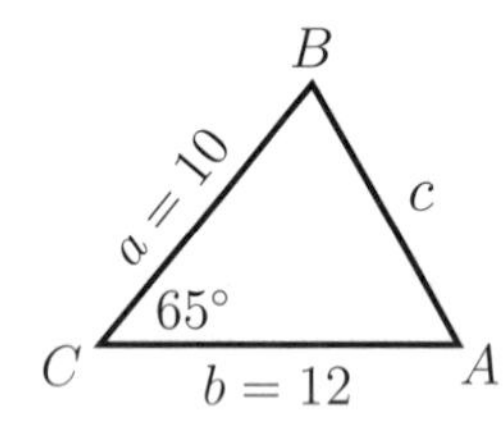

Since triangle ABC is a non-right triangle and is a SAS triangle, use the Law of Cosines to find the side c.

$$c^2 = 10^2 + 12^2 - 2(10)(12)\cos 65° \quad \Longrightarrow \quad c = 11.94$$

Therefore, the magnitude of the resultant vector is 11.94.

25. (B)

Tip $\log_a x - \log_a y = \log_a \frac{x}{y}$

$$\begin{aligned} \ln(x^2+x-2) - \ln(x-1) &= \ln\frac{x^2+x-2}{x-1} && \text{Factor} \\ &= \ln\frac{(x+2)(x-1)}{(x-1)} && \text{Cancel } (x-1) \text{ out} \\ &= \ln(x+2) \end{aligned}$$

Therefore, the logarithmic expression $\ln(x^2+x-2) - \ln(x-1)$ is equivalent to $\ln(x+2)$.

26. (D)

Divide each side of the equation $x^2 + 4y^2 = 4$ by 4. We find that $\frac{x^2}{2^2} + \frac{y^2}{1^2} = 1$.

$$e = \sqrt{1 - \frac{b^2}{a^2}} = \sqrt{1 - \frac{1^2}{2^2}} = 0.87$$

Therefore, the eccentricity of the ellipse $x^2 + 4y^2 = 4$ is 0.87.

27. (A)

> **Tip** To add (or subtract) two rational expressions with unlike denominators, find the least common denominator of the rational expressions. Then, rewrite each expression as an equivalent expression using the least common denominator. Finally, add (or subtract) their numerators and put the result over the least common denominator. For instance,
>
> $$\frac{3}{x+1}+\frac{2}{x-2}=\frac{3(x-2)}{(x+1)(x-2)}+\frac{2(x+1)}{(x+1)(x-2)}=\frac{5x-4}{(x+1)(x-2)}$$

Since the two rational expressions have unlike denominators, find the least common denominator of the rational expressions, which is $2(x-1)(x+2)$.

$$\begin{aligned}\frac{x}{x-1}+\frac{x}{2x+4}&=\frac{2x(x+2)}{2(x-1)(x+2)}+\frac{x(x-1)}{2(x-1)(x+2)}\\&=\frac{2x^2+4x+x^2-x}{2(x-1)(x+2)}\\&=\frac{3x^2+3x}{2(x-1)(x+2)}\\&=\frac{3x(x+1)}{2(x-1)(x+2)}\end{aligned}$$

Therefore, $\frac{x}{x-1}+\frac{x}{2x+4}=\frac{3x(x+1)}{2(x-1)(x+2)}$.

28. (D)

> **Tip** Solving $x^2+7x-18<0$ means finding the x-values for which the graph of $y=x^2+7x-18$ lies below the x-axis.

Subtract 18 from each side of the inequality $x^2+7x<18$. We find that $x^2+7x-18<0$. Let's solve the quadratic inequality $x^2+7x-18<0$ graphically. Find the x-intercepts of $y=x^2+7x-18$ first. Since $y=x^2+7x-18=(x+9)(x-2)$, the x-intercepts are -9 and 2 as shown in Figure A.

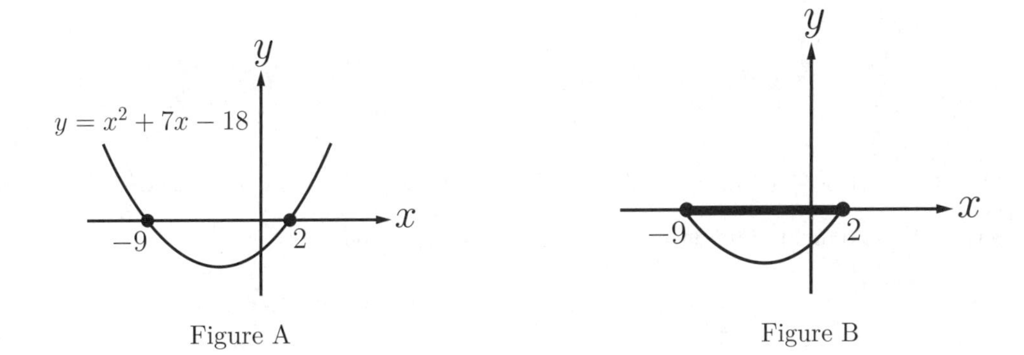

Figure A

Figure B

The graph of $y=x^2+7x-18$ lies below the x-axis when $-9<x<2$ as shown in Figure B. Therefore, the solution to $x^2+7x<18$ is $-9<x<2$.

29. (A)

In order to find the vertical asymptote, set the denominator equal to zero and solve for x. Thus, the vertical asymptote of $f(x) = \dfrac{-x+3}{x-1}$ is $x = 1$. In addition, both the numerator and the denominator of $f(x) = \dfrac{-x+3}{x-1}$ are first degree polynomials. Thus, the horizontal asymptote of $f(x) = \dfrac{-x+3}{x-1}$ is the ratio of the leading coefficients, or $y = -1$. Since $g(x) = f(x-1)+2$, the vertical and horizontal asymptotes of $g(x)$ is obtained by moving the vertical and horizontal asymptotes of $f(x)$ 1 unit to the right and 2 units up. Therefore, the vertical and horizontal asymptotes of $g(x)$ are $x = 2$ and $y = 1$, respectively.

30. (C)

A cylindrical tank with a radius of 1 foot and a height of 20 feet is half full. Thus, the water level in the beginning is 10 feet. We need to raise the water level by 5 feet so that the water level is 15 feet from the bottom. The total amount of water needed to raise the water level by 5 feet is $\pi(1\,\text{ft})^2(5\,\text{ft}) = 5\pi\ \text{ft}^3$. Since the water is being pumped into the cylindrical tank at a rate of $\dfrac{\pi}{6}\ \text{ft}^3$ per minute, $5\pi\ \text{ft}^3$ of water would be pumped into the cylindrical tank in 30 minutes. Therefore, (C) is the correct answer.

31. (B)

In order to solve this problem, it is not necessary to determine the other terms. In a set of five terms of an arithmetic sequence, the third term is both the mean and the median. For instance, 4, 7, 10, 13, and 16 is an arithmetic sequence in which the third term, 10, is both the mean and the median. Since 5 is added to each element in the set T, the new mean and the median of set T are both 15. Therefore, the sum of the new mean and median of set T is $15 + 15 = 30$.

32. (A)

Set the angle mode to Radians in your calculator.

$$2\cos\theta + 0.36 = 1$$

$$\cos\theta = 0.32 \qquad \text{Take the inverse cosine of both sides}$$

$$\theta = \cos^{-1}(0.32) = 1.25$$

Therefore, (A) is the correct answer.

33. (E)

Tip The nth partial sum for an arithmetic series: $S_n = \dfrac{n}{2}(a_1 + a_n)$

$\displaystyle\sum_{n=10}^{21}(3n+1) = 31 + 34 + 37 + \cdots + 64$, which is the sum of the first 12 terms of the arithmetic sequence with a common difference of 3, $a_1 = 31$, and $a_{12} = 64$.

$$S_n = \frac{n}{2}(a_1 + a_n) \quad \Longrightarrow \quad S_{12} = \frac{12}{2}(31 + 64) = 570$$

Therefore, $\displaystyle\sum_{n=10}^{21}(3n+1) = 570$.

34. (E)

> **Tip** Solutions to a system of nonlinear equations are ordered pairs (x, y) that satisfy all equations in the system. In other words, solutions to a system of nonlinear equations are intersection points that lie on all graphs.

$(x-2)^2+(y-2)^2=4$ represents a circle with radius 2 as shown in the figure below. The center of the circle is $(2, 2)$.

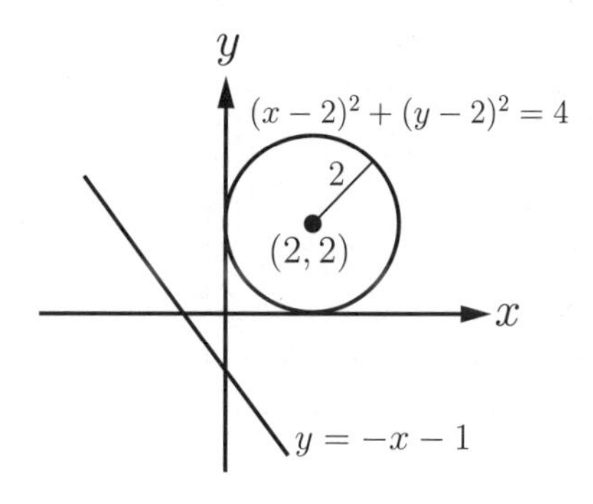

$y=-x-1$ represents a line that does not intersect the circle $(x-2)^2+(y-2)^2=4$. Since there are no intersection points that lie on the two graphs, there are no solutions to a system of nonlinear equations.

35. (B)

Below shows the graphs of functions in answer choices.

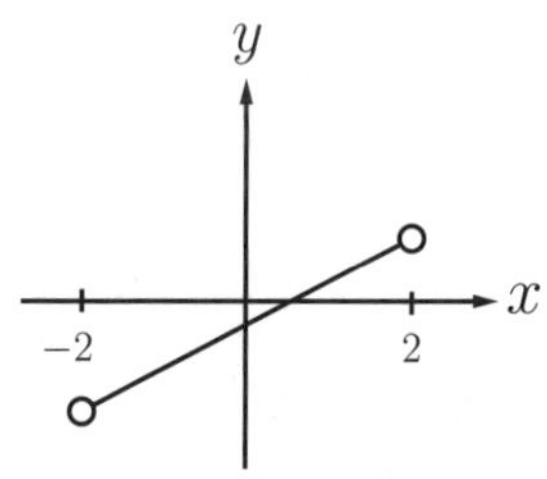

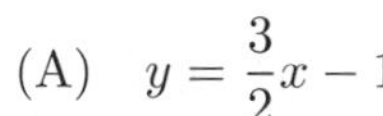
(A) $y=\frac{3}{2}x-1$

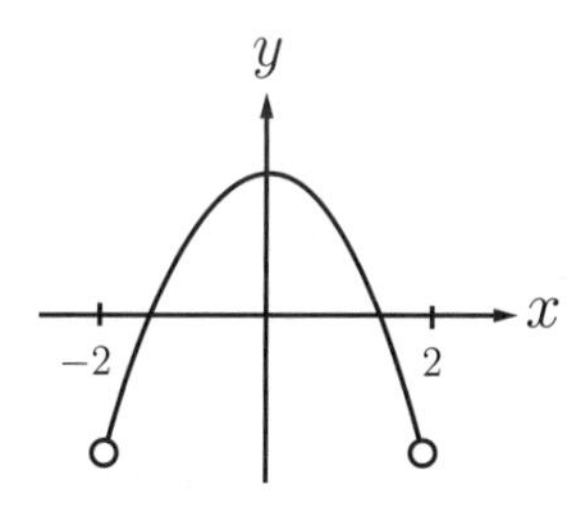

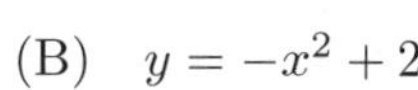
(B) $y=-x^2+2$

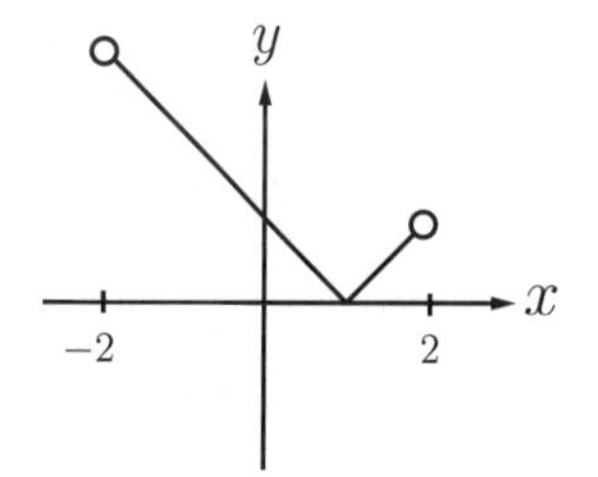

(C) $y=|x-1|$

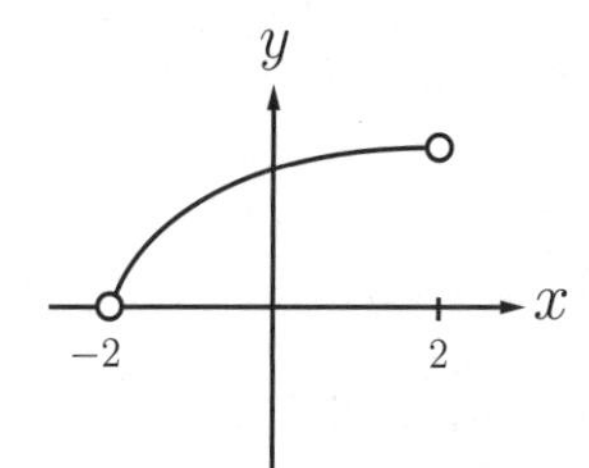

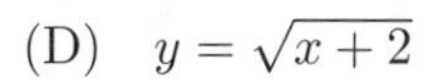
(D) $y=\sqrt{x+2}$

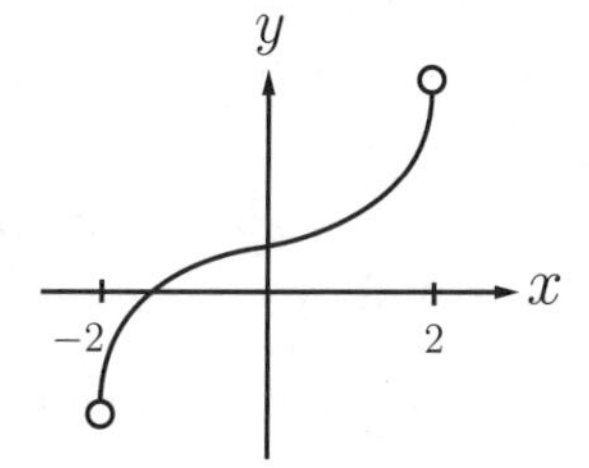

(E) $y=x^3+1$

Only function that has a maximum point in the interval $-2<x<2$ is $y=-x^2+2$. Therefore, (B) is the correct answer.

36. (D)

Tip
1. $(a \pm b)^2 = a^2 \pm 2ab + b^2$
2. $\sin^2\theta + \cos^2\theta = 1$

$$\begin{aligned}(\sin\theta+\cos\theta)^2-(\sin\theta-\cos\theta)^2 &= \sin^2\theta+2\sin\theta\cos\theta+\cos^2\theta-(\sin^2\theta-2\sin\theta\cos\theta+\cos^2\theta)\\ &= 1+2\sin\theta\cos\theta-(1-2\sin\theta\cos\theta)\\ &= 4\sin\theta\cos\theta\end{aligned}$$

Therefore, (D) is the correct answer.

37. (B)

The surface area A of the rectangular box in terms of x and y is $A = 2x^2 + 4xy$ as shown below.

Top	Bottom	Left	Right	Front	Back	Surface area
x^2	x^2	xy	xy	xy	xy	$2x^2+4xy$

The volume V of the rectangular box, $V = x^2y$, is equal to 1000; that is $x^2y = 1000$. Thus, $y = \frac{1000}{x^2}$. In order to write the surface area of the box in terms of x, substitute $\frac{1000}{x^2}$ for y in $A = 2x^2 + 4xy$.

$$\begin{aligned}A &= 2x^2+4xy\\ &= 2x^2+4x\left(\frac{1000}{x^2}\right)\\ &= 2x^2+\frac{4000}{x}\end{aligned}$$

Therefore, the surface area of the box in terms of x is $2x^2 + \frac{4000}{x}$.

38. (A)

Tip
1. $\log_a x - \log_a y = \log_a \frac{x}{y}$
2. $\log_a y = x \iff y = a^x$

$\ln y = -2x + \ln 3$	Subtract $\ln 3$ from each side
$\ln y - \ln 3 = -2x$	Simplify
$\ln \frac{y}{3} = -2x$	Convert the equation to an exponential equation
$\frac{y}{3} = e^{-2x}$	Multiply each side by 3
$y = 3e^{-2x}$	

Therefore, y as a function of x is $y = 3e^{-2x}$.

39. (D)

Tip

If point $A(x_1, y_1)$ is the initial point and point $B(x_2, y_2)$ is the terminal point, a vector, denoted by **AB**, is defined as

$$\mathbf{AB} = \langle x_2 - x_1, y_2 - y_1 \rangle$$

and the magnitude of the vector, denoted by $|\mathbf{AB}|$, is defined as

$$|\mathbf{AB}| = \sqrt{(x_2 - x_1)^2 + (y_2 - y_1)^2}$$

The initial point is $(-2, -3)$. If the terminal point is $(6, 3)$, vector $\mathbf{V} = \langle 8, 6 \rangle$ such that $|\mathbf{V}| = 10$. Therefore, the terminal point of vector $\mathbf{V}$ is $(6, 3)$.

40. (B)

Eliminate answer choices (C), and (E) because both $\sec x$ and $\csc x$ have a period of 2π. Since $\tan \pi = 0$, eliminate answer choice (A). Substitute π for x in both answer choices (B) and (D). Since $\tan\left(\pi - \frac{\pi}{2}\right) = \tan\frac{\pi}{2} =$ undefined, (B) is the correct answer.

41. (E)

Tip

Solving $\cos x < \tan x$ means finding the x-values for which the graph of $\tan x$ lies above the graph of $\cos x$.

Graph both $\cos x$ and $\tan x$ for $\frac{\pi}{2} < x < \frac{3\pi}{2}$ as shown in the figure below.

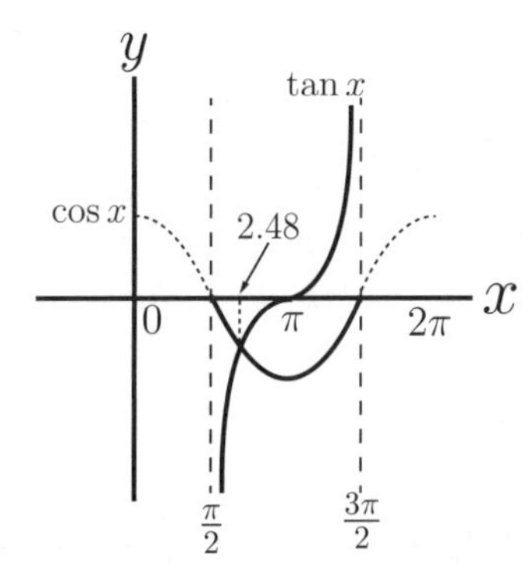

$\cos x$ and $\tan x$ intersect at $x = 2.48$. For $2.48 < x < \frac{3\pi}{2}$, the graph of $\tan x$ lies above the graph of $\cos x$. Therefore, (E) is the correct answer.

42. (D)

Tip | Factor Theorem: If the remainder $r = f(k) = 0$, then k is a zero of $f(x)$ and $x - k$ is a factor of $f(x)$.

When $x = -1$, $x = 2$, and $x = 5$, $y = 0$; that is, $f(-1) = 0$, $f(2) = 0$, and $f(5) = 0$. Thus, the polynomial function f has factors of $x + 1$, $x - 2$ and $x - 5$. Since the leading coefficient of f is unknown, let $f(x) = a(x+1)(x-2)(x-5)$. When $x = 0$, $y = -20$. Substitute 0 for x, and -20 for $f(x)$ to solve for a.

$$f(x) = a(x+1)(x-2)(x-5) \qquad \text{Substitute 0 for } x\text{, and } -20 \text{ for } f(x)$$

$$-20 = a(0+1)(0-2)(0-5) \qquad \text{Solve for } a$$

$$a = -2$$

Therefore, the polynomial function f is $f(x) = -2(x+1)(x-2)(x-5)$.

43. (A)

Tip

1. $(a+b)(a-b) = a^2 - b^2$
2. In order to rationalize the denominator, multiply the numerator and the denominator by the denominator's conjugate. For instance,

$$\frac{1}{1+\sqrt{5}} = \frac{1}{(1+\sqrt{5})} \cdot \frac{(1-\sqrt{5})}{(1-\sqrt{5})} = \frac{1-\sqrt{5}}{1-5} = \frac{1-\sqrt{5}}{-4}$$

Rationalize R as shown below.

$$R = \frac{2}{1+\sqrt{5}} = \frac{2}{1+\sqrt{5}} \cdot \frac{1-\sqrt{5}}{1-\sqrt{5}} = \frac{2(1-\sqrt{5})}{-4} = \frac{\sqrt{5}-1}{2}$$

Evaluate R^2 as shown below.

$$R^2 = \left(\frac{\sqrt{5}-1}{2}\right)^2 = \frac{5-2\sqrt{5}+1}{4} = \frac{6-2\sqrt{5}}{4} = \frac{3-\sqrt{5}}{2}$$

Therefore,

$$R^2 + \frac{1}{R} = \frac{3-\sqrt{5}}{2} + \frac{1+\sqrt{5}}{2} = \frac{4}{2} = 2$$

44. (E)

Tip | The domain of any exponential function is all real numbers.

$f(x) = \frac{1}{2}x^2 - 2x$ and $g(x) = e^x$. Thus, $g(f(x)) = e^{\frac{1}{2}x^2 - 2x}$. Since $g(f(x))$ is an exponential function, the domain of $g(f(x))$ is all real numbers.

45. (B)

In the figure below, the area of the shaded region equals the area of the triangle. Since the y-intercept of $y = 2x + 8$ is 8, the length of the base of the triangle is 8.

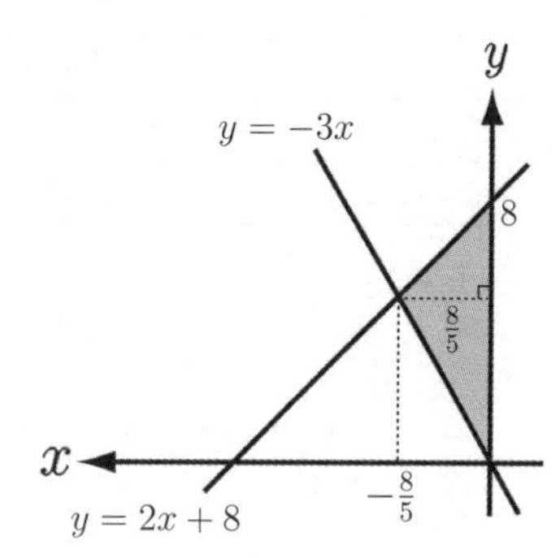

Two lines $y = 2x + 8$ and $y = -3x$ intersect. Set each equation equal to each other. Solve for x, which determines the height of the triangle.

$$\begin{aligned} 2x + 8 &= -3x \\ 5x &= -8 \\ x &= -\frac{8}{5} \end{aligned}$$

Since the x-coordinate of the intersection point is $-\frac{8}{5}$, the height of the triangle is $\frac{8}{5}$. Therefore, the area of the shaded region is $\frac{1}{2} \times 8 \times \frac{8}{5} = \frac{32}{5} = 6.4$.

46. (B)

In order to find the time at which the temperature of the pizza reaches $85°F$, substitute 85 for $P(t)$ and solve for t.

$P(t) = 72 + (450 - 72)e^{-0.4t}$	Substitute 85 for $P(t)$
$85 = 72 + (450 - 72)e^{-0.4t}$	Subtract 72 from each side
$378e^{-0.4t} = 13$	Divide 378 from each side
$e^{-0.4t} = \frac{13}{378}$	Convert the equation to a logarithmic equation
$-0.4t = \ln\left(\frac{13}{378}\right)$	Divide each side by -0.4
$t = 8.42$	

Therefore, after the pizza is cooled at room temperature for 8.42 minutes, the temperature of the pizza reaches $85°F$.

47. (A)

Define x as Jason's present age. Then, the age of Joshua is $3x$. In six years, Jason will be $x+6$ years old and Joshua will be $3x+6$ years old. Below is the summary of the defined variables.

	Jason	Joshua
Age at present	x	$3x$
Age in 6 years	$x+6$	$3x+6$

Since Joshua will be twice as old as Jason in 6 years, set up an equation in terms of age in six years and solve for x.

$$3x+6=2(x+6)$$
$$3x+6=2x+12$$
$$x=6$$

Thus, Jason is 6 years old and Joshua is $3x=18$ years old at present. Define y as the number of years from present at which Jason will be $\frac{2}{3}$ as old as Joshua.

	Jason	Joshua
Age at present	6	18
Age in y years	$6+y$	$18+y$

Since Jason will be $\frac{2}{3}$ as old as Joshua in y years, set up an equation in terms of age in y years and solve for y.

$$6+y=\frac{2}{3}(18+y) \qquad \text{Multiply each side by 3}$$
$$3(6+y)=2(18+y) \qquad \text{Expand each side}$$
$$3y+18=2y+36 \qquad \text{Solve for } y$$
$$y=18$$

Therefore, Jason will be $\frac{2}{3}$ as old as Joshua in 18 years.

48. (B)

The figure below shows the two lines. Let's find the equation of the first line that passes through the two points $(3,3)$ and $(0,-3)$ when $x \leq 3$. The slope of the line that passes through the two points is

$$\text{Slope} = \frac{-3-3}{0-3} = 2$$

Start with the slope-intercept form, $y = mx+b = 2x+b$. Since the point $(0,-3)$ is the y-intercept, $b = -3$. Thus, the equation of the first line that passes through the two points $(3,3)$ and $(0,-3)$ is $y = 2x-3$, $x \leq 3$.

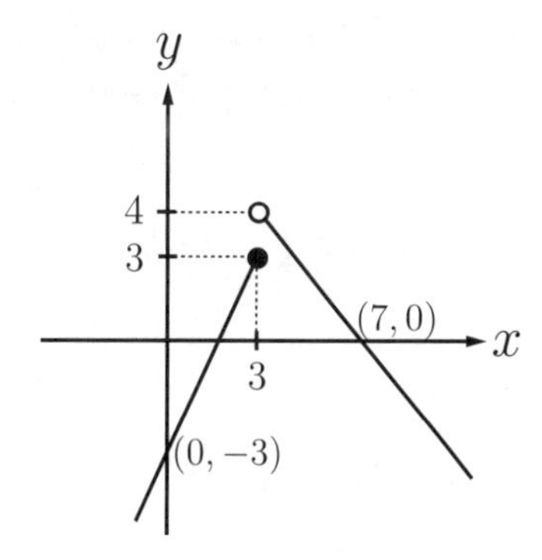

Furthermore, find the equation of the second line that passes through the two points $(3,4)$ and $(7,0)$ when $x > 3$. The slope of the line that passes through the two points is

$$\text{Slope} = \frac{0-4}{7-3} = -1$$

Start with the slope-intercept form, $y = mx+b = -x+b$. Since the point $(7,0)$ is on the line, $(7,0)$ is a solution to the equation $y = -x+b$. Substitute 7 for x and 0 for y in the equation and then solve for b.

$y = -x + b$	Substitute 7 for x and 0 for y
$0 = -7 + b$	Solve for b
$b = 7$	

Thus, the equation of the second line that passes through the two points $(3,4)$ and $(7,0)$ is $y = -x+7$, $x > 3$. Therefore, the piecewise function that represents the figure above is as follows:

$$f(x) = \begin{cases} 2x-3, & x \leq 3 \\ -x+7, & x > 3 \end{cases}$$

49. (D)

> **Tip** To solve a radical equation, square both sides of the equation to eliminate the square root. Then, solve for the variable. Once you get the solution of the radical equation, you need to substitute the solution into the original equation to check the solution. If the solution doesn't make the equation true, it is called an extraneous solution and is disregarded.

$x-1=\sqrt{2x+22}$	Square both sides
$(x-1)^2=2x+22$	Use the binomial expansion formula
$x^2-2x+1=2x+22$	Subtract $2x+22$ from each side
$x^2-4x-21=0$	Factor the quadratic expression
$(x-7)(x+3)=0$	Use zero product property: If $ab=0$, then $a=0$ or $b=0$.
$x=7$ or $x=-3$	

Substitute 7 and -3 for x in the original equation to check the solutions.

$$7-1=\sqrt{2(7)+22} \qquad\qquad -3-1=\sqrt{2(-3)+22}$$

$$6=6 \quad \checkmark\ \text{(Solution)} \qquad\qquad -4=4 \quad \text{(Not a solution)}$$

Therefore, the solution to $x-1=\sqrt{2x+22}$ is $x=7$.

50. (D)

The frequency f of a vibrating guitar string can be written as $f=\frac{1}{2}\frac{\sqrt{T}}{L\sqrt{\mu}}$, which means that f varies directly with $\sqrt{T}$, inversely with L, and inversely with $\sqrt{\mu}$. The statements in answer choice (A), (B), (C), and (E) are correct. However, the statement in (D) is wrong because the larger the linear density μ, the lower the frequency f. Therefore, (D) is the correct answer.

SAT II MATH LEVEL 2 TEST 6

Directions: Among the given answer choices, choose the BEST answer for each problem. If the exact numerical value is not within the given answer choices, select the answer that best approximates this value. Afterwards, fill in the corresponding oval on the answer sheet.

Notes:

1. A calculator may be required to answer some of the questions. Scientific and graphing calculators are allowed during this test.
2. For some questions, it is up to you to decide whether the calculator should be in degree mode or radian mode.
3. Provided figures for questions are drawn as accurately as possible UNLESS otherwise specified by the problem. Unless otherwise indicated, all figures are assumed to lie in a plane.
4. Unless otherwise specified, it can be assumed that the domain of any function f is to be the set of all real numbers x for which $f(x)$ is a real number.

Reference Information: Use the following information and formulas as a reference in answering questions on this test.

1. If the radius and height of a right circular cone are r and h respectively, then the Volume V of the cone is $V = \frac{1}{3}\pi r^2 h$.
2. If the circumference and slant height of a right circular cone are c and ℓ respectively, then the Lateral Area A of the cone is $A = \frac{1}{2}c\ell$.
3. At any given radius r, the Volume V of a sphere is $V = \frac{4}{3}\pi r^3$.
4. At any given radius r, the Surface Area A of a sphere is $A = 4\pi r^2$.
5. The Volume V of a pyramid is $V = \frac{1}{3}Bh$; given that B and h represent the base area and height of the pyramid respectively.

USE THIS SPACE FOR SCRATCH WORK

1. What are the rectangular coordinates (x, y) that correspond with the polar coordinates $(10, 135°)$?

(A) $(-5, -5)$

(B) $(-5, 8.66)$

(C) $(-7.07, 7.07)$

(D) $(7.07, -7.07)$

(E) $(8.66, -5)$

2. The probability that it will rain on Monday is $\frac{7}{10}$. The probability that it will snow on Tuesday is $\frac{1}{5}$. What is the probability that it will not rain on Monday and it will snow on Tuesday? (Assume that rain on Monday and snow on Tuesday are independent events.)

(A) $\frac{1}{50}$

(B) $\frac{3}{50}$

(C) $\frac{7}{50}$

(D) $\frac{7}{25}$

(E) $\frac{9}{10}$

3. $\sqrt[3]{27^2} =$

(A) 9

(B) 12

(C) 15

(D) 18

(E) 24

USE THIS SPACE FOR SCRATCH WORK

4. Solve the inequality: $\frac{1}{2}(x+3) < \frac{1}{3}(x+4)$

   (A) $x < 6$

   (B) $x > 1$

   (C) $x < 1$

   (D) $x > -1$

   (E) $x < -1$

5. Solve: $5x^2 + 7 = -3$

   (A) $x = i$ or $x = -i$

   (B) $x = i\sqrt{2}$ or $x = -i\sqrt{2}$

   (C) $x = 2i$ or $x = -2i$

   (D) $x = \sqrt{2}$ or $x = -\sqrt{2}$

   (E) $x = 2$ or $x = -2$

6. The equation $x^2 + y^2 + 4x - 6y = 3$ represents a circle. What is the radius of the circle?

   (A) 4

   (B) 6

   (C) 8

   (D) 12

   (E) 16

USE THIS SPACE FOR SCRATCH WORK

7. $\sec^2 x(1 - \sin^2 x) =$

(A) $\tan^2 x$

(B) $\cot^2 x$

(C) $\sin x \cos x$

(D) $\csc x \sec x$

(E) 1

8. If -2 and 3 are the x-intercepts of the quadratic function $y = x^2 + bx + c$, what is value of $b + c$?

(A) -8

(B) -7

(C) -6

(D) -5

(E) -4

9. Mr. Rhee traveled 120 miles at 60 miles per hour to visit his friend. On the way home, he was caught in heavy traffic so that the average speed for the entire trip was 40 miles per hour. How fast did Mr. Rhee travel on the way home in miles per hour?

(A) 50 miles per hour

(B) 40 miles per hour

(C) 30 miles per hour

(D) 20 miles per hour

(E) 10 miles per hour

USE THIS SPACE FOR SCRATCH WORK

10. If $31\cos\theta = -7,\ 0 \le \theta \le 180°$, what is the degree measure of angle θ ?

(A) 72°

(B) 85°

(C) 94°

(D) 103°

(E) 111°

11. If $\cos 2x = \frac{1}{3}$, what is the value of $\frac{1}{1-2\sin^2 x}$?

(A) 6

(B) 3

(C) 1

(D) $\frac{2}{3}$

(E) $\frac{1}{3}$

12. A cone and a cylinder have the same radius and the same height as shown in Figure 1. If the cone is put inside the cylinder whose volume is $\frac{2}{3}V$, what is the volume outside the cone but inside the cylinder in terms of V ?

(A) $\frac{4}{9}V$

(B) $\frac{5}{9}V$

(C) $\frac{2}{3}V$

(D) $\frac{3}{4}V$

(E) $\frac{7}{9}V$

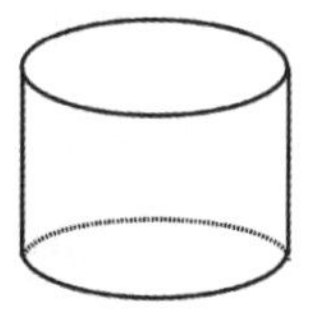

Figure 1

USE THIS SPACE FOR SCRATCH WORK

13. If $4^x = \frac{1}{5}$, what is the value of 4^{2-x} ?

(A) 16

(B) 48

(C) 60

(D) 64

(E) 80

14. A set S has four numbers which form an arithmetic sequence. If the sum of the first three terms of the sequence is 49, and the sum of the last three terms is 74, what is the range of set S ?

(A) 13

(B) 17

(C) 22

(D) 25

(E) 29

15. If the equation of the sphere is $x^2+y^2+(z-3)^2 = 25$, which of the following ordered triple (x, y, z) is on the surface of the sphere?

(A) $(4, 3, 0)$

(B) $(4, 3, 3)$

(C) $(4, 4, 3)$

(D) $(3, 4, 5)$

(E) $(3, 3, 3)$

USE THIS SPACE FOR SCRATCH WORK

16. $P(x) = -\frac{1}{32}x^2 + 4x - 100$ represents the production cost P of manufacturing x units of USB flash drive, where P is in dollars, and $x \geq 45$. How many units of the USB flash drive does a company need to manufacture to get a production cost of \$10 ?

(A) 48 units

(B) 60 units

(C) 72 units

(D) 88 units

(E) 96 units

17. In $\triangle ABC$, $m\angle C = 90°$ as shown in Figure 2. If $\cos A = \frac{1}{x}$, what is the value of $\sin A$?

(A) $\frac{x}{\sqrt{x^2 - 1}}$

(B) $\frac{x}{x^2 + 1}$

(C) $\frac{\sqrt{x^2 - 1}}{x}$

(D) $\frac{\sqrt{x^2 + 1}}{x}$

(E) $\frac{x^2 - 1}{x}$

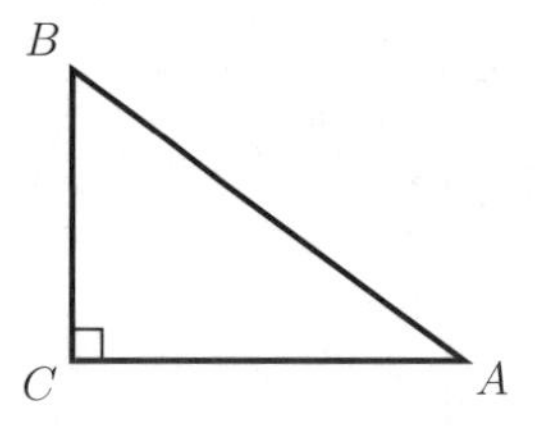

Figure 2

18. Which of the following rational function has a slant asymptote?

(A) $y = \frac{1}{x}$

(B) $y = \frac{1}{x^2}$

(C) $y = \frac{x^2 - x - 2}{x + 3}$

(D) $y = \frac{x^2 - 3x - 1}{x^2 + 1}$

(E) $y = \frac{x}{x^2 - 1}$

USE THIS SPACE FOR SCRATCH WORK

19. If $a < 0$, $b > 0$, and $|a| > |b|$, which of the following expression must be true?

(A) $\frac{a+b}{a} > 0$

(B) $\frac{ab}{b} > 0$

(C) $\frac{ab}{a+b} < 0$

(D) $a(a+b) < 0$

(E) $a^2b < 0$

20. For $x > 1$, what is the value of x for which $e^x - 3 = \ln x$?

(A) 3.35

(B) 2.78

(C) 2.12

(D) 1.73

(E) 1.14

21. What is the number of solutions to the following system of nonlinear equations?

$$2x^2 + y^2 = 9$$
$$x^2 + y^2 = 5$$

(A) 1

(B) 2

(C) 3

(D) 4

(E) 5

USE THIS SPACE FOR SCRATCH WORK

22. In $\triangle ABC$, $BC = 5$, $AC = 6$, and $m\angle C = 55°$ as shown in Figure 3. What is the area of $\triangle ABC$?

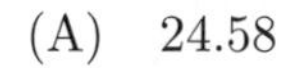

(A) 24.58

(B) 18.82

(C) 12.29

(D) 8.36

(E) 6.15

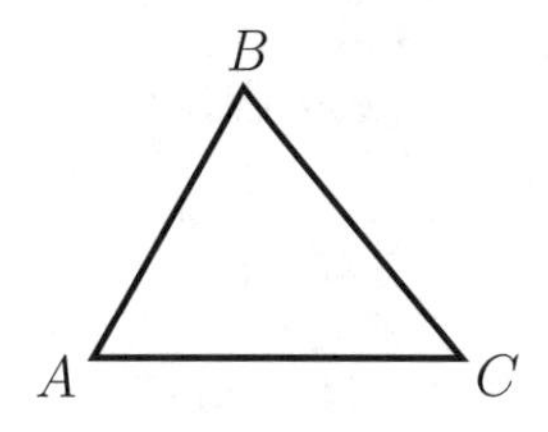

Figure 3

23. What is the sum of the following series ?

$$\frac{1}{2} + \frac{1}{2^2} + \frac{1}{2^3} + \cdots + \frac{1}{2^9}$$

(A) $\frac{255}{256}$

(B) $\frac{511}{512}$

(C) $\frac{1023}{1024}$

(D) 1

(E) $\frac{3}{2}$

24. What is the amplitude of the trigonometric function $y = -4\cos(\pi x + 1) - 2$?

(A) -4

(B) -2

(C) 1

(D) 2

(E) 4

USE THIS SPACE FOR SCRATCH WORK

25. Suppose vector $\mathbf{V} = \langle a, b \rangle$, where a and b are the horizontal and vertical components of $\mathbf{V}$. If the horizontal component of $\mathbf{V}$ is three times the vertical component of $\mathbf{V}$, and the magnitude of $\mathbf{V}$ is $\sqrt{40}$, which of the following vector can be vector $\mathbf{V}$?

    (A) $\langle -6, -2 \rangle$

    (B) $\langle -3, -1 \rangle$

    (C) $\langle 2, -6 \rangle$

    (D) $\langle 3, -1 \rangle$

    (E) $\langle 9, 3 \rangle$

26. If $\sec\theta + \tan\theta = \frac{1}{2}$, what is the value of $\sec\theta - \tan\theta$?

    (A) 4

    (B) 2

    (C) 1

    (D) $-\frac{1}{2}$

    (E) $-\frac{1}{4}$

27. The ratio of chocolate chip cookies to oatmeal cookies in a jar is 1:2. After Joshua ate four chocolate chip cookies and half of the oatmeal cookies in the jar, the ratio of chocolate chip cookies to oatmeal cookies is 3:4. How many chocolate chip cookies are left in the jar?

    (A) 10

    (B) 12

    (C) 14

    (D) 16

    (E) 18

USE THIS SPACE FOR SCRATCH WORK

28. In triangle ABC, $m\angle C = 40°$ as shown in Figure 4. If the length of $\overline{AC}$ is 20% longer than the length of $\overline{AB}$, which of the following could be the degree measure of angle B ?

(A) 50.47°

(B) 54.26°

(C) 59.49°

(D) 62.33°

(E) 65.82°

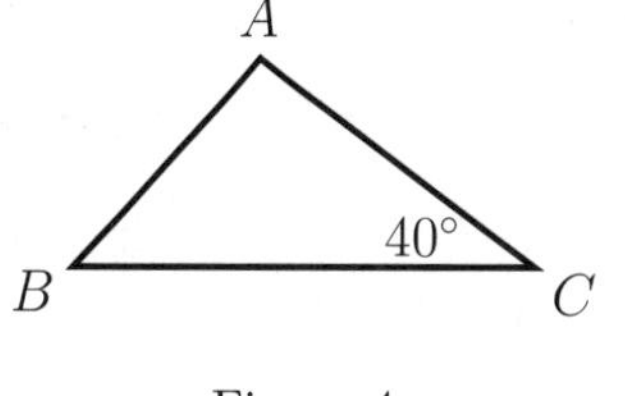

Figure 4

29. An initial amount of \$1000 is deposited into a savings account. The interest rate is compunded quaterly. If the amount after 10 years is \$3000, what is the interest rate?

(A) 11.14%

(B) 12.56%

(C) 14.09%

(D) 15.23%

(E) 16.74%

30. Solve the equation: $1 - \cos\theta = \sin^2\theta,\ 0 \leq \theta < 2\pi$.

(A) $\theta = 0,\ \frac{\pi}{2},\ \frac{3\pi}{2}$

(B) $\theta = 0,\ \pi,\ 2\pi$

(C) $\theta = \frac{\pi}{2},\ \pi,\ \frac{3\pi}{2}$

(D) $\theta = \frac{\pi}{2},\ \frac{7\pi}{6},\ \frac{3\pi}{2}$

(E) $\theta = \frac{5\pi}{6},\ \frac{7\pi}{6},\ \frac{5\pi}{3}$

USE THIS SPACE FOR SCRATCH WORK

31. For all numbers a and b, a linear function, $f(x)$, satisfies $f(a) > f(b)$ if $a < b$. Which of the following statement must be true about the graph of the linear function $f(x)$ on the xy-coordinate plane?

(A) A line with undefined slope.

(B) A line with zero slope.

(C) A line with positive slope.

(D) A line with negative slope.

(E) A line that passes through the origin.

32. $\lim_{x \to 4} \frac{\sqrt{x} - 2}{x - 4} =$

(A) $\frac{1}{6}$

(B) $\frac{1}{4}$

(C) $\frac{1}{3}$

(D) $\frac{1}{2}$

(E) $\frac{3}{2}$

33. Which of the following statement must be true?

(A) If $f(x) = x^3 + 3x + 1$, the graph of f rises to the right and falls to the left.

(B) Some polynomial functions are NOT continuous.

(C) If f and g are inverse functions, then the range of g is the range of f.

(D) If $f(x) = x^2 + 1$, the domain of f is $(-\infty, \infty)$ and the range is $(-\infty, \infty)$.

(E) If 2 is a zero of a polynomial function f, -2 is also zero of f.

USE THIS SPACE FOR SCRATCH WORK

34. Which of the following trigonometric function has an amplitude of 4, a period of π, and a vertical shift of 1 down ?

(A) $y = -2\cos 2x + 1$

(B) $y = 2\cos\left(\frac{1}{2}x\right) + 1$

(C) $y = -4\sin 2x - 1$

(D) $y = 4\sin\left(\frac{1}{2}x\right) - 1$

(E) $y = 4\tan x - 1$

35. What are the real zeros of $f(x) = 2x^3 - x^2 - 8x + 4$?

(A) $-1,\ \frac{1}{4},\ 2$

(B) $-2,\ \frac{1}{2},\ 2$

(C) $\frac{1}{2},\ 1,\ 2$

(D) $\frac{1}{4},\ \frac{1}{2},\ 2$

(E) $\frac{1}{2},\ 2,\ 4$

36. Figure 5 shows a rectangle that has one corner on the graph of $f(x) = 9 - x^2$, another on the positive x-axis, a third at the origin, and the fourth on the positive y-axis. What is the area A of the rectangle as a function of x ?

(A) $A(x) = -9 + x^2$

(B) $A(x) = -9x + x^2$

(C) $A(x) = 9 + x - x^2$

(D) $A(x) = 9x^2 - x^3$

(E) $A(x) = 9x - x^3$

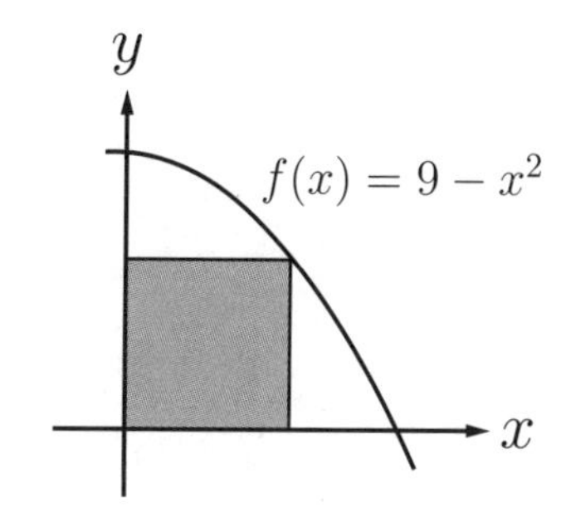

Figure 5

USE THIS SPACE FOR SCRATCH WORK

37. If $\dfrac{\sqrt{2}\cdot\sqrt[3]{4}}{\sqrt[4]{8}} = 2^k$, what is the value of k ?

(A) $\dfrac{1}{4}$

(B) $\dfrac{1}{3}$

(C) $\dfrac{5}{12}$

(D) $\dfrac{1}{2}$

(E) $\dfrac{7}{12}$

38. In quadrilateral ABCD as shown in Figure 6, $AB = BC = \sqrt{2}$, $m\angle BCD = 110°$, and $CD = 3$. What is AD ?

(A) 2.45

(B) 2.82

(C) 3.19

(D) 3.67

(E) 4.03

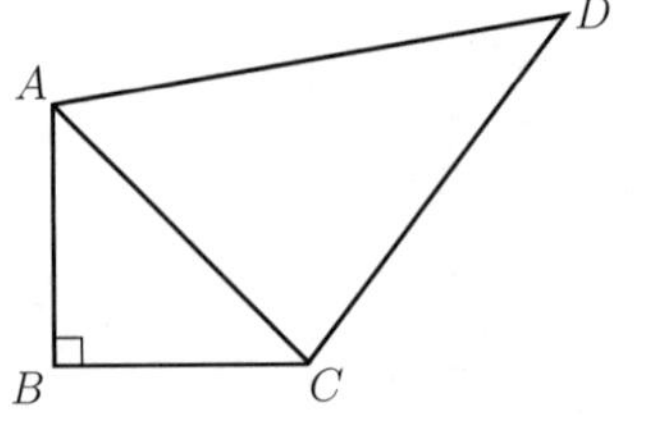

Figure 6

39. Solve: $\sqrt{x+1} - \sqrt{x-1} = 1$

(A) $\dfrac{1}{8}$

(B) $\dfrac{1}{4}$

(C) $\dfrac{1}{2}$

(D) $\dfrac{3}{4}$

(E) $\dfrac{5}{4}$

USE THIS SPACE FOR SCRATCH WORK

40. A big water bottle in the cooler was full on Monday. Students drank one-fourth the water on Tuesday, one-third of the remaining water on Wednesday and one-half of the remaining water on Thursday. What fractional part of the water would still be remaining in the water bottle?

(A) $\frac{1}{3}$

(B) $\frac{1}{4}$

(C) $\frac{1}{5}$

(D) $\frac{1}{6}$

(E) $\frac{1}{8}$

41. Solve the inequality: $-(x-1)(x-2)(x-3) < 0$

(A) $x < 1$ or $x > 3$

(B) $1 < x < 3$

(C) $x < 1$ or $2 < x < 3$

(D) $1 < x < 2$ or $x > 3$

(E) $x < 2$ or $x > 3$

42. Three points $A(x, y)$, $B(0, 0)$, and $C(2, -1)$ are on the coordinates plane. Point $A(x, y)$ lies in the first quadrant. Two segments $\overline{AB}$ and $\overline{CB}$ are perpendicular to each other. If the length of $\overline{AB}$ is $5\sqrt{5}$, what is the value of $x + y$?

(A) 3

(B) 7

(C) 10

(D) 15

(E) 20

USE THIS SPACE FOR SCRATCH WORK

43. If $f(x) = 2\log_7 x$, what is the value of $f^{-1}(0.3)$?

(A) 1.34

(B) 2.11

(C) 3.08

(D) 4.52

(E) 5.05

44. Half-life is the amount of time required for a radioactive substance to decrease to half its value as measured at the beginning of the time period. The amount A of plutonium-243 at time t is given by $A(t) = A_0e^{kt}$, where A_0 is the original amount and k is a negative number that represents the rate of decay. If the half-life of plutonium-243 is 4.956 hours, what is the value of k ?

(A) -0.098

(B) -0.107

(C) -0.119

(D) -0.128

(E) -0.140

45. If $f(x) = \dfrac{x+1}{x-2}$ and $g(x) = \dfrac{1}{x}$, what is $f(g(x))$?

(A) $\dfrac{1-2x}{1+x}$

(B) $\dfrac{1+x}{1-2x}$

(C) $\dfrac{2x-1}{x+1}$

(D) $\dfrac{x+1}{2x-1}$

(E) $\dfrac{1+2x}{1-x}$

USE THIS SPACE FOR SCRATCH WORK

46. The equation $x^2 - 4y^2 - 8x + 8y + 8 = 0$ represents a hyperbola. Which of the following equation can be the asymptote of the hyperbola?

(A) $y + 1 = \frac{1}{2}(x + 4)$

(B) $y - 1 = \frac{1}{2}(x - 4)$

(C) $y + 1 = 2(x + 4)$

(D) $y - 1 = 2(x - 4)$

(E) $y - 4 = 2(x - 1)$

47. What is the minimum value of the trigonometric function $y = -5\sin(2\pi x + 4) + 3$ for $-3 < x < 3$?

(A) -8

(B) -6

(C) -5

(D) -3

(E) -2

48. If the sequence is defined as shown below, what is the 8th term?

$$a_n = \frac{a_{n-1} \cdot a_{n-2}}{2},\ a_1 = 1, a_2 = 2,\ \text{for } n \geq 3$$

(A) $\frac{1}{128}$

(B) $\frac{1}{64}$

(C) $\frac{1}{32}$

(D) $\frac{1}{16}$

(E) $\frac{1}{8}$

USE THIS SPACE FOR SCRATCH WORK

49. There are nine points on the circumference of a circle. If three points are randomly selected to form a triangle, how many different triangles can be formed?

(A) 56

(B) 64

(C) 84

(D) 100

(E) 120

50. Figure 7 shows a circle whose center is located at $(1, 0)$. If a line is tangent to the circle at $(4, 4)$, what is the equation of the tangent line?

(A) $4x + 3y = 28$

(B) $4x - 3y = 4$

(C) $3x - 4y = -4$

(D) $3x + 4y = 28$

(E) $3x + 4y = -4$

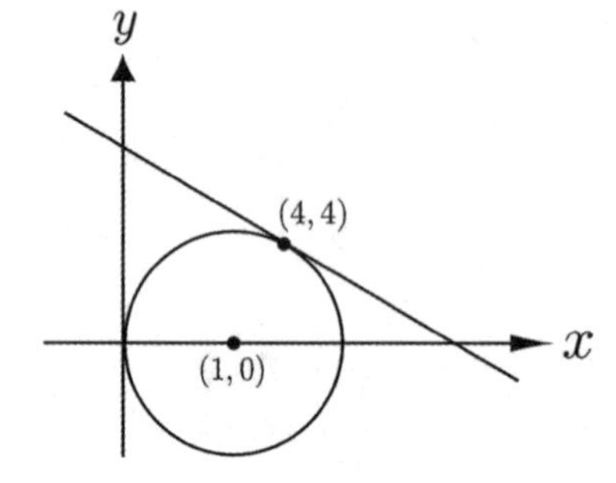

Figure 7

STOP

Mathematics Scoring Worksheet

Directions: In order to calculate your score correctly, fill out the table below. After calculating your raw score, round the raw score to the nearest whole number. The scaled score can be determined using the "Math Test Score Conversion Table".

Mathematics Score			
A. Number Correct		**B.** Number Incorrect ÷ 4	
Total Unrounded Raw Score $A - B$		**Total Rounded Raw Score** Round to nearest whole number	

Math Test Score Conversion Table					
Raw Score	Scaled Score	Raw Score	Scaled Score	Raw Score	Scaled Score
50	800	29	660	8	490
49	800	28	650	7	480
48	800	27	640	6	480
47	800	26	630	5	470
46	800	25	630	4	460
45	800	24	620	3	450
44	800	23	610	2	440
43	800	22	600	1	430
42	790	21	590	0	410
41	780	20	580	−1	390
40	770	19	570	−2	370
39	760	18	560	−3	360
38	750	17	560	−4	340
37	740	16	550	−5	340
36	730	15	540	−6	330
35	720	14	530	−7	320
34	710	13	530	−8	320
33	700	12	520	−9	320
32	690	11	510	−10	320
31	680	10	500	−11	310
30	670	9	500	−12	310

ANSWERS AND SOLUTIONS

Answers

1. C	11. B	21. D	31. D	41. D
2. B	12. A	22. C	32. B	42. D
3. A	13. E	23. B	33. A	43. A
4. E	14. D	24. E	34. C	44. E
5. B	15. B	25. A	35. B	45. B
6. A	16. D	26. B	36. E	46. B
7. E	17. C	27. B	37. C	47. E
8. B	18. C	28. A	38. B	48. A
9. C	19. A	29. A	39. E	49. C
10. D	20. E	30. A	40. B	50. D

Solutions

1. (C)

Tip

If the polar coordinates of a point is (r, θ), the rectangular coordinates of the point (x, y) are given by

$$x = r\cos\theta, \qquad y = r\sin\theta$$

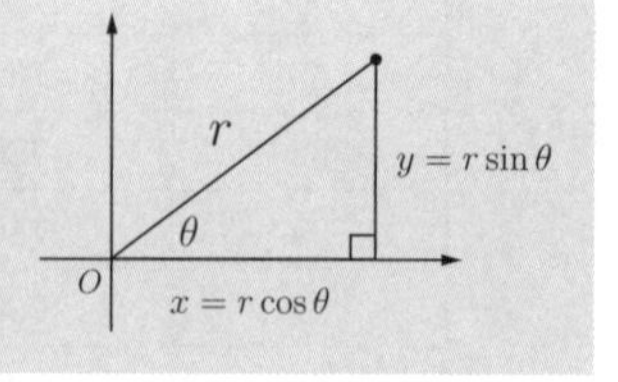

$$x = r\cos\theta \implies x = 10\cos 135^\circ = 10\left(-\frac{\sqrt{2}}{2}\right) = -7.07$$

$$y = r\sin\theta \implies y = 10\sin 135^\circ = 10\left(\frac{\sqrt{2}}{2}\right) = 7.07$$

Therefore, the rectangular coordinates (x, y) that correspond with the polar coordinates $(10, 135^\circ)$ are $(-7.07, 7.07)$.

2. (B)

> Tip
>
> Two events, A and B, are said to be independent if the outcome of A does not affect the outcome of B. If two events, A and B, are independent, the probability of both occurring is as follows:
>
> $$\text{P}(A \text{ and } B) = P(A) \times P(B)$$

The probability that it will rain on Monday is $\frac{7}{10}$. Thus, the probability that it will not rain on Monday is $1 - \frac{7}{10} = \frac{3}{10}$. The probability that it will snow on Tuesday is $\frac{1}{5}$. Since rain on Monday and snow on Tuesday are independent events, the probability that it will not rain on Monday and it will snow on Tuesday is $\frac{3}{10} \times \frac{1}{5} = \frac{3}{50}$.

3. (A)

> Tip
>
> 1. $\sqrt[n]{a^m} = a^{\frac{m}{n}}$
> 2. $(a^m)^n = a^{mn}$

$$\sqrt[3]{27^2} = (27)^{\frac{2}{3}} = (3^3)^{\frac{2}{3}} = 3^2 = 9$$

4. (E)

Multiply each side of the inequality by 6 and solve the inequality.

$\frac{1}{2}(x+3) < \frac{1}{3}(x+4)$	Multiply each side by 6
$3(x+3) < 2(x+4)$	Expand
$3x + 9 < 2x + 8$	Solve
$x < -1$	

Therefore, the solution to $\frac{1}{2}(x+3) < \frac{1}{3}(x+4)$ is $x < -1$.

5. (B)

> Tip
>
> $i^2 = -1$

$5x^2 + 7 = -3$	Subtract 7 from each side
$5x^2 = -10$	Divide each side by 5
$x^2 = -2$	Convert -2 to $2i^2$
$x^2 = 2i^2$	Take the square root of both sides
$x = \pm i\sqrt{2}$	

Therefore, the solutions to $5x^2 + 7 = -3$ are $i\sqrt{2}$ or $-i\sqrt{2}$.

6. (A)

> **Tip** The general equation of a circle is given by $(x-h)^2+(y-k)^2=r^2$, where (h,k) is the center of the circle and r is the radius of the circle.

$x^2+y^2+4x-6y=3$	Rearrange the terms
$x^2+4x+y^2-6y=3$	Add 4 to each side to complete squares in x
$(x+2)^2+y^2-6y=7$	Add 9 to each side to complete squares in y
$(x+2)^2+(y-3)^2=4^2$	

Therefore, the radius of the circle $x^2+y^2+4x-6y=3$ is 4.

7. (E)

> **Tip**
> 1. $\sec^2 x = \dfrac{1}{\cos^2 x}$
> 2. $\sin^2 x + \cos^2 x = 1 \implies \cos^2 x = 1-\sin^2 x$

$$\sec^2 x(1-\sin^2 x) = \frac{1}{\cos^2 x}\times\cos^2 x = 1$$

Therefore, $\sec^2 x(1-\sin^2 x)=1$.

8. (B)

> **Tip** Vieta's formulas relate the coefficients of a polynomial to the sum and product of its zeros and are described below. For a quadratic function $f(x)=x^2+bx+c$, let z_1 and z_2 be the zeros of f.
>
> $z_1+z_2=-b$ Sum of zeros equals the opposite of the coefficient of x
>
> $z_1z_2=c$ Product of zeros equals the constant term

Since -2 and 3 are the zeros of $y=x^2+bx+c$, use Vieta's formula to find the value of b and c.

Sum of zeros: $-2+3=-b \implies b=-1$

Product of zeros: $-2(3)=c \implies c=-6$

Therefore, the value of $b+c$ is $-1+-6=-7$.

9. (C)

Mr. Rhee traveled 120 miles at 60 miles per hour to visit his friend. It took him $\frac{120\text{ miles}}{60\text{mph}}$ or 2 hours to drive to his friend's house. The total distance for the entire trip is 2×120 miles or 240 miles. Since the average speed for the entire trip is 40 miles per hour, the total time for the entire trip is $\frac{240\text{ miles}}{40\text{mph}}$ or 6 hours. This implies that it took Mr. Rhee $6-2=4$ hours to drive back to his home. Therefore, the rate at which Mr. Rhee traveled on the way home is $\frac{120\text{ miles}}{4\text{ hours}}$ or 30 miles per hour.

10. (D)

Set the angle mode to Degrees in your calculator.

$$\begin{aligned} 31\cos\theta &= -7 && \text{Divide each side by 31} \\ \cos\theta &= -\frac{7}{31} && \text{Solve for } \theta \\ \theta &= \cos^{-1}\left(-\frac{7}{31}\right) = 103^\circ \end{aligned}$$

Therefore, the degree measure of angle θ for which $31\cos\theta = -7$ is 103°.

11. (B)

Tip $\cos 2x = \cos^2 x - \sin^2 x = 2\cos^2 x - 1 = 1 - 2\sin^2 x$

Since $\cos 2x = \frac{1}{3}$, $\frac{1}{1-2\sin^2 x} = \frac{1}{\cos 2x} = \frac{1}{\frac{1}{3}} = 3$. Therefore, (B) is the correct answer.

12. (A)

The cone is put inside the cylinder as shown below. The volume of the cylinder is $\pi r^2 h$ and the volume of the cone is $\frac{1}{3}\pi r^2 h$. This means that the volume of the cone is $\frac{1}{3}$ of the volume of the cylinder and the volume outside the cone but inside the cylinder is $\frac{2}{3}$ of the volume of the cylinder.

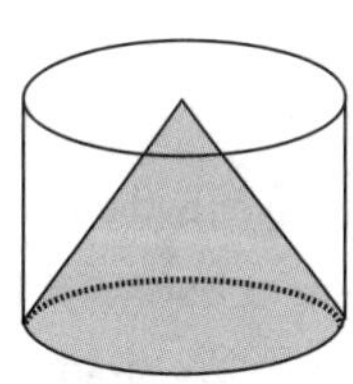

Since the volume of the cylinder is given as $\frac{2}{3}V$, the volume outside the cone but inside the cylinder is $\frac{2}{3} \times \frac{2}{3}V = \frac{4}{9}V$.

13. (E)

Tip $a^{m-n} = \frac{a^m}{a^n}$

$$4^{2-x} = \frac{4^2}{4^x} = \frac{16}{\frac{1}{5}} = 80$$

Therefore, the value of 4^{2-x} is 80.

14. (D)

Let a_1, a_2, a_3 and a_4 be the first term, the second term, the third term, and the fourth term of the arithmetic sequence, respectively, and $a_1 < a_2 < a_3 < a_4$. The sum of the first three terms of the sequence is 49, which can be expressed as $a_1+a_2+a_3 = 49$. The sum of the last three terms is 74, which can be expressed as $a_2+a_3+a_4 = 74$. Furthermore, the range of set S can be expressed as $a_4 - a_1$. In order to find the range of set S, subtract the first equation $a_1 + a_2 + a_3 = 49$ from the second equation $a_2 + a_3 + a_4 = 74$ as shown below.

$$\begin{array}{rl} a_2 + a_3 + a_4 = 74 & \\ a_1 + a_2 + a_3 = 49 & \text{Subtract the two equations} \\ \hline a_4 - a_1 \quad\quad = 25 & \end{array}$$

Therefore, the range of set S, $a_4 - a_1$, is 25.

15. (B)

Tip: If an ordered triple triple (x, y, z) is on the surface of the sphere, the ordered triple satisfies the equation of the sphere $x^2 + y^2 + (z-3)^2 = 25$.

The ordered triple $(4, 3, 3)$ satisfies the equation of the sphere $x^2 + y^2 + (z-3)^2 = 25$ as shown below.

$$x^2 + y^2 + (z-3)^2 = 25 \qquad \text{Substitute } x = 4,\ y = 3, \text{ and } z = 3$$
$$4^2 + 3^2 + (3-3)^2 = 25 \qquad \text{(True)} \checkmark$$

Therefore, (B) is the correct answer.

16. (D)

In order to find the number of units of the USB flash drive to get a production cost of \$10, graph the two functions $P(x) = -\frac{1}{32}x^2 + 4x - 100$ and $y = 10$ in your graphing calculator and find the intersection points as shown below in the figure below.

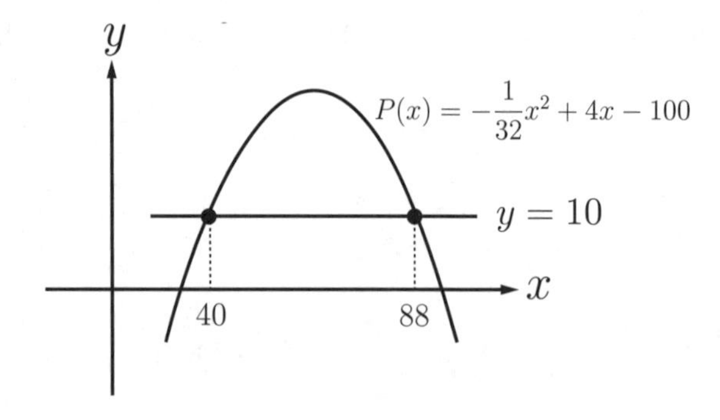

The graph of $y = 10$ intersects the graph of $P(x) = -\frac{1}{32}x^2 + 4x - 100$ at $x = 88$ for $x \geq 45$. Therefore, the company needs to manufacture 88 units of the USB flash drive to get a production cost of \$10.

17. (C)

> Tip $\cos\theta = \dfrac{\text{adjacent side}}{\text{hypotenuse}}, \qquad \sin\theta = \dfrac{\text{opposite side}}{\text{hypotenuse}}$

Since $\cos A = \dfrac{\text{adjacent side}}{\text{hypotenuse}} = \dfrac{1}{x}$, $AC = 1$, and $AB = x$ as shown in the figure below.

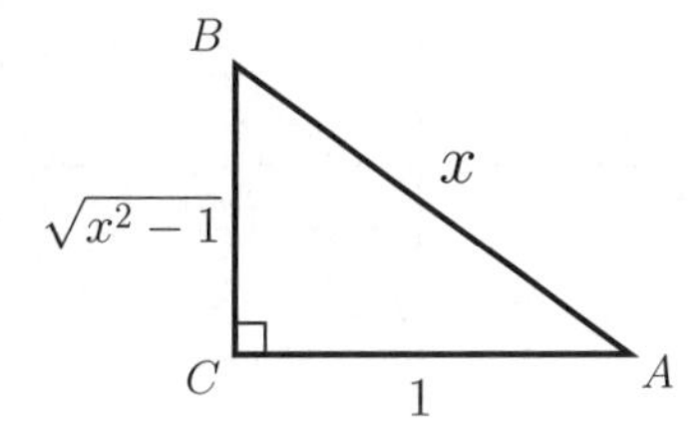

In order to find BC, use the Pythagorean theorem: $x^2 = BC^2 + 1^2$. Thus, $BC = \sqrt{x^2-1}$. Therefore, $\sin A = \dfrac{\text{opposite side}}{\text{hypotenuse}} = \dfrac{\sqrt{x^2-1}}{x}$.

18. (C)

> Tip For the rational function
>
> $$f(x) = \frac{p(x)}{q(x)} = \frac{ax^m + \cdots}{bx^n + \cdots}$$
>
> where m is the degree of the numerator and n is the degree of the denominator, f has a slant asymptote if $n < m$.

For the rational function $y = \dfrac{x^2-x-2}{x+3}$ in answer choice (C), the numerator is a second degree polynomial ($m = 2$) and the denominator is a first degree polynomial ($n = 1$). Since $n < m$, the rational function $y = \dfrac{x^2-x-2}{x+3}$ has a slant asymptote. Therefore, (C) is the correct answer.

19. (A)

$a < 0$, $b > 0$, and $|a| > |b|$ implies that $a + b < 0$ and $ab < 0$. For instance, if $a = -3$ and $b = 2$, $|a| > |b|$, $a + b < 0$, and $ab < 0$. Thus, $\dfrac{a+b}{a} > 0$. Therefore, (A) is the correct answer.

20. (E)

Graph the two functions $y = e^x - 3$ and $y = \ln x$ in your graphing calculator and find the intersection point for $x > 1$ as shown below in the figure below.

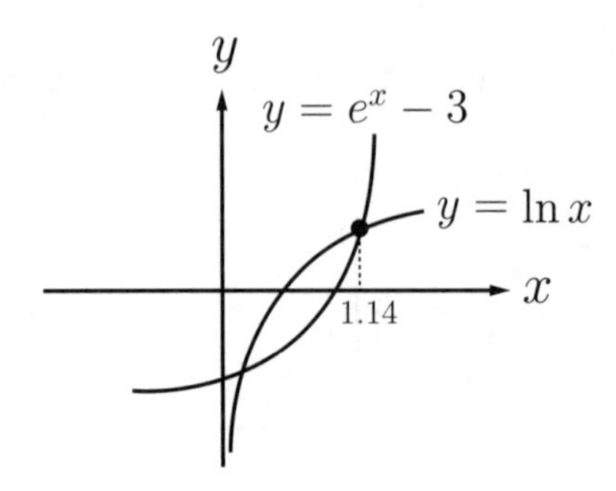

Since the graph of $y = e^x - 3$ intersects the graph of $y = \ln x$ at $x = 1.14$, the value of x for which $e^x - 3 = \ln x$ is 1.14.

21. (D)

Subtract the second equation $x^2 + y^2 = 5$ from the first equation $2x^2 + y^2 = 9$ as shown below.

$$\begin{aligned} 2x^2 + y^2 &= 9 \\ x^2 + y^2 &= 5 \quad \text{Subtract the two equations} \\ \hline x^2 \quad &= 4 \\ x \quad &= \pm 2 \end{aligned}$$

In order to find the y-values, substitute $x = 2$ and $x = -2$ into the second equation $x^2 + y^2 = 5$. When $x = 2$, $y = \pm 1$. When $x = -2$, $y = \pm 1$. Thus, the solutions to the system of nonlinear equations are $(2, 1)$, $(2, -1)$, $(-2, 1)$, and $(-2, -1)$. Therefore, there are 4 solutions to the system of nonlinear equations.

22. (C)

Tip

If a triangle shown at the right is a SAS triangle (a, b, and $m\angle C$ are known), the area of the triangle is as follows:

$$A = \frac{1}{2}ab\sin C$$

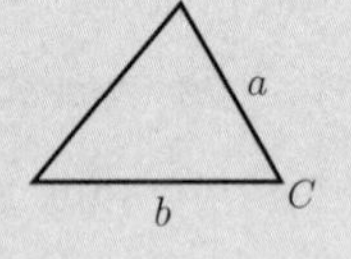

In $\triangle ABC$, $BC = 5$, $AC = 6$, and $m\angle C = 55°$ as shown in the figure below.

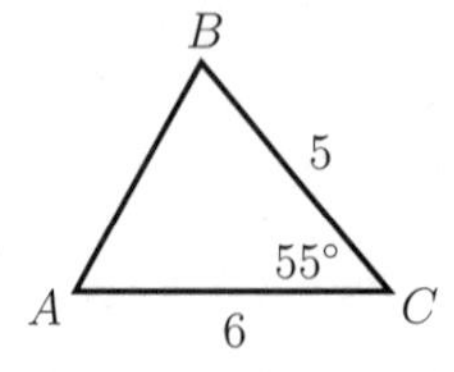

Therefore, the area A of $\triangle ABC$ is $A = \frac{1}{2}ab\sin C = \frac{1}{2}(5)(6)\sin 55° = 12.29$.

23. (B)

Tip nth partial sum of a geometric series: $S_n = \dfrac{a_1(1-r^n)}{1-r}$

$\dfrac{1}{2}+\dfrac{1}{2^2}+\dfrac{1}{2^3}+\cdots+\dfrac{1}{2^9}$ is the 9th partial sum S_9 of a geometric series with $r=\dfrac{1}{2}$.

$$S_9 = \frac{\frac{1}{2}\left(1-\left(\frac{1}{2}\right)^9\right)}{1-\frac{1}{2}} = 1-\left(\frac{1}{2}\right)^9 = 1-\frac{1}{512} = \frac{511}{512}$$

Therefore, $\dfrac{1}{2}+\dfrac{1}{2^2}+\dfrac{1}{2^3}+\cdots+\dfrac{1}{2^9}=\dfrac{511}{512}$.

24. (E)

Tip For the general form of the cosine function $y = A\cos\big(B(x-C)\big)+D$, A affects the amplitude. The amplitude is half the distance between the maximum and minimum values of the function and is $|A|$.

The amplitude of the trigonometric function $y=-4\cos(\pi x+1)-2$ is $|-4|=4$. Therefore, (E) is the correct answer.

25. (A)

Tip If vector $\mathbf{V}=\langle a,b\rangle$, the magnitude of the vector $\mathbf{V}$ is $|\mathbf{V}|=\sqrt{a^2+b^2}$

Vector $\mathbf{V}$ is given as $\langle a,b\rangle$. The horizontal component of $\mathbf{V}$ is three times the vertical component of $\mathbf{V}$. Thus, $a=3b$. Since the magnitude of $\mathbf{V}$ is $\sqrt{40}$,

$\lvert\mathbf{V}\rvert=\sqrt{a^2+b^2}$	Substitute $3b$ for a and $\sqrt{40}$ for $\lvert\mathbf{V}\rvert$
$\sqrt{40}=\sqrt{(3b)^2+b^2}$	Square both sides
$10b^2=40$	Divide each side by 10
$b^2=4$	Solve for b
$b=\pm 2$	

When $b=2$, $a=3b=6$. When $b=-2$, $a=-6$. Thus, vector $\mathbf{V}$ can be either $\langle 6,2\rangle$ or $\langle -6,-2\rangle$. Therefore, (A) is the correct answer.

26. (B)

Tip
1. $\sec^2\theta-\tan^2\theta=1$
2. $\sec^2\theta-\tan^2\theta=(\sec\theta+\tan\theta)(\sec\theta-\tan\theta)$

$\sec^2\theta-\tan^2\theta=1$	Factor
$(\sec\theta+\tan\theta)(\sec\theta-\tan\theta)=1$	Substitute $\frac{1}{2}$ for $\sec\theta+\tan\theta$
$\frac{1}{2}(\sec\theta-\tan\theta)=1$	Multiply each side by 2
$\sec\theta-\tan\theta=2$	

Therefore, the value of $\sec\theta-\tan\theta$ is 2.

27. (B)

Since the ratio of chocolate chip cookies to oatmeal cookies is $1:2$, let x and $2x$ be the number of chocolate chip cookies and oatmeal cookies, respectively. Joshua ate four chocolate chip cookies and half of the oatmeal cookies. Thus, the remaining chocolate chip cookies and oatmeal cookies are $x-4$ and x, respectively. The table below summarizes the number of the chocolate and oatmeal cookies in terms of x.

	Chocolate chip	Oatmeal
Number of cookies in the beginning	x	$2x$
How many Joshua ate	4	$\frac{1}{2}(2x)=x$
How many cookies are left	$x-4$	x

The new ratio of chocolate chip cookies to oatmeal is $3:4$. Set up an equation and solve for x as shown below.

$$\frac{x-4}{x}=\frac{3}{4} \qquad \text{Cross multiply}$$

$$4(x-4)=3x \qquad \text{Expand}$$

$$4x-16=3x \qquad \text{Solve for } x$$

$$x=16$$

Therefore, the number of chocolate chip cookies that are left in the jar is $x-4=12$.

28. (A)

Tip

The Law of Sines: If a, b, and c are the lengths of the sides of a triangle, and A, B, and C are the opposite angles, then

$$\frac{a}{\sin A} = \frac{b}{\sin B} = \frac{c}{\sin C}$$

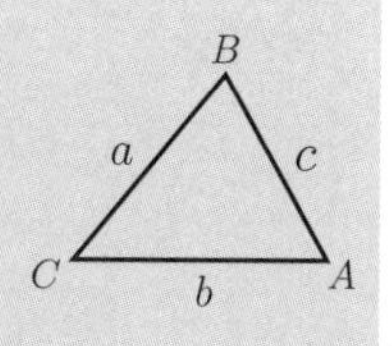

Let x be the length of $\overline{AB}$. Since the length of $\overline{AC}$ is 20% longer than the length of $\overline{AB}$, it can be expressed as $1.2x$ as shown in the figure below.

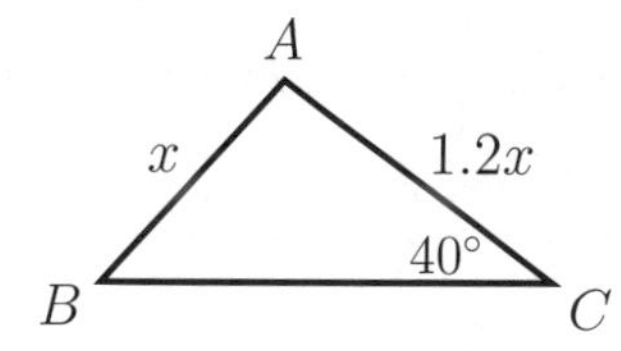

In order to find the degree measure of angle B, use the Law of Sines.

$$\frac{x}{\sin 40°} = \frac{1.2x}{\sin B}$$ Cross multiply

$$x \sin B = 1.2x \sin 40°$$ Divide each side by x

$$\sin B = 1.2 \sin 40°$$ Solve for B

$$B = \sin^{-1}(1.2 \sin 40°) = 50.47°$$

Therefore, the degree measure of angle B could be 50.47°.

29. (A)

Tip

If an initial amount P is invested at an annual interest rate (expressed as a decimal) r compounded n times per year, the amount of money A accumulated in t years is as follows:

$$A = P\left(1 + \frac{r}{n}\right)^{nt}$$

The interest rate is compunded quaterly. Thus, $n = 4$.

$$A = P\left(1 + \frac{r}{n}\right)^{nt}$$ $A = 3000$, $P = 1000$, $n = 4$ and $t = 10$

$$3000 = 1000\left(1 + \frac{r}{4}\right)^{4(10)}$$ Divide each side by 1000

$$\left(1 + \frac{r}{4}\right)^{40} = 3$$ Raise each side of the equation to the power of $\frac{1}{40}$

$$1 + \frac{r}{4} = 3^{\frac{1}{40}}$$ Subtract 1 from each side

$$\frac{r}{4} = 0.02785$$ Multiply each side by 4

$$r = 0.1114$$

Since $r = 0.1114$, the interest rate is 11.14%.

30. (A)

Tip $\sin^2\theta + \cos^2\theta = 1 \implies \cos^2\theta = 1 - \sin^2\theta$

$1 - \cos\theta = \sin^2\theta$	Subtract $\sin^2\theta$ from each side
$1 - \sin^2\theta - \cos\theta = 0$	Replace $1 - \sin^2\theta$ with $\cos^2\theta$
$\cos^2\theta - \cos\theta = 0$	Factor
$\cos\theta(\cos\theta - 1) = 0$	Solve
$\cos\theta = 0 \quad \text{or} \quad \cos\theta = 1$	

The solutions to $\cos\theta = 0$, where $0 \le \theta < 2\pi$ are $\frac{\pi}{2}$ and $\frac{3\pi}{2}$. Futhermore, the solution to $\cos\theta = 1$, where $0 \le \theta < 2\pi$ is 0. Therefore, the solutions to $1 - \cos\theta = \sin^2\theta$ are 0, $\frac{\pi}{2}$ and $\frac{3\pi}{2}$.

31. (D)

In the figure below, the linear function $f(x)$ represents a line.

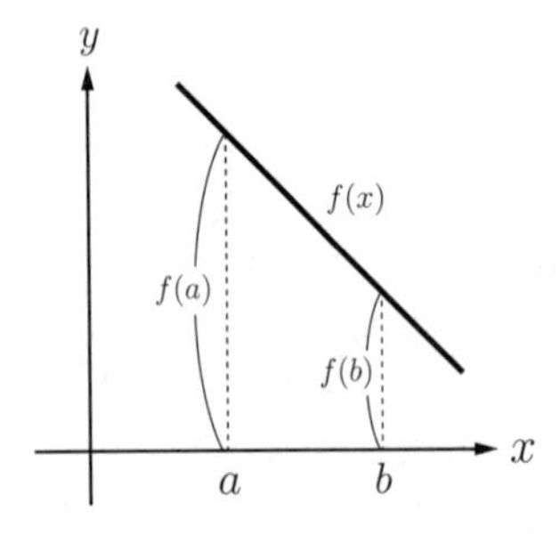

Thus, $f(a)$ and $f(b)$ represent the function's values at $x = a$ and $x = b$, respectively. Since $a < b$ and $f(a) > f(b)$, the linear function $f(x)$ must have a negative slope. Therefore, (D) is the correct answer.

32. (B)

$$\begin{aligned}
\lim_{x\to4}\frac{\sqrt{x}-2}{x-4} &= \lim_{x\to4}\frac{\sqrt{x}-2}{x-4}\cdot\frac{\sqrt{x}+2}{\sqrt{x}+2} \\
&= \lim_{x\to4}\frac{(\sqrt{x}-2)(\sqrt{x}+2)}{(x-4)(\sqrt{x}+2)} \\
&= \lim_{x\to4}\frac{x-4}{(x-4)(\sqrt{x}+2)} \\
&= \lim_{x\to4}\frac{1}{\sqrt{x}+2)} \\
&= \frac{1}{4}
\end{aligned}$$

Therefore, $\lim_{x\to4}\frac{\sqrt{x}-2}{x-4} = \frac{1}{4}$.

33. (A)

All polynomial functions are continuous on $(-\infty, \infty)$. So eliminate answer choice (B). If f and g are inverse functions, then the range of g is the domain of f. So eliminate answer choice (C). If $f(x) = x^2 + 1$, the domain of f is $(-\infty, \infty)$ and the range is $[1, \infty)$. So eliminate answer choice (D). If 2 is a zero of a polynomial function f, -2 is NOT always zero of f. So eliminate answer choice (E). Therefore, (A) is the correct answer.

34. (C)

Compare $y = -4\sin 2x - 1$ in answer choice (C) to $y = A\sin\big(B(x - C)\big) + D$. We found that $A = -4$, $B = 2$, $C = 0$, and $D = -1$. Thus, the amplitude is $|-4| = 4$. The period P is $P = \frac{2\pi}{B} = \frac{2\pi}{2} = \pi$. The vertical shift is 1 down. Therefore, (C) is the correct answer.

35. (B)

Substitute 0 for y and solve for x.

$$2x^3 - x^2 - 8x + 4 = 0 \qquad \text{Group the first two terms and last two terms}$$
$$(2x^3 - x^2) + (-8x + 4) = 0 \qquad \text{Factor}$$
$$x^2(2x - 1) - 4(2x - 1) = 0$$
$$(2x - 1)(x^2 - 4) = 0$$
$$(2x - 1)(x + 2)(x - 2) = 0$$
$$x = -2, \ \frac{1}{2}, \ 2$$

Therefore, the real zeros of $f(x) = 2x^3 - x^2 - 8x + 4$ are -2, $\frac{1}{2}$, and 2.

36. (E)

Let the x and y-coordinates of the corner of the rectangle be (x, y) as shown in the figure below.

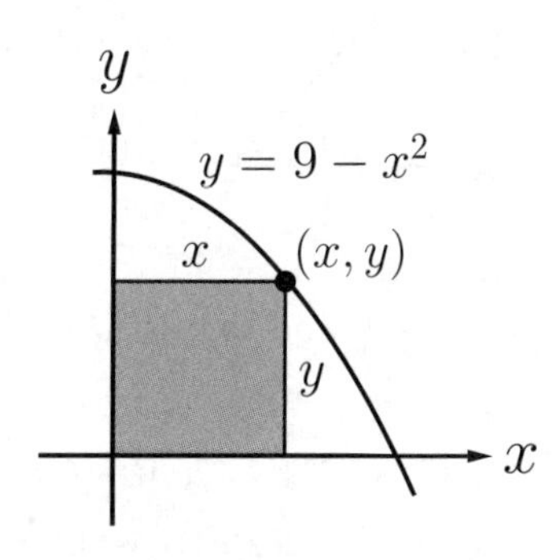

The length and the width of the rectangle are x, y, respectively. Thus, the area A of the rectangle is $A = xy$. Since $y = 9 - x^2$, the area A of the rectangle as a function of x is $A = xy = x(9 - x^2) = 9x - x^3$.

37. (C)

Tip

1. $a^m \cdot a^n = a^{m+n}$
2. $\frac{a^m}{a^n} = a^{m-n}$
3. $\sqrt[n]{a^m} = a^{\frac{m}{n}}$

Change $\sqrt{2}$ to $2^{\frac{1}{2}}$, $\sqrt[3]{4} = \sqrt[3]{2^2} = 2^{\frac{2}{3}}$, and $\sqrt[4]{8} = \sqrt[4]{2^3} = 2^{\frac{3}{4}}$.

$$\frac{\sqrt{2} \cdot \sqrt[3]{4}}{\sqrt[4]{8}} = \frac{2^{\frac{1}{2}} \cdot 2^{\frac{2}{3}}}{2^{\frac{3}{4}}} = 2^{\frac{1}{2}+\frac{2}{3}-\frac{3}{4}} = 2^{\frac{5}{12}}$$

Therefore, the value of k is $\frac{5}{12}$.

38. (B)

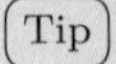

If triangle ABC shown to the right is a SAS triangle (a, b, and $m\angle C$ are known), side c can be calculated by the Law of Cosines.

$$c^2 = a^2 + b^2 - 2ab\cos C$$

Note that side c is opposite angle C.

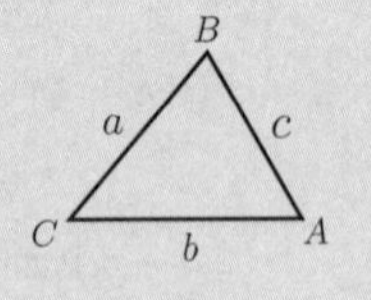

$\triangle ABC$ is a 45°-45°-90° right triangle with $AB = BC = \sqrt{2}$ as shown in the figure below. Use the Phygorean theorem: $AB^2 = (\sqrt{2})^2 + (\sqrt{2})^2$ or use the ratio $1 : 1 : \sqrt{2}$ such that the length of the hypotenuse is $\sqrt{2}$ times the length of each leg. Thus, $AB = 2$. Since $m\angle BCD = 110°$ and $m\angle BCA = 45°$, $m\angle ACD = 65°$.

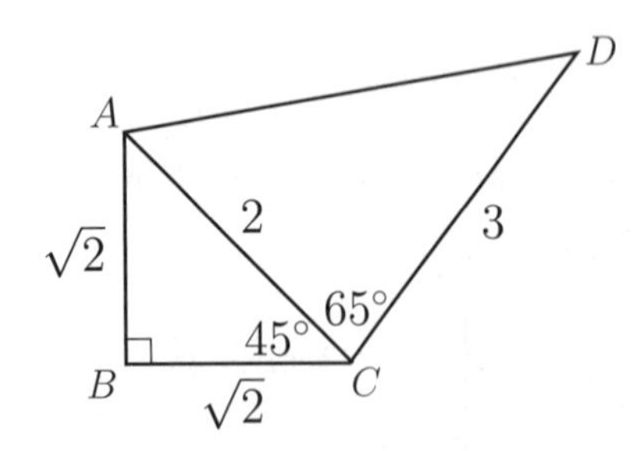

Since $\triangle ACD$ is a SAS triangle (AC, CD, and $m\angle ACD$ are known), use the Law of Cosines to find AD.

$$AD^2 = AC^2 + CD^2 - 2(AC)(CD)\cos 65°$$
$$AD^2 = 2^2 + 3^2 - 2(2)(3)\cos 65° = 7.929$$
$$AD = 2.82$$

Therefore, $AD = 2.82$.

39. (E)

Tip $(a-b)^2 = a^2 - 2ab + b^2$

$$\begin{aligned} \sqrt{x+1} - \sqrt{x-1} &= 1 && \text{Add } \sqrt{x+1} \text{ to each side} \\ \sqrt{x+1} &= 1 + \sqrt{x-1} && \text{Square both sides} \\ x+1 &= 1 + 2\sqrt{x-1} + x - 1 && \text{Subtract } x \text{ from each side} \\ 1 &= 2\sqrt{x-1} && \text{Divide each side by 2} \\ \sqrt{x-1} &= \frac{1}{2} && \text{Square both sides} \\ x-1 &= \frac{1}{4} && \text{Add 1 to each side} \\ x &= \frac{5}{4} \end{aligned}$$

Substitute $\frac{5}{4}$ for x in the original equation to check the solution.

$$\begin{aligned} \sqrt{\frac{5}{4}+1} - \sqrt{\frac{5}{4}-1} &= 1 \\ \sqrt{\frac{9}{4}} - \sqrt{\frac{1}{4}} &= 1 \\ \frac{3}{2} - \frac{1}{2} &= 1 \\ 1 &= 1 \quad \checkmark \text{ (Solution)} \end{aligned}$$

Therefore, the solution to $\sqrt{x+1} - \sqrt{x-1} = 1$ is $x = \frac{5}{4}$.

40. (B)

On Tuesday, students drank $\frac{1}{4}$ of the water. Thus, $\frac{3}{4}$ of the water would be remaining on Tuesday. Students drank $\frac{1}{3}$ of the remaining water on Wednesday. Thus, $\frac{2}{3}$ of the remaining water would be remaining on Wednesday. This means that $\frac{2}{3} \times \frac{3}{4} = \frac{1}{2}$ of the water would be remaining on Wednesday. Students drank $\frac{1}{2}$ of the remaining water on Thursday. Thus, $\frac{1}{2}$ of the remaining water would be remaining. Therefore, $\frac{1}{2} \times \frac{1}{2} = \frac{1}{4}$ of the water would be remaining on Thursday.

41. (D)

Tip: Solving $-(x-1)(x-2)(x-3) < 0$ means finding the x-values for which the graph of $f(x) = -(x-1)(x-2)(x-3)$ lies below the x-axis.

First, graph the polynomial function $f(x) = -(x-1)(x-2)(x-3)$ with the leading coefficient 1. Since 1, 2, and 3 are zeros of multiplicity 1, the graph of f crosses the x-axis at $x = 1$, $x = 2$, and $x = 3$ as shown in Figure A.

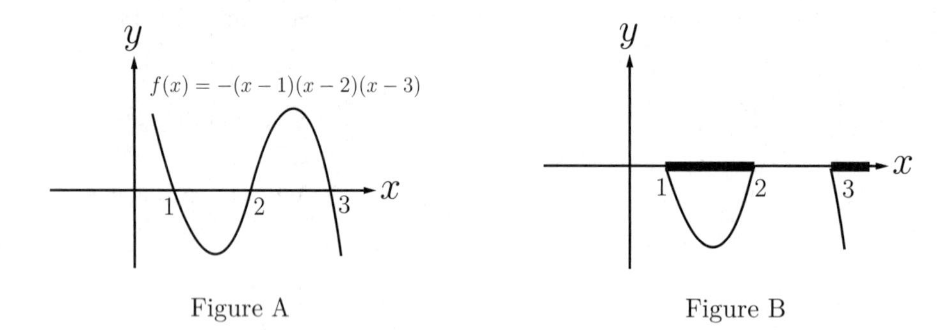

Figure A　　　　Figure B

As shown in Figure B, the graph of f lies below the x-axis when $1 < x < 2$ or $x > 3$. Therefore, the solution to $(x-1)(x-2)(x-3) > 0$ is $1 < x < 2$ or $x > 3$.

42. (D)

Three points $A(x, y)$, $B(0, 0)$, and $C(2, -1)$ are on the coordinate plane as shown below. Point $A(x, y)$ lies in the first quadrant.

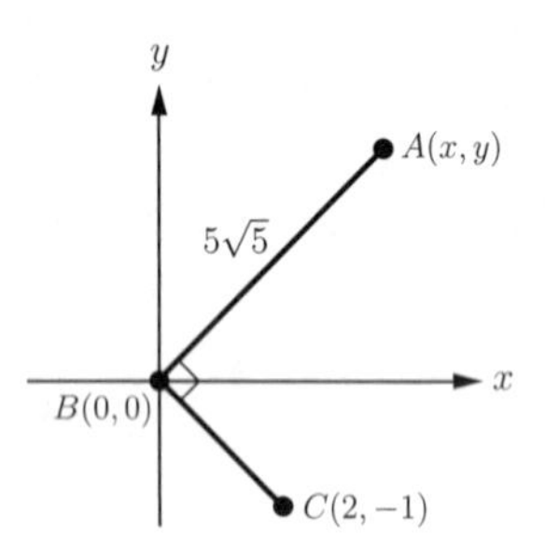

The slope of $\overline{CB}$ is $\frac{-1-0}{2-0} = -\frac{1}{2}$. Since $\overline{AB}$ and $\overline{CB}$ are perpendicular to each other, the slope of $\overline{AB}$ is the negative reciprocal of $-\frac{1}{2}$ or 2. $\overline{AB}$ passes through the origin, which means that the y-intercept of $\overline{AB}$ is 0. Thus, the equation of $\overline{AB}$ is $y = 2x$. The length $\overline{AB}$ is $5\sqrt{5}$. Use the distance formula, $d = \sqrt{(x_1 - x_2)^2 + (y_1 - y_2)^2}$, to write AB in terms of x and y.

$AB = \sqrt{(x-0)^2 + (y-0)^2}$	Substitute $5\sqrt{5}$ for AB and $2x$ for y
$5\sqrt{5} = \sqrt{(x)^2 + (2x)^2}$	Square both sides
$5x^2 = 125$	Divide each side by 5
$x^2 = 25$	Solve for x
$x = 5$ or $x = -5$	

Since point $A(x, y)$ lies in the first quadrant, $x > 0$ and $y > 0$. Thus, $x = 5$ and $y = 2x = 10$. Therefore, the value of $x + y$ is 15.

43. (A)

Tip $\log_a y = x \iff y = a^x$

In order to find the inverse function of $y = 2\log_7 x$, switch the x and y variables and solve for y.

$$y = 2\log_7 x \qquad \text{Switch the } x \text{ and } y \text{ variables}$$
$$x = 2\log_7 y \qquad \text{Divide each side by 2}$$
$$\log_7 y = \frac{x}{2} \qquad \text{Convert the equation to an exponential equation}$$
$$y = 7^{\frac{x}{2}}$$

Thus, the inverse function is $f^{-1}(x) = 7^{\frac{x}{2}}$. Therefore, the value of $f^{-1}(0.3) = 7^{\frac{0.3}{2}} = 1.34$

44. (E)

Tip
1. $a^x = y \iff x = \log_a y$
2. $\log_e x = \ln x$

The original amount of plutonium-243 at $t = 0$ is A_0. Since the half-life of plutonium-243 is 4.956 hours, the amount A of plutonium-243 at time $t = 4.956$ is $A = \frac{A_0}{2}$. In order to find the value of k, substitute 4.956 for t and $\frac{A_0}{2}$ for $A(t)$.

$$A(t) = A_0 e^{kt} \qquad \text{Substitute 4.956 for } t \text{ and } \frac{A_0}{2} \text{ for } A(t)$$
$$A_0 e^{4.956k} = \frac{A_0}{2} \qquad \text{Divide each side by } A_0$$
$$e^{4.956k} = \frac{1}{2} \qquad \text{Convert the equation to a logarithmic equation}$$
$$4.956k = \log_e \frac{1}{2} \qquad \text{Divide each side by 4.956}$$
$$k = \frac{1}{4.956}\ln\frac{1}{2} = -0.140$$

Therefore, the value of k is -0.140.

45. (B)

Since $f(x) = \frac{x+1}{x-2}$ and $g(x) = \frac{1}{x}$,

$$f(g(x)) = f\left(\frac{1}{x}\right) = \frac{\frac{1}{x}+1}{\frac{1}{x}-2} = \frac{\frac{1+x}{x}}{\frac{1-2x}{x}} = \frac{1+x}{1-2x}$$

Therefore, $f(g(x)) = \frac{1+x}{1-2x}$.

46. (B)

> Tip The general equation of a hyperbola with a horizontal transverse axis is $\dfrac{(x-h)^2}{a^2} - \dfrac{(y-k)^2}{b^2} = 1$, where the center of the hyperbola is (h,k). The equations of asymptotes are $y-k = \pm\frac{b}{a}(x-h)$

In order to write a general equation of the hyperbola $\frac{(x-h)^2}{a^2} - \frac{(y-k)^2}{b^2} = 1$, complete the squares in x and in y.

$x^2 - 4y^2 - 8x + 8y + 8 = 0$	Subtract 8 from each side
$x^2 - 4y^2 - 8x + 8y = -8$	Rearrange the terms
$(x^2 - 8x) - 4(y^2 - 2y) = -8$	Add 16 to each side to complete the squares in x
$(x^2 - 8x + 16) - 4(y^2 - 2y) = 8$	Subtract 4 from each side to complete the squares in y
$(x-4)^2 - 4(y^2 - 2y + 1) = 4$	Divide each side by 4
$\dfrac{(x-4)^2}{2^2} - \dfrac{(y-1)^2}{1^2} = 1$	

In order to find the center of the hyperbola, set $x - 4 = 0$ and $y - 1 = 0$ and solve for x and y. Thus, the center of the hyperbola is $(4,1)$ as shown in the figure below.

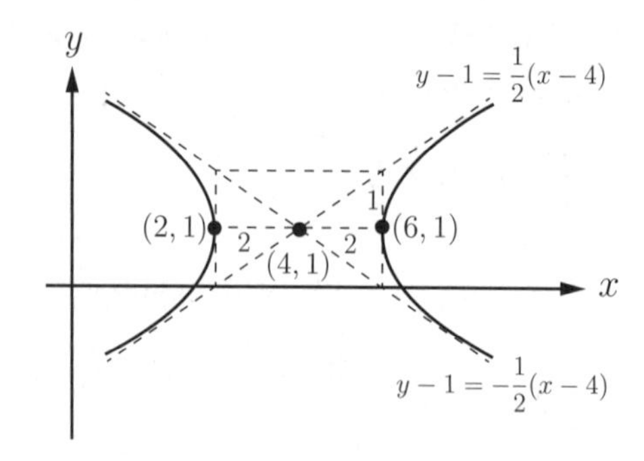

Since the vertices are horizontally 2 units from the center, the vertices are at $(6,1)$ and $(2,1)$. The slopes of the asymptotes are $\pm\dfrac{1}{2}$ and the asymptotes pass through the center $(4,1)$. Thus, the equations of the asymptotes are $y - 1 = \dfrac{1}{2}(x-4)$ and $y - 1 = -\dfrac{1}{2}(x-4)$. Therefore, (B) is the correct answer.

47. (E)

For any angle x, the range of a sine function is $[-1,1]$. For instance, both sine functions, $\sin x$ and $\sin(2x-1)$ have the range of $[-1,1]$.

$-1 \le \sin(2\pi x + 4) \le 1$	Multiply each side of inequality by -5
$-5 \le -5\sin(2\pi x + 4) \le 5$	Add 3 to each side of inequality
$-2 \le -5\sin(2\pi x + 4) + 3 \le 8$	

Since the range of $y = -5\sin(2\pi x + 4) + 3$ is $[-2, 8]$, the minimun value of $y = -5\sin(2\pi x + 4) + 3$ is -2.

48. (A)

In order to evaluate the 8th term, we need to find the previous seven terms as shown below.

$a_n = \dfrac{a_{n-1} \cdot a_{n-2}}{2}$, $a_1 = 1, a_2 = 2,$	Recursive formula with $a_1 = 1$ and $a_2 = 2$
$a_3 = \dfrac{a_2 \cdot a_1}{2} = \dfrac{2 \cdot 1}{2} = 1$	Substitute 3 for n to find a_3
$a_4 = \dfrac{a_3 \cdot a_2}{2} = \dfrac{1 \cdot 2}{2} = 1$	Substitute 4 for n to find a_4
$a_5 = \dfrac{a_4 \cdot a_3}{2} = \dfrac{1 \cdot 1}{2} = \dfrac{1}{2}$	Substitute 5 for n to find a_5
$a_6 = \dfrac{a_5 \cdot a_4}{2} = \dfrac{\frac{1}{2} \cdot 1}{2} = \dfrac{1}{4}$	Substitute 6 for n to find a_6
$a_7 = \dfrac{a_6 \cdot a_5}{2} = \dfrac{\frac{1}{4} \cdot \frac{1}{2}}{2} = \dfrac{1}{16}$	Substitute 7 for n to find a_7
$a_8 = \dfrac{a_7 \cdot a_6}{2} = \dfrac{\frac{1}{16} \cdot \frac{1}{4}}{2} = \dfrac{1}{128}$	Substitute 8 for n to find a_8

Therefore, the value of the 8th term is $\dfrac{1}{128}$.

49. (C)

Since the nine points on the circumference of a circle are indistinguishable, the order is not important. Thus, this is a combination problem. Three points are randomly selected to form a triangle. Therefore, the number of different triangles that can be formed is $\dbinom{9}{3} = \dfrac{9!}{3! \cdot 6!} = 84$.

50. (D)

In the figure below, the tangent line is perpendicular to the radius drawn from the center $(1, 0)$ to the point of tangency $(4, 4)$. Since the slope of the radius is $\frac{4-0}{4-1} = \frac{4}{3}$, the slope of the tangent line is the negative reciprocal, or $-\frac{3}{4}$.

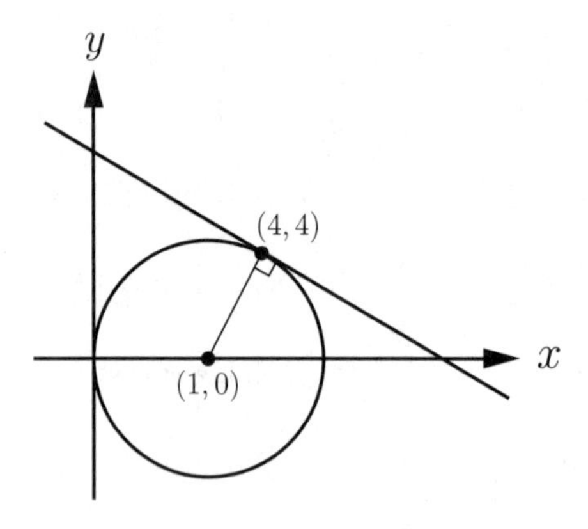

The equation of the tangent line is $y = -\frac{3}{4}x + b$, where b is the y-intercept. Since the tangent line passes through the point $(4, 4)$, substitute 4 for x and 4 for y to find the y-intercept.

$$y = -\frac{3}{4}x + b \qquad \text{Substitute 4 for } x \text{ and 4 for } y$$

$$4 = -\frac{3}{4}(4) + b \qquad \text{Solve for } b$$

$$b = 7$$

Thus, the equation of the tangent line is $y = -\frac{3}{4}x + 7$. Multiply each side of the equation of the tangent line by 4 to write the equation in standard form. Therefore, the equation of the tangent line is $3x + 4y = 28$.

Memo

SAT SUBJECT TEST MATH LEVEL 2

발 행 2020년 12월 4일 초판 1쇄
저 자 이연욱
발행인 최영민
발행처 헤르몬하우스
주 소 경기도 파주시 신촌2로 24
전 화 031 - 8071 - 0088
팩 스 031 - 942 - 8688
전자우편 hermonh@naver.com
출판등록 2015년 3월 27일
등록번호 제406 - 2015 - 31호
정 가 25,000원

ISBN 979-11-91188-05-9 (53410)

• 헤르몬하우스는 피앤피북의 임프린트 출판사입니다.